World Regional Geography Book Series

Series Editor
E. F. J. De Mulder, DANS, NARCIS, Utrecht, Noord-Holland, The Netherlands

What does Finland mean to a Finn, Sichuan to a Sichuanian, and California to a Californian? How are physical and human geographical factors reflected in their present-day inhabitants? And how are these factors interrelated? How does history, culture, socio-economy, language and demography impact and characterize and identify an average person in such regions today? How does that determine her or his well-being, behaviour, ambitions and perspectives for the future? These are the type of questions that are central to The World Regional Geography Book Series, where physically and socially coherent regions are being characterized by their roots and future perspectives described through a wide variety of scientific disciplines. The Book Series presents a dynamic overall and in-depth picture of specific regions and their people. In times of globalization renewed interest emerges for the region as an entity, its people, its landscapes and their roots. Books in this Series will also provide insight in how people from different regions in the world will anticipate on and adapt to global challenges as climate change and to supra-regional mitigation measures. This, in turn, will contribute to the ambitions of the International Year of Global Understanding to link the local with the global, to be proclaimed by the United Nations as a UN-Year for 2016, as initiated by the International Geographical Union. Submissions to the Book Series are also invited on the theme 'The Geography of…', with a relevant subtitle of the authors/editors choice. Proposals for the series will be considered by the Series Editor and International Editorial Board. An author/editor questionnaire and instructions for authors can be obtained from the Publisher.

This book series is published in cooperation with the International Geographical Union (IGU). The IGU is an international, non-governmental, professional organization devoted to the development of the discipline of Geography. The purposes of the IGU are primarily to promote Geography through initiating and coordinating geographical research and teaching in all countries of the world.

More information about this series at https://link.springer.com/bookseries/13179

Dhimitër Doka • Perikli Qiriazi

The Geography of Albania

Problems and Perspectives

 Springer

Dhimitër Doka
Department of Geography
University of Tirana
Tirana, Albania

Perikli Qiriazi
Department of History and Geography
University of Elbasan
Elbasan, Albania

ISSN 2363-9083   ISSN 2363-9091   (electronic)
World Regional Geography Book Series
ISBN 978-3-030-85553-6   ISBN 978-3-030-85551-2   (eBook)
https://doi.org/10.1007/978-3-030-85551-2

This Springer imprint is published by the registered company Springer Nature Switzerland AG
The registered company address is: Gewerbestrasse 11, 6330 Cham, Switzerland

Preface

The science of geography is as old as it is new, as complex and generalizing as it is specific and concrete. Its object has continuously evolved from the general description of various phenomena occurring in the world to the analysis of many changes occurring in certain geographical spaces because of the confrontation between nature and its resources on the one hand and the human activity on the other hand.

On this basis, besides the general geography (physical and human), the need arose for the development of regional geography, which has as its object the treatment of the potentials, alternatives, problems, and perspectives that appear in different regions. The concept of "the region" covers spaces of different sizes: from continents or parts of continents to countries or states, or even parts within the same country. This book, *The Geography of Albania,* is included within the framework of the prestigious Springer Nature publications focusing on different countries and regions of the world.

Although a small country located in Southeast Europe and little known around the world, Albania is very interesting due to its diversity and peculiarities related to its nature, population, history, culture, and economic activities. Positioned in the belt of young mountains and the Mediterranean climate, filled with water resources and greenery, located on the intersection of roads between the East and the West, the present-day Albanian territory bears the historical marks of different cultures, from the Illyrian and the antiquity to the Byzantine and Venetian, to the medieval and Ottoman, but also the traces of the history of an independent country to the present day.

It is precisely this diverse natural and cultural mosaic that underlies the geographical treatment in this book. The analysis of the abundant potential and natural resources of the country; the problems and perspectives in its geopolitical, demographic, social, and economic developments; and the treatment of its specific regions are the subjects of this publication.

Based on this rationale, the book has been organized into five main parts, which are further divided into chapters and sub-chapters.

The first part focuses on the introduction of the current problems that the country is facing and its general natural, demographic, and economic features. An important part here is the historical and geopolitical analysis of the territory at different time periods, with their respective reverberations in many directions, especially in the ethnic and religious structure.

The second part, focusing on the landscape of Albania, deals with the physical, morphodynamic, and bioclimatic basis of the geographical space of the country, the geographical laws, and the role of human activity in the geographical environment. On this basis, the extraordinary diversity of geographical landscapes, their current state, evolution, and trends have been treated. Albania's natural resources, their use, and environmental consequences have also been assessed here. This is the longest part and consists of six subdivisions. The geological building is the object of the first chapter, which is about the geological position of the Albanian territory and its role in the great variety of lithological composition and tectonic structures, the long and complex geological and neo-tectonic evolution, the marked seismic activity, and the mineral resources. The second chapter is about the morphological features and genetic types of relief, the karst ecosystems, the desertification of landscapes, and the crisis of beaches. Climatic features, assets, and problems, and also climatic zones and current problems of global warming and its consequences in Albania are addressed in the third chapter. Waters and their general flow characteristics, river floods, water resources, and water crisis are the subjects of the fourth chapter. The fifth chapter deals with soils, diversity and vertical zonation, and land assets and their damage. The flora and fauna, the extraordinary diversity, the geographical distribution of the vertical zonation, and the damage and protection of biodiversity are analyzed in the sixth chapter. The seventh chapter deals with the Albanian world and national natural and cultural heritage, its degree of damage, and its management.

The third part focuses on the demographic developments of the Albanian population, by analyzing a series of indicators, from the change in the total population number to its geographical distribution, the progress of the natural increase, the significant role of external migration and domestic movement, and the situation in the population structure by age, gender, and residence, but also social and cultural indicators. From the analysis of all these indicators, what becomes evident is the ancient origin and the continuity of the population of the Albanian territory, but also the variety of problems that have historically arisen for this population to the present day.

The fourth part is dedicated to the economy of Albania and it analyzes both the resources and opportunities that the country has for its development. It also deals with the history of this development and the structural changes in the economy. Besides the problems that the economic development of the country faces, the opportunities and alternatives for a better future development are also addressed here.

The fifth and last part deals with the treatment of the four geographical regions of Albania. The complex connection between nature, population, and economy is analyzed for each of these four regions. This enables, besides the overall assessment of the natural and human conditions of the country, the comparison of these conditions at regional and local level. From this perspective, besides the similarities, there are also many differences and inequalities in the socio-economic development of the different regions. The further deepening of these differences could upset the balances of regional development; therefore, interventions through inter-regional development strategies and projects are required.

The book includes a wealth of illustrations and references to make it as scientific and understandable as possible.

This is the first book on the type of regional geography of Albania, where in one single book are treated many aspects of Albanian nature and society. It is also the first one to be published in this form in the English language by one of the most internationally well-known publishers like Springer. Therefore, we believe that this book will fill the gap that exists in the country and abroad in this regard and it will be valuable for many scientists, scholars, students, and various readers who have an interest in knowing or studying Albania.

Besides the voluminous work of the two authors, several institutions and persons have helped with their valuable information, advice, and contribution in the realization of this book. Among them, we would like to single out and particularly thank the English translator of the material Dr. Albert Sheqi; Mr. Ledjo Serferkolli and Ervin Shameti for the cartographic works of many maps and graphs; our geographer colleagues, Prof. Dr. Skënder Sala, Pal Nikolli, and Bilal Draçi for their opinions and oppositions; the geologists, Prof. Dr. Shyqyri Aliaj, Prof. Dr. Vangjel Melo, and Prof. Dr. Romeo Eftimi; the biologists, Prof. Dr. Jani Vangjeli and Prof. Dr. Petrit Hodo; and the pedologists, Prof. Dr. Fran Gjoka and Prof. Dr. Ferdi Brahushi.

Although we have tried to do our best, any evaluative or critical suggestions are welcome from the authors of this book.

Tirana, Albania   Dhimitër Doka

Elbasan, Albania   Perikli Qiriazi
May 2021

Contents

List of Figures

List of Photos

List of Tables

About the Authors

Dhimitër Doka was born in Elbasan on September 05, 1963.

1982–1986 he studied at the Tirana University, Geography Department.

Since 1986 lecturer in the Department of Geography of the University of Tirana. Main areas of study and teaching: Human and Regional Geography of Albania and Geography of Tourism.

1992–1994 – first specialization abroad at the University of Bamberg in Germany.

1995 receiving the title of Doctor of Science (Ph.D.) on the topic: Urban tourism and tourist values of Albanian cities.

1994–2003 – Head of the Geography Department at the University of Tirana.

1997–1998 Qualifications in the University of Potsdam in Germany and Vienna in Austria.

1999 receiving the scientific title of Associate Professor.

1999–2001 – Vice/Dean of the Faculty of History and Filology at the University of Tirana.

2004–2005 research stay at the University of Potsdam Germany in the Institute of Geography, supported by the Humboldt Scholarship.

2007 receiving the scientific title Full Professor.

2013 research stay at the University of Bamberg Germany in the Institute of Geography, supported by the Humboldt Scholarship.

2014 research stay at the Humboldt University of Berlin in the Institute of Geography, supported by the Humboldt Scholarship.

He has published two monographs, four books, atlases, maps, guides, and many scientific articles in Albania and internationally.

Since 2000, he has been a participant and leader in many projects, realized especially with German colleagues and students.

Perikli Qiriazi was born in the village of Podë on March 22, 1944. He completed his higher education studies in the Faculty of History and Geography at the University of Tirana in 1968. From 1970 until 2011, he worked as a lecturer in the Department of Geography at the University of Tirana, then continued at the University of Elbasan. After his retirement, he continues to be a part-time lecturer at several universities in Albania.

He defended his thesis "Morphology and morphogenesis of the Albanian Southeastern Pits (Plains) and their surrounding mountains." He received his scientific degree "Candidate of Sciences" (in 1983), "Doctor of Geographical Sciences" (in 1993), and the title "Professor" (in 1995). He made scientific specializations at the University of Bucharest (in 1975), University of Paris VII (in 1987), and University of Anger (in 1998). He has conducted studies and delivered lectures in several fields, such as: physical and regional geography of Albania, natural heritage of Albania, and degradation and desertification of landscapes.

Prof. Qiriazi has chaired educational and scientific structures in the national university: head of the Department of Geography at the University of Tirana (1985–1994), dean of the Faculty of History and Philology at the University of Tirana (1997–2000), and member of the University Senate (1997–2003).

He has extensive scientific and publishing activities: author and co-author of 7 monographs, the *Albanian Encyclopedic Dictionary*, 80 scientific articles, 15 books on the geography of Albania, and 14 pre-university textbooks. He has delivered papers in 65 national and international congresses and conferences, and has managed environmental projects.

Part I

Introduction, Political Geography, Ethnic Groups and Religions

Introduction

Dhimitër Doka

Abstract

Albania is a small European country, little known in the world. Although a country with an ancient population and history, with important natural resources and favorable geographical position, the numerous wars, political uncertainty, lack of stability, and governance instability have not allowed the country to have a normal development or play its part in the region and Europe. The prolonged self-isolation (1945–1990) further alienated the country from Europe and the world.

Following the collapse of communism and the opening of the country after 1990, the interest in getting to know Albania has increased, and there has been significant progress in this regard. Thus, over 6 million foreign tourists visit the country every year, compared to no more than 10,000 until 1990. The country is open and connected to Europe and the rest of the world by land, sea, and air. The country has already become part of important European organizations and is an EU candidate country.

However, regardless of these important developments have taken place in the last three decades, Albania is still little known and studied in the world. From this perspective, the general treatment of the country is necessary first, including such issues as follows: What is Albania's geographical position and its role in the country's development? What do the natural resources offer for the country? What are the most important events in the country's history? How has the political system changed? How many inhabitants live in Albania and abroad? What is the current state of economic development? etc.

Keywords

Geographical position · Natural resources · Important events in the history · Population · Economy

1.1 Albania's Current "Catastrophes" and Their Geographical Relevance

At the time of writing, Albania was facing two major problems, on top of the global pandemic caused by COVID-19 virus. These problems have pushed the country on the brink of disaster.

The first problem is social, and it has to do with the ***mass migration of Albanians, abandoning their country***. They no longer envisage their future in Albania. This phenomenon began immediately after the collapse of communism but has continued unabated ever since. Consequently, in the last 30 years, over 1.6 million Albanians have left their country, which means that almost half of the population that lived in Albania until 1990 is no longer residing here. The decreasing natural population growth, on the one hand, coupled with the huge uninterrupted departure of the population, on the other hand, have reduced the population of Albania by about 30,000 inhabitants every year, most of them at a young age.

This massive abandonment is accompanied by a series of socioeconomic problems and is putting the country's future in question. In recent years, the cause is not only the difficult economic situation and unemployment but also the disillusionment and frustration caused by politics and lack of justice. There is still some meager hope which lies in the assistance and intervention of the international community.

However, even this hope seems to have faded when EU member states did not approve for several years the opening of negotiations for Albania's EU integration. This event has started a great debate not only within the country but also among the EU countries. For many Albanians, this was another major disappointment and one more reason to leave their country. In this sense, it can be said that the wrong policy, disappointment, and insecurity are leading the country toward the catastrophe of mass abandonment. The effects of this social catastrophe are manifold, and they will be analyzed in detail in the chapter on demographics.

D. Doka, P. Qiriazi, *The Geography of Albania*, World Regional Geography Book Series,
https://doi.org/10.1007/978-3-030-85551-2_1

The second catastrophe was natural. On November 26, 2019, at 03:54 local time, Albania was hit by a *devastating earthquake* (magnitude 6.3 on the Richter scale and intensity IX). It had been 40 years since the country was devastated by such a catastrophe with tragic consequences: 51 dead, about 2,000 injured, over 200,000 people were affected, while over 10,000 inhabitants were left homeless; many family dwellings, health-care center, education facilities, hotels, and the like were reduced to rubble. The whole country was shocked by the natural calamity.

However, unlike the last equally strong earthquake of 1979 when the country was under communist isolation and did not accept any outside help, the November 2019 earthquake happened in the time of the Internet and electronic media, when news travels around the world within minutes. Thus, at the dawn of November 26, the earthquake in Albania would make it into the headline of the world media, and condolences came from all over the world, as well as the willingness to help. This help was started by the Albanians themselves, living in Albania and abroad, and it would gradually include assistance from different countries and peoples around the world. This generous support reached its peak on February 17, 2020, with the Donors' Conference held in Brussels, where about 100 countries that took part expressed their willingness to help Albania and donations from around the world amounting to 1.15 billion Euros were made to overcome the consequences of this earthquake.

To add to these two Albanian problems in 2020, the *great global situation of the pandemic caused by the COVID-19 virus* has put a huge strain on the general socioeconomic situation of the country. From the onset of the pandemic on March 9, 2020, the country entered an extreme three-month lockdown (quarantine), which brought about not only deeper economic and financial problems but also social and psychological troubles for many Albanians and their families. Even after the three-month-long quarantine, the pandemic situation might continue for a long time, causing more difficulties for the economy and disrupting people's normal lives.

1.1.1   What Do These Events Have to Do with Our Geographical Treatment?

Geography is precisely the science that can connect and explain many phenomena that occur in space by human interaction with nature. Herein lies the connection with the catastrophes analyzed above. They are related to the space where we live. Although at first glance they look quite different from each other, as two of them occur on the surface of the earth while the earthquake originates underground, they are based on the very interaction and coexistence between man and nature.

The first one, the abandonment of the country, although it is a mainly a social phenomenon, it reverberates with many effects in space, and it also highlights the many problems related to the nonfunctioning of various sociopolitical structures in utilizing of the country's abundant natural resources for the benefit of its economic and human development, thus offering better opportunities and generating more hope for people to stay in their country.

The second one, the earthquake, is indeed a genuine natural phenomenon resulting from geotectonic processes that occur underground, and as such, it cannot be predicted. However, the manifestations of its effects on the earth's surface are multifaceted. In the first place, the earthquake has put people's lives in danger, but it has also shown how informed and prepared the people are to face such natural disasters.

As a result, such a catastrophe tests how the state structures work, as well as how state policy on territory regulation and management functions in such a case. It is also an important moment of reflection on what needs to be done further in this necessary and obligatory nature-human coexistence and interaction. All this was best proved in the case of Albania.

The first task that presented itself for geographers, along with other specialists (seismologists, geologists, and the like), was the need for information so that people can understand and explain how and why such natural phenomena occur. Why is Albania one of the countries with a high level of seismic risk? What are the most endangered areas? How strong can the earthquake be, and how should we react in such cases? The analysis of the consequences brought about by the earthquake on the earth's surface was equally important. Why did so many buildings collapse within seconds, claiming many lives and leaving many others homeless? Were these buildings built in the right place and with the proper quality? What were the nature-human relations, and how were they tested in this case?

The third one, the pandemic, although a global phenomenon that has swept the entire world, showed that particularly small countries like Albania, with meager economic and financial power, were quickly exposed and unprepared to cope with such an unforeseen situation. This has put a severe strain on the country with numerous negative effects on the economic, medical, and social sphere.

The unprecedented situation also raised the need for geographers to conduct various analyses, such as the country's ability to cope with such situations, the social stratification of the population based on income distribution and the opportunities for individuals and families to react and survive such crises, the rapid and large concentration of the population in a few cities of the country and the impact of their response to the pandemic crisis, the readiness of health centers in central and suburban areas to serve the people, the preparedness of the education system to function even in pandemic conditions and isolation, the functioning of the economic and financial structures of the country to compensate and support especially those economic activities and areas that were most affected by the pandemic, etc.

Finally, all these issues highlighted the proper functioning or nonfunctioning of local and central government structures in managing such human and natural disasters. On this occasion, it became clear how important the cooperation between politics, economy, and science is for the study these phenomena and, based on this cooperation, the design of programs and strategies are aiming at achieving a more sustainable development.

Another very important dimension is social and human. Although a society still traumatized by communist period lasting 45 years (1954–1990) and going through various transition problems, Albanians proved that when disasters strike, they become united and stay strong together. In everyday Albanian language, there is a saying *(ra ky mort dhe u pamë)* *"this death (calamity) came about, and we met each other."* However, it would be good for the Albanians not to become united together only for calamities but also for many other things that can be done united.

All these arguments, along with many others like these, highlight the importance of knowing and studying different geographical areas, which in our case is called Albania. The geographical perspective of this area, with its specific features, is one of the main directions of comprehensive study. Also, as the examples above have shown, time has increasingly underscored the need for a close and inseparable treatment of natural and human phenomena.

Therefore, today's science of geography is no longer divided into physical and human but mostly aims at an integration of them toward a single geography. This is best proved in the complex regional treatment of different spaces and countries, including Albania, which is precisely the object of this publication.

1.2 General Geographical Features of Albania

1.2.1 Geographical Position, Neighboring Countries, Borders, and Size

Albania is located (Qiriazi 2019) in the southwestern part of the Balkans, on the eastern shores of the Adriatic and Ionian Seas. It lies between geographical coordinates from 39°38′ to 42°39′ in the north latitude and from 19°16′ to 21°4′ in the east longitude. The difference between the northernmost and southernmost edge is 3°1′ or 335 km, while between the westernmost and easternmost edge 1° 48′, or 150 km. The area of the country is 28,748 km². Albania borders the Republic of Montenegro, Kosovo, North Macedonia, and

Greece, while the Strait of Otranto separates it from the Republic of Italy. The state maritime border goes 12 nautical miles from the Albanian coastline of the Adriatic and Ionian Seas (Fig. 1.1).

Albania, with its mainly hilly-mountainous relief, is part of the rugged Alpine region and the Mediterranean Alpine belt, which is distinguished for its tectonic instability, its high amplitude of differentiating neotectonic movements, its high volcanic and seismic activity, its great variety and contrast in relief, and its rapid morpho-dynamics (Qiriazi 2019).

Having access to two Mediterranean Seas (the Adriatic and Ionian Seas), Albania is part of the European Mediterranean region. Therefore, it has dominant Mediterranean features, varied relief with sharp contrasts, Mediterranean climate and hydrological regime, Mediterranean features of the soil, and Mediterranean flora and fauna (Vangjeli 2015). These Mediterranean features are also found in human aspects: ancient population, inhabited centers, and social and economic activities related to the Mediterranean nature.

The country's location on the animal migration routes from north to south and vice versa has enriched biodiversity, diversity of habitats, ecosystems, landscapes, and the like.

With no obvious cultural, linguistic, and religious similarities with the Slavs, Greeks, and Italians, the Albanian civilization and identity has a special place in the general, cultural, and social composition of the Southeastern European Region. Albania lies in the Western Balkans, which is identified as a subregion with common problems, development strategies, and integration projects.

Albania is at the crossroads between the East and the West, but it is more oriented toward the West than its neighbors (Çabej 2006). This position, and the firm perseverance to preserve its ethnic nature, has forever endowed Albania with its special features, which even the long Ottoman rule found impossible to erase. These factors and other common ones formed and ensured the continuity of the identity and unity among the provinces and all the Albanian territories, which find expression in way of life, history, religious beliefs, customs, folk, and artistic poetry of the Albanian people.

With its Eastern and Western traditions and its Muslim, Orthodox, and Catholic religions, Albania is the epitome of the complexity of the Balkans, as a region of clashes between the Eastern and Western interests; between Catholic, Orthodox, and Muslim civilizations; and between Europe, Russia, and the USA.

Albania's geographical position offers great opportunities for economic development, thanks to its access to the Mediterranean Sea, which has been the cradle of the birth and development of European civilization, and its position at the intersection of the shortest and largest international roads between the East and the West and vice versa and between the

Fig. 1.1 Albania's position in Europe
Cartography: Ledjo Seferkolli, Mrs. Elda Pineti

North and the South and vice versa. Utilizing the advantage of geographical position and relief (transverse mountain valleys, the only ones in the Western Balkans, which have connected the coast with its interior since ancient times), international roads were built, passing through the Albanian territory. These roads strongly influenced the early population of Albanian territories and the development of Illyrian civilization. Large civic centers flourished in Illyria like Buthrotum (Butrint), Dyrrachium (Durrës), Apollonia, Lisus (Lezha), Scutari (Shkodra), Antipatrea, Bylis, Amantia, etc. (Çabej 2006).

These roads proved invaluable in all subsequent times. Even now, their value is great. Some of them are turning into major corridors like Durrësi-Prishtina (completed), Dibra (under construction), the Eight Corridor, the Adriatic-Ionian Project, etc. Energy arteries of regional, European, and wider importance are of great value. Currently, the large Trans-Atlantic Pipeline (TAP) is being completed, while others are in the design phase.

1.2.1.1 Has Albania Made Good Use of Its Geographical Position for the Development of the Country?

Despite the opportunities offered by the favorable geographical position, Albania has not always used it adequately in the benefit of the country's socioeconomic development, which has historically been associated with either an increase or a decrease in the importance of this position. For instance, since the time of Illyria, the Illyrians were distinguished not only as skillful sailors, but also as powerful merchants, who made good use of the favorable conditions provided by their geographical position, with its wide access to the Adriatic and Ionian Seas (Doka 2015).

During antiquity, it is also worth mentioning the so-called Via Egnatia (Egnatia Road, as a continuation of Via Api in Italy), which followed with its two branches in Durrësi and Apollonia that joined together near Elbasan and then continued to Ohrid, Thessaloniki, and onto Constantinople (modern-day Istanbul). This was the most important trade route of the time because it connected the two centers of the world of that time (Rome and Constantinople). Since a significant part of Via Egnatia passed through Illyria (modern-day Albania), it had a significant impact on the country, as it strengthened some cities of special importance (Dyrrachium, Apollonia) and facilitated the movement of goods and people, etc. (Historia e Popullit Shqiptar 2002).

During the Middle Ages and the long Ottoman occupation, the country's main direction would be oriented more toward the East, thus diluting both the role and the traditional effects of the geographical position and its connections to the West.

The Declaration of Independence in 1912 and especially the period under King Zog's reign would restore the importance of the geographical position and the country's orienta-tion toward the West. A series of political, economic, and infrastructural connections (by sea and air) with important countries such as Italy, Austria, etc., would significantly increase the role of geographical position in the development of the country. The role of the geographical position was reduced to its minimum level during the 45-year-long period of the communist regime because of the extreme isolation of the country (Historia e Popullit Shqiptar 2002).

In the wake of the political changes in 1990–1991 that put an end to the extreme isolation, the situation changed significantly. After the opening of the country to the outside world, many Albanians used the favorable geographical position to emigrate by sea and land to the neighboring countries (Italy and Greece, both members of the European Union).

The role of geographical position would be used further for the establishment of various economic and trade links, for the movement of people and the attraction of foreign investors. Thus, the role and importance of the geographical position have constantly been on the increase in the last three decades. The infrastructure improvement has had a significant impact in this aspect, such as the expansion of connections by sea and air, the construction of several road access to the North, East, and South, which have had a great impact not only on developments within the country but also on the strengthening of the role of Albania as a transit country with important connections to different countries in the region (Malaj 2009).

Based on the current developments, the role and importance of the geographical position and other related natural resources of Albania remain a great opportunity to be further used in the future to the benefit of socioeconomic development of the country. The implementation of the major infrastructure projects mentioned above would significantly increase the role of the geographical position in the further development of the country and play a part in the integration and stability of Albania and the region of Southeast Europe.

1.2.2   Rich and Diverse Nature

With its abundant and diverse natural resources, Albania is considered a privileged country. The country's small area abounds in a variety of natural resources. This is shown in all components of the geographical environment, such as in geological construction, relief forms, climatic conditions, water resources, as well as its plant and animal diversity.

The *geological evolution* of the Albanian territories started (Qiriazi 2019) about 500 million years ago, and it is still ongoing. The history of geological development has been complex and accompanied by several folding phases, volcanic activity, sea advances and retreats, horizontal displacement, as well as numerous tectonic subsidence or uplift. The great variety of rocks and composition structures is precisely related with the geological position of the land and its complex geological evolution. Therefore, we distinguish different and diverse types of rocks, such as different magmatic types, sedimentary (limestone, terrigenous, etc.), and metamorphic types. Sixty-one types of minerals have been discovered in the composition of these rocks, most of them of industrial importance, such as chromium, copper, iron-nickel, oil, natural gas, etc.

The diverse geological construction is also reflected in *the relief forms*, (Qiriazi 2019) which are mainly distinguished for the hilly-mountainous character and sharp contrasts. Albania is dominated by mountains and hills, which are separated by numerous river valleys. Thus, about 80% of the country's surface area is above 200 m from the sea level, while the average relief height is 708 m, which means almost twice the average height of Europe. The flat plain relief occupies a smaller area and is dominant mainly in the western part of the country, while there are many plains, plateaus, and river valleys between the mountains (Qiriazi 2019).

Albania has Mediterranean climate which is distinguished by its typical seasonal character of the solar radiation regime (winter season amount to less than 1/3 of summer radiation), and by atmospheric circulation, consequently, the seasonal character appears in all climatic elements. In general, winter is mild and humid, whereas summer is hot and dry. The factual average annual temperatures range from 7°C to 17–18°C. January temperatures range from about -3°C to around 10°C, and July temperatures range from 15.9°C to 25.8°C. The average annual rainfall ranges from 620 mm to about 3100 mm. From 60% to about 80% of the annual rainfall falls during the cold half of the year, while from about 4% to 15% of the annual rainfall occurs during the summer (Instituti Hidrometeorologjik: Klima e Shqiperisë 1975).

The country's climate, being Mediterranean, is distinguished by great fluctuations of weather and the values of climatic elements in their perennial average. The extreme values of temperature vary from around -27°C (in mountainous areas) to 43.9°C (in western lowland area), while the recorded extreme values of the annual rainfall vary from 5352 mm (in the Albanian Alps, in 1958) to 423 mm (in the Korça Plain, in 1961). The irregular character of the precipitation regime is expressed even in the high values of their intensity, up to about 420 mm/24 hours (Instituti Hidrometeorologjik: Klima e Shqiperisë 1975). This irregular rainfall regime is associated with summer drought and river floods, which inundate fields and residential areas. The climate is also distinct for its great diversity from one area to another. Rarely can one find another country in the world, with the same area as Albania, with so much climate alteration from one province to another.

Besides the two seas, *Albania's water wealth* consists of many lakes, rivers, streams, and natural resources. In terms of economic and tourist importance, it is worth mentioning the natural lakes (Shkodra, Ohrid, Prespa, etc.), the artificial lakes (Koman, Fierze, Vau i Dejes, etc.), the rivers, (Drini, Mati, Shkumbini, Semani, Vjosa), and numerous natural resources, such as Blue Eye Spring, Sotira Springs, etc. (Instituti Hidrometeorologjik: Klima e Shqiperisë 1975).

The *flora and fauna* of Albania is quite rich with many species of plants and animals. There are four plant layers (levels), starting with the Mediterranean shrubs, and then with the oak, coniferous beech, and Alpine pastures. There are a total of 3,651 different plant species, of which about 1% are endemic and about 4% subendemic (Vangjeli 2015). This wealth of plants is home to a rich animal world, represented by different species of animals and birds.

The natural heritage of the country is also rich and diverse. There are three world heritage sites, two strictly protected reserves, fourteen national parks, many nature monuments, managed nature reserves, protected landscapes, managed resource areas, and municipal natural parks. All these areas enjoy the legal status of natural protected areas, which, together with other very beautiful natural landscapes, form the basis for the development of ecological tourism and agritourism.

The great diversity in the geological construction, relief features, climate, hydrography, vegetation, and animal diversity has brought about the diversity of numerous natural landscapes, which are divided into *four large natural units*, (Qiriazi 2019) the Northern Mountainous Province, the Central Mountainous Province, the Southern Mountainous Province, and the Western Lowlands with its hilly regions (Fig. 1.2).

The Northern Mountainous Province consists of the Albanian Alps and the Highlands of Hasi and Gjakova. One characteristic of the Albanian Alps is the sharp contrast between the high mountains and the deep valleys. About 30 mountain peaks are 2000 m above sea level. Some of the valleys are the Valbona, the Shala valleys, etc., while the Highlands of Hasi and Gjakova have lower relief and smaller contrasts (Qiriazi 2019).

The Central Mountainous Province has a very complex and diverse nature. The northern and southern parts are distinguished in this province. The first consists of mountain ranges, highlands, and large valleys, which are separated by the Korabi range; the (Drinii Zi) Black Drini Valley; the Puka-Mirdita Highlands; the Lura Ranges; the Golloborda, Martanesh, and Çermenika Highlands; the Mati valley; and the Skanderbeg Ranges. In its southern part, there are large depression (field), plains, and valleys separated by mountains, mountain ranges, and highlands (Qiriazi 2019).

The Southern Mountainous Province extends as far as the Ionian Sea to the South and is also divided into several smaller geographical units. However, this province has a more regular relief consisting mainly of mountain ranges and valleys which intertwine. The Albanian Riviera, which stretches from the bay of Vlora to the bay of Ftelia in the South, has a special place in this province.

Fig. 1.2 Physical map of Albania
Cartography: Ledjo Seferkolli

Along the Adriatic coast in the west of the country lies the Province of the Western Lowlands and its hilly regions, with a length of about 200 km from north to south and a width of up to 50 km from west to east (Qiriazi 2019). This lowland is the largest flat plain area of the western part of the Balkans, and it is the most economically developed region of Albania.

1.2.3 Ancient History

Albania is distinguished for its very early history. Numerous archeological evidences prove that the Albanian territories have been inhabited since the Middle Paleolithic period (100,000 to 30,000 years B.C.) (Historia e Popullit Shqiptar 2002).

Present-day Albanians are descendants of the Illyrians, who, together with the ancient Greeks and Thracians, are one of the oldest peoples of the Balkans and Europe. The Illyrian population in present-day Albanian territories was organized in several Illyrian tribes, such as the Enchelei (Enkelejdët), the Taulantis (Taulantët), the Dalmatae (Dalmatët), the Dardanis (Dardanët), etc. In the second century B.C., Illyria was conquered by the Roman Empire.

The evidence found on the period of antiquity is proof of Illyrian ethnocultural continuity, as well as of the existence of a considerable population. The characteristic element of this period was the continuous increase in the number of cities, with about 20. Among these cities, several were particularly important, such as Scutari (Skodra), Lisus, Dyrrachium, Apolonia, Buthroti, Bylis, Antigonea, etc.

After the division of the Roman Empire into two parts in 395, the Illyrian territories became part of the Byzantine Empire, as the Eastern Roman Empire was called after that date. The peripheral position on the border with Italy, with which Byzantine interests remained closely linked, determined the extraordinary role of the Illyrian-Albanian territories within the framework of the Byzantine Empire. They became a communication hub between the East and the West and vice versa. In the seventh and eighth centuries, the medieval homeland of the Albanians was known as Albania-Arbëria, which remained essentially unchanged in the later centuries (Historia e Popullit Shqiptar 2002).

The medieval period marked a further important development for the country. It is estimated that during this period, the Albanian population reached about 1.5 million inhabitants.

An important period in the history of Albania is the overlong Ottoman occupation. From the middle of the fifteenth century until the beginning of the twentieth century, Albania became part of the Ottoman Empire for almost 500 years.

The Declaration of Independence on November 28, 1912, and the creation of an independent Albanian state would be the greatest victory in the Albanian history. Despite the difficulties of that time and the two world wars, Albania managed to survive. Moreover, it was during this time that the country became a kingdom for 11 years (1928–1939), a period which laid the foundations of a consolidated state with all the relevant institutions (Historia e Popullit Shqiptar 2002).

After the Second World War, Albania became part of Eastern Bloc countries for 45 years (1945–1990). During this period, the country applied the communist, even Stalinist, model with staunch fanaticism.

Following the collapse of communism in 1990–1991, a new historical period would begin for Albania, marked by the transition from a totalitarian system and centralized economy to a democracy and market economy. This process of transformation is still ongoing in the country.

Some of the most important events in the history of Albania from the earliest periods to the present are listed in Table 1.1.

Table 1.1 Some of the most important events in the history of Albania

Historical periods	Historical events
The Illyrian-Roman period	1200–168 BC: The Albanian territories are an important part of Illyria. 229–167 BC: The period of Roman-Illyrian wars 168 BC: Rome conquer Illyria 395 AD: The division of the Roman Empire and the creation of the Byzantine Empire
The early middle ages period	395–1430 AD: The Illyrian territories of present-day Albania become part of the Byzantine Empire. Ninth to eleventh centuries: The Albanian territories became part of the Bulgarian Kingdom. Twelfth to fifteenth centuries: The establishment and functioning of principalities. The country was known as *Arbëria*, and the inhabitants were called *Arbër*. Fourteenth to fifteenth centuries: The Venetians rule the main centers of the Albanian coasts.
The Ottoman period	1430–1912: The Albanian territories become part of the Ottoman Empire 1443–1468: The Albanian resistance against the Ottomans under the leadership of National Hero Gjergj Kastrioti (Scanderbeg) Late nineteenth century to beginning of the twentieth century: The national movement called "Renaissance" and the country's preparation for the Declaration of Independence
The period from the Declaration of Independence until 1990	November 28, 1912: The Declaration of Independence from the Ottoman Empire 1913: Determination by the great powers of the current political borders of Albania 1914: For 6 months, Albania under the leadership of German Prince Wilhelm von Wied 1914–1918: First World War. Albania divided between Austro-Hungarians, Italians, and French. 1920: The Congress of Lushnje and the establishment of the Albanian Government. Tirana is declared the capital of the country. 1928–1939: Albania becomes a kingdom under King Zogu I. 1939: Occupation of the country by fascist Italy 1943: Occupation of the country by the German army November 29, 1944: Liberation of the country. End of Second World War 1945–1990: Albania under communist rule
The period from 1990 to the present day	1990: The Democratic Movement and the Fall of Communism 1992: The first free democratic elections since 1923. Opposition parties come to power 1997: The collapse of pyramid schemes. Albania in chaos. Early parliamentary elections 1997–2005: The Socialist Party in power 2005–2013: The Democratic Party in power Sins 2013–…: The Socialist Party in power

Work by Dhimitër Doka

All this ancient and rich history of Albania and the Albanians has not been easy. It has often been accompanied by great difficulties and problems. Due to its transit position, the country has had to "navigate the rough waves" that came sometimes from the East and sometimes from the West. After the important developments during the Illyrian and Roman periods, the country was put to the first test of its survival after the division of the Roman Empire in 395, when it became a peripheral part of the Byzantine Empire. Even later, Albania would find no peace, due to its position between the Slavic and Greek interests. Likewise, the overlong Ottoman occupation oriented the country mainly toward the East, separating it from its early tradition with the West and significantly influencing many features of political, economic, social, and cultural development. The consequences of this situation would be reflected even after the declaration of independence, where the country would find it difficult to have the peace and support needed for adequate development (Doka 2015).

After the Second World War, Albania experimented with worst model of communism, which isolated the country from the rest of the world and reduced Albanians into survival conditions. These earlier developments have made the country's transition into a democratic state and market economy very difficult and with many problems in achieving the necessary political stability, proper economic development, and the necessary social and cultural cohesion.

1.2.4   The Political System

Albania is a parliamentary republic. This means that the Parliament (Assembly) is the highest state body in the country, and it consists of 140 deputies elected every 4 years directly by the people. The political party, or coalition of parties, that wins the majority of the seats in parliament enjoys the right to form the government, which governs the country for a period of 4 years. The President of the Republic is elected by the parliament for a 5-year term. The president has mostly an honorary role.

Albania does not have a long-established tradition and experience with its political system. As underscored above, for almost 500 years, Albania was part of the Ottoman Empire and functioned within its framework without developing any political system of its own (Historia e Popullit Shqiptar 2002).

After the declaration of independence in 1912 and the formation of the independent Albanian state, it became possible for the country to establish the first institutions of its political system. The most important result of these developments was the creation of the government of Vlora. Although functioning for only 14 months, this government laid the foundations of the independent Albanian state. However, from the resignation of the government of Vlora until 1945, the political situation in the country became unclear and volatile, and the governance of the country changed hands many times. During this unstable era, the period of the Albanian Kingdom (1928–1939) stands out as a time when genuine political structures were established and functioned, such as parliament, government, ministries, etc.

During the period 1945–1990, Albania was a dictatorial communist country; therefore, the political organization and system suited that regime. Although during this period there were bodies such as parliament or government, they were commanded by a single man and rubber-stamped the dictator's will. Human rights and freedoms were out of the question (Historia e Popullit Shqiptar 2002).

The situation changed after 1990, marking the fall of the communist system, and since then, Albania has officially entered the phase of creating and strengthening its democratic political system.

1.2.5   The Population

According to the last general census of 2011, there are 2,831,000 inhabitants living in Albania. However, it is difficult to provide an accurate figure, as the population of Albania is a migrant population and more than 1.5 million Albanians have left the country in the last 30 years. The thorough changes that took place after 1990 in Albania included not only the politics and economy but also the entire Albanian society (INSTAT 2011).

All these changes, among other things, influenced the decline of the total population of Albania, especially because of two main factors: massive external emigration and the reduction of the birth rate. This decrease is clear from the data of the general census of Albania of 2001, when the population of Albania was 3,069,275 inhabitants, about 130,000 or almost 4% lower than in 1989 (3.2 million). Even more so compared to the data of the last census of Albania conducted in October 2011, when this population decline has continued even further, reducing the total population of the country to only 2,831,000 inhabitants (INSTAT 2001 and 2011).

The latter is the number of Albanians living in the country today, while the number of Albanians in the neighboring countries, Europe, and the world has grown much larger. Although accurate data are lacking, various estimates put the total number of Albanians in the world at around 9 million, of whom about 6 million live in the region (Doka 2015). Thus, besides Albania, many Albanians live in the neighboring countries, primarily in Kosovo, but also in North Macedonia, Montenegro, Greece, Serbia, etc. Albania is one of the few countries that is surrounded by other Albanians on its land borders. The Albanian diaspora, living in many European countries and around the world, also plays an important part. It is worth mentioning the special role played by Albanians living in Italy, the USA, Canada, England, etc.

The current geographical distribution of the population in Albania has undergone significant changes, with over 2/3 of the population concentrated only in the western region, especially in its central part, or what is called the region around the capital (Doka 2015). The peripheral regions have been characterized by mass departures, and there is currently very little population left.

Similarly, important changes have been observed in the population structure, such as age. Due to the decrease in the birth rate, the young age group has been shrinking, whereas the senior age group has expended. At the same time, most of the country's population, about 55%, lives in the cities today, whereas that figure did not exceed 35% in 1990 (Draçi et al. 2018).

Great changes have also taken place in the human, cultural, and educational development of the population. There changes are the natural concomitant of the opening of the country and the new opportunities that have been created for Albanians domestically and abroad.

1.2.6   The Economy

In terms of economic structure, Albania has traditionally been an agricultural-industrial country. This means that agriculture has played a primary role and has provided most of the national income in the economic development of the country. However, each stage of the historical development of the country has been characterized by its own specific features, both in the importance of specific sectors of the economy and in the structure of their products.

Until the end of the Second World War, Albania was considered an economically backward agrarian country. Over 80% of the population was engaged in agriculture and livestock, and the agricultural economy provided about 92% of the total national gross domestic product (Doka 2015). Although agriculture was the main sector of the economy, it was characterized by very low level of development, with primitive working conditions and very low yields in agricultural produce and livestock products. While the industry was underdeveloped and it was mainly represented by some small factories processing agricultural produce and livestock products, as well as by some small industrial centers for the

extraction of minerals, oil, gas, and bitumen, the development of transport, trade, and services was also in a difficult situation.

During the period 1945–1990, the industry, including the heavy industry, became the priority in the economic development of Albania. Industry was the main objective of the economy in all areas of the country. However, even during that period, Albania was considered the most backward country in Europe. The average per capita income did not exceed US $800 per year. Regardless of Albania's claim that it was an industrially well-developed country at the time, Albania was mainly oriented toward the production of grain and small livestock. At the time, every region of the country, down to the village level, had to provide its own agricultural products.

Despite the advances made in the field of agriculture, compared to the period before the Second World War, no significant improvements in increasing the general welfare of the population or their way of life were made during that time. In contrast, the revenues from the agricultural economy were used inefficiently in investments for the industrialization and militarization of the country, while the third sector of the economy (services, hotel, trade, tourism, etc.) played a negligible role in the Albanian economy, amounting to only about 5% of the total gross domestic product of the country. (INSTAT 2019)

The transition period that began after 1990 in Albania saw major changes in the development of the economy. This was primarily the result of the key change from the state into private ownership in economy. Today, over 100,000 enterprises operate in the private sector. No such enterprises existed until 1990. Most of these enterprises were established in Tirana and the largest cities of the country. Likewise, the agricultural economy is currently almost completely privatized, with over 400,000 small private farmsteads in the villages (INSTAT 2019).

Rapid developments have been made in recent years in the sectors of construction, transport, trade, tourism, etc. Currently (see Table 1.2), the services sector plays a key role in Albania's economy. This sector provides about 60% of the

Table 1.2 Different sectors of the Albanian economy

Sectors of economy	Contributions to GDP (in %)	Contributions in employment (in %)
Agriculture	21.4	45.8
Industry	19.4	15.0
Service	59.2	39.2

Source: INSTAT, Bulletin of Statistics 2019

national revenue and employs about 40% of the country's workforce.

The important natural and human potentials of the country, the new economic and trade ties with regional countries and beyond, and the involvement in important European and world organizations are important stepping-stones for a better perspective in the future development of Albania.

Among the many opportunities, tourism stands out as a priority for the country's economy, and its importance is expected to increase in the future. In the meantime, the country should benefit much more from the production and trade of numerous agricultural products, the export of minerals, and the production of energy, generated not only from hydropower plants but also from the solar panels and wind turbines.

References

Çabej E (2006) Shqiptarët midis perëndimit dhe lindjes

Doka D (2015) Gjeografia e Shqipërisë (Popullsia dhe Ekonomia). Tekst Universitar

Draçi B, Doka D, Qiriazi P (2018) Gjeografia e Problemeve Globale. Tirana

Historia e Popullit Shqiptar (2002) Botimi Akademisë së Shkencave të Shqipërisë. "Toena", Tiranë

INSTAT (2001, 2011, 2019) Botime periodike të Fletores Zyrtare

Insttut i Hidrometeorologjik: Klima e Shqipërisë, 1975.

Malaj A (2009) Integrimi i Ballkanit Perëndimor në BE, një persepektvë ekonomike

Qiriazi P (2019) Gjeografia fizike e Shqipërisë, Shtëpia Botuese Mediaprint

Vangjeli J (2015) Excursinflora of Albania. Koenigstein, Germany

Historical and Political Geography

2

Dhimitër Doka

Abstract

In the last 2500 years, different peoples and cultures have lived in the territory of Albania where they have left their cultural traces, starting with the Illyrians-Arbëri-Albanian, Greek, Roman, Byzantine, Venetian, Ottoman, and up to modern history. All these elements together have created cultural and natural landscapes which have left their special features during the history of the transformation in the territory of Albania.

These events are related to the fact that the territory of Albania has often been part of various great empires or kingdoms dominant in history, such as the Roman, Byzantine, Bulgarian, and Ottoman Empires. The territory got the name "Albania" only in the late Middle Ages.

All these historical developments have influenced the geopolitical role that the country has played in different historical periods. This role has been dependent on many historical events, economic developments, and its influence and power in the region and beyond. The question arises, How has Albania used its geographical position and specific natural and cultural conditions in favor of its geopolitical role?

Keywords

History of the territories · Administration of the territory · Geopolitical problems

2.1 Early History of the Territories

After two Roman-Illyrian wars (229–167 BC), the present-day Albanian territories became part of the Roman Empire. After the administrative reform undertaken by Diocletian, these territories became part of the new province of Preval with Scutari (Shkodra) as its capital, the New Epirus with its capital Dyrrachium (Durrës), and the Old Epirus with its capital Nikopolis (located in present-day Preveza, Greece) (Albanien 2003).

Following the division of the Roman Empire in 395, the present-day Albanian territories remained in the eastern half, known as the Byzantine Empire. During the first to fourth centuries, while the northern part of present-day Albanian territories were included in the process of Romanization, the southern parts, especially the coastal areas, came under the influence of the Hellenic Byzantines.

Under the Byzantine Empire, present-day Albanian territories were divided into administrative-military divisions called "themes" (*thémata*). There were two themes, Durrësi and Kosovo, which had a wide territorial extent and were divided into smaller units (Johan Georg von Hahn 1854).

The Slavic invasions in the sixth and seventh centuries to the south had consequences in Albanian territories, and these consequences increased by the beginning of the ninth century, when the expansion of the Bulgarian Empire toward these territories began. The interior part of the territory, with such important centers as Berati and Durrësi, fell under Bulgarian rule, which lasted for about two centuries.

Although all these events bear no direct connection with the later history of Albania, their traces are still present today. The clearest evidence of this period is the many Slavic names and denominations that exist, especially in the central and southern part of the country (Albanien 2003).

2.1.1 The Albanian Territories from the Beginning of the Twelfth Century Until the Declaration of Independence

Although the name "Albania" is mentioned early, at least since Ptolemy in the second century AD, it derives from the

Illyrian tribe of "Albanoi," who lived in the present-day area between Kruja and Tirana. Due to many invasions by different empires, the history of present-day Albania and the Albanians begins with the fall of the Bulgarian Empire in the early twelfth century. Initially, the name for the country was "Arbëria" and "Arbër" for the people, while in the neighboring countries, the names "Arvanitas" (in Greece) and "Albanesi" (in Italy) were used (Peter Jordan 2003).

The most important process that took place in the Arbëri territories during that period was the division of the country into several principalities and the consolidation of these principalities headed by the most powerful and the wealthiest lords of the time, known by their feudal names. Three of the largest feudal families of that time were the Ballshaj in the north, the Topia in the central part, and the Muzaka in the south (see Fig. 2.1). These feudal families dominated the Albanian territories at the time.

The Venetians also left important traces in these territories, especially during the fourteenth century, when they dominated the coastal areas with important city centers such as Durrësi and Shkodra. The evidence of this period is still present today in these city centers such as the Venetian Wall in Durrësi or the ruins and edifices in the city of Shkodra (Albania 2003).

This was the situation in the Albanian territories prior to the overlong Ottoman occupation.

The Ottoman idea and intention to penetrate the west and the south was conceived in Skopje in 1391, and within a few years, they occupied important centers like Berati (in 1417), Ioannina (in 1430), and Arta (in 1449). Meanwhile, after the Battle of Kosovo (in 1389), the Ottoman occupation would extend to Kosovo and the other Albanian territories of the northern and central part. In 1430, the Albanian territories were divided between the Ottoman-ruled ones, mainly in the central and southern part, and the ones under the Venetians, mainly in the coastal area, while the highlands and the isolated parts of the country were still under the rule of the powerful Albanian families (Albania 2001).

Most of the Albanian territories after 1431 would function within the large Ottoman administrative unit called the "Albanian Sanjak," which consisted of 10 *vilayets* (provinces). However, with the administrative reform of 1864–1868, the vilayet was established as the largest Ottoman local administrative unit, which was further divided into *sanjaks, kazas,* and *nahiyas* (see Fig. 2.2). Based on this administrative and organizational reform, the Albanian territories were divided into four vilayets: Shkodra, Kosovo (with Skopje as its main center), Bitola, and Ioannina (Herbert Louis 1925).

Each of these vilayets consisted of several sanjaks, kazas, and nahiyas. This organization, with minor changes, continued to function as such until the end of the Ottoman occupation and the creation of the independent Albanian state.

2.1.2 The Albanian Territories from the Declaration of Independence Until the Liberation of the Country

The Declaration of the Independence of Albania in 1912 created the possibility for the Albanian territories to be administered by Albanians themselves through their own independent state for the first time (Albania 2001). However, numerous uncertainties were left behind after the fall of the Ottoman Empire. Lack of internal unity among Albanians themselves and different interests of the neighboring countries but also of the great powers made it impossible to bring all Albanian territories under Albanian administration and to determine the political boundaries. Under these conditions, the political map of the Albanian territories was determined at the London Conference in 1913, when the great powers determined the current political border of the Republic of Albania, leaving out of this map many Albanian territories which became part of the neighboring countries. This has made Albania one of the few countries in the world that is surrounded by Albanian territories and populations in its entire land border, such as Kosovo, Albanians in North Macedonia, Montenegro, and Greece (Grothusen 1993). The 1913 border division was also one of the causes of various conflicts in the region, the latest being the War of Kosovo in 1999 that ended with the country's secession from Serbia and the formation of the independent Republic of Kosovo.

During the First World War (1914–1918), the territory of Albania was under separate administration among the Austro-Hungarians, Italians, and French. The period of King Zog enabled the governance of the entire territory, and the administration of the country was divided into prefectures, subprefectures, municipalities, and communes. Therefore, whole country was divided into 10 prefectures, 30 subprefectures, 160 communes, and 2551 villages (Doka 2015). This administrative structure, with minor addition or subtraction of small administrative units, was maintained until the beginning of Second World War.

During the Second World War (1939–1944), the country was occupied initially by Italy. During this time, the same administrative organization was maintained, introducing the element of municipality for some cities. Thus, in 1940, the administrative-territorial organization of Albania consisted of 10 prefectures, 30 subprefectures, 22 municipalities, 145 communes, and 2551 villages (Doka 2015).

2.1.3 Administration of the Territory During the Communist Period and Until Today

During the period of communist rule (1945–1990), the country was isolated, and the administration of the territory went

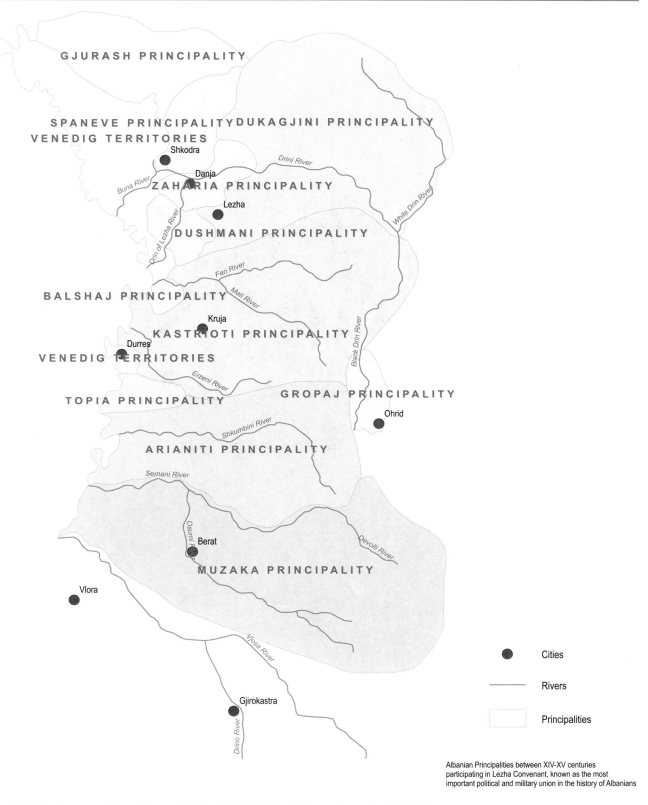

Fig. 2.1 Albanian principalities. (Source: Draft by Dhimitër Doka; Cartography: Ledjo Seferkolli)

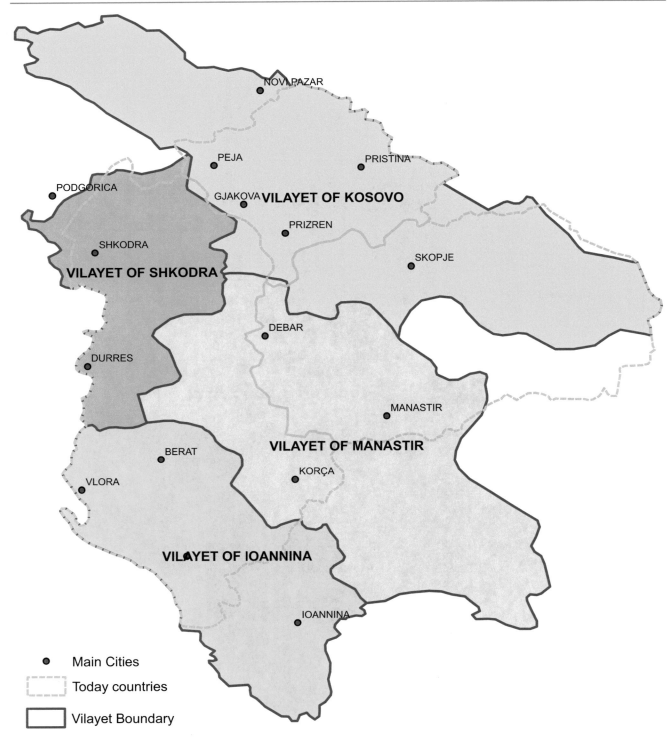

Fig. 2.2 Vilayets. (Source: Draft by Dhimitër Doka; Cartography: Ledjo Seferkolli)

through numerous transformations. In 1946, the new administrative division of Albania was approved, and its territory was divided into 10 prefectures, 39 subprefectures, 116 communes, and several local units. However, in 1953, the district (*qarku*) was determined as the largest administrative unit,

which was divided into counties (*rrethe*) and further into local units.

In 1958, the districts ceased to exist, and the country was divided into 26 large administrative units called counties. In 1969, with the consolidation of the internal organization of counties, Albania was reorganized into 26 counties, 104 local units, 56 cities, and 2860 villages. In 1979, the local units ceased to exist, but all other divisions were maintained until 1992, when the first new administrative division of Albania took place following the change of the political system (Doka 2015).

Under the new administrative division of 1992, the number of districts was increased to 36 and the prefectures, municipalities, and communes were reestablished as units of administrative organization. Following these changes, the territory of the country was reorganized in 12 prefectures, 36 districts, 44 municipalities, and 313 communes (Berxholi et al. 2003).

In 2000, the prefecture was superseded by the district, as the largest unit of territorial organization, while the role of the prefecture was limited to monitoring the proper implementation of laws adopted by the central government. In 2000, the functions of the county were repealed, as a unit of administrative organization, defining the district as the largest unit, and its municipalities and communes.

In 2014, the government undertook the reform of the new administrative organization of the country. The result of this reform was the limitation of the number of first-level administrative units (municipalities and communes) from 373 to only 61 municipalities throughout the country, thus eliminating the communes (see Fig. 2.3). While the district would remain as an administrative unit, the second level and the largest administrative-territorial organization of the country.

Though this reform was necessary and welcomed, it was mostly of a mechanical nature, not considering the many natural, economic, and human criteria on which such organizations must be based.

This makes the implementation of such a reform difficult, and it leaves many questions unanswered, related to the further progress of the administrative and regional organization of the country, such as:

- Can there be a new administrative organization without initially conducting the regionalization of the country based on natural and human potentials and the development priorities that each region should have?
- How appropriate is this new administrative organization with the neighboring countries of the region to implement joint development projects, especially in cross-border areas?
- How is this new administrative organization harmonized with the criteria of the European Union so that the country meets an important condition for its EU integration?

All these questions, and many others, leave the topic of administrative-territorial organization of the country still open. This topic will continue to be significant and important for the future socioeconomic and regional development of the country.

Based on this analysis of the changes in the functioning and organization of Albanian territories in various periods, we can conclude that the creation of the cultural landscape of Albania has been influenced by various factors, which have left their marks to this day. From the earliest periods until the onset of communism, the cultural landscape of Albania was dominated by vast natural spaces, few settlements, but also important Illyrian and Roman centers and ruins, as well as medieval castles and citadels. There are also architectural constructions of modern western civilization (like Italian, French, Austrian ones) including the period of fascism.

During the communist period, the changes in the landscape occurred not so much through settlements but through the transformation of nature for agricultural development (drying of swamps, opening of new arable lands, construction of terraces for agricultural cultivation, and the like) and industry (construction of many large-scale industrial facilities in the shape of the manufacturing factories).

Since 1990, the main agent that has played a major role in the transformation of the Albanian space has been the great boom of the construction industry, especially in the western and southern part of the country. Another factor that has played a part is the abandonment of many peripheral and mountainous areas. Such a situation has created numerous environmental problems, such as excessive pollution being one of the biggest issues requiring special attention in the country.

2.2 Geopolitical Problems

2.2.1 The Geopolitical Role of Geographical Position

Although evaluated for its favorable geographical position in the region, as a transit country between Southern Europe (Italy) and Southeastern European countries, and onto Asia Minor (Turkey and beyond), Albania has not always had the opportunity to properly make use of its geographical position to its geopolitical advantage. This has happened because Albanian territories have seen long periods of numerous invasions by different empires and powers. These invasions have deprived Albanians from the opportunity to make their own decisions on their geopolitical role in the region. Thus, the geopolitical influence of Albania's geographical position has mostly been conditioned by macro political factors and the political interests of other countries. This has been the cause for the changing influence and role, sometimes shifting to the West and sometimes to the East, depending on the time of dominance of the respective powers (Grothusen 1993).

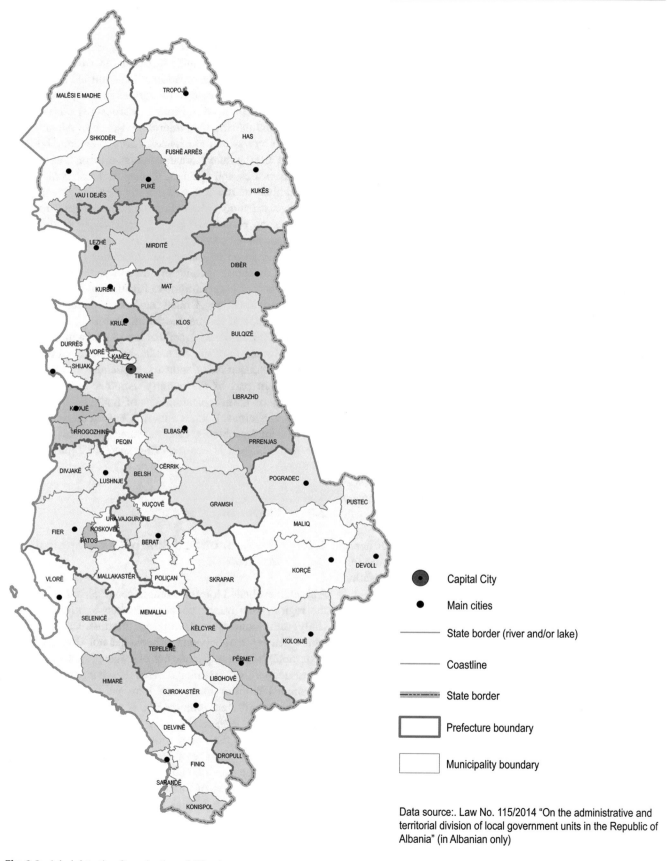

Fig. 2.3 Administrative Organization of Albania

The use of Albania's geographical position as a bridge connecting the West has been apparent in several cases. The first case is related to the Roman Empire's interests and the needs for expansion to the East, especially during the first and second centuries AD, and the Albanian territory with its wide access to the Adriatic and Ionian Seas and its transit position played a significant geopolitical role as a gateway or connecting bridge between the two most important world centers of the time, Rome and Constantinople (modern-day Istanbul), through the so-called Via Egnatia (Egnatia Road). This important role and influence were materialized in the rapid development and strengthening of the important civic and commercial centers, such as Dyrrachium (Durrës), Apollonia, Butrinti (Buthrotum), and other cities (Peter Jordan 2003). The Albanian territory played an equally influential role during the Venetian period in the fourteenth and fifteenth century, especially with its main coastal centers Durrësi, Shkodra, and Vlora.

During the First World War, Italy occupied Vlora (in 1915–1920) with the aim of using its favorable geographical position as a beachhead and springboard to other parts of the region. Italy had the same objective in terms of its penetration toward the Albanian territories and the rest of the region even during the 1930s and the Second World War (Albanien 2003).

Another case of geopolitical use of Albania's geographical position presented itself during the War of Kosovo in 1999, when the country's territory was used as a NATO military base to conduct military operations against Serbia for the liberation of Kosovo.

Since 1991, following the opening of the country after the communist self-isolation, the role and influence of the West is mostly manifested through the influx of Italian investments, which amount to almost half of all foreign investments in Albania. At the same time, the strengthening of Rinas airport and the Albanian ports (especially Durrësi) and the construction of road corridors to the east and north have significantly increased the flow of goods and the movement of people toward Italy and the other Southwestern European countries in the last 30 years (Albanien 2003).

However, the Albanian territory with its geographical position has often been separated from the western geopolitical influence, shifting toward the East. An early case of this Eastern orientation was first observed at the time of the division of the Roman Empire in 395 and the creation of the Byzantine Empire (Peter Jordan 2003). After this division, Albanian territories were also divided, and the north of the country remained under the Roman influence whereas the southern part under the Byzantine influence. In the meantime, the role and importance of Via Egnatia (Egnatia Road), besides serving as the border between the two parts, was significantly increased. In addition to its traditional use from the west to east, the east to west direction was also strengthened. The power-shifting tendency, along with the strengthening of the various cultural and religious elements, brought about the orientation of Albanian territories more and more toward the East. The Eastern orientation of Albanian territories was further deepened during the period of the Bulgarian Empire between the ninth and twelfth centuries, and even more so during the Ottoman occupation (1466–1912). This overlong period had two main effects on the country: its separation from Western influence and its peripheral position within the Ottoman Empire. The introduction of many oriental elements to the life and activity of Albanians was the concomitant of the period of the Ottoman rule (Hatschikjan and Troebst 1999).

The change of the geopolitical role of Albania in different periods is given in the following scheme (see Fig. 2.4).

2.2.2 The Role of Natural Factors

The first role is related to Albania's wide access to the Adriatic and Ionian Seas. However, this wide access is closed by the high mountains in the east, thus limiting this advantage "only to Albania itself." The movement to the east is only possible through some river valleys and mountain passes, "forcing" the neighboring countries (like Kosovo, North Macedonia, Serbia) to make use of the Albanian territory for their sea access (Peter Jordan 2003).

The Adriatic lowland plain of Albania is the largest in the entire Western Balkans, with a length of about 50 km. This feature makes it ideal for use, not only by Albania but the whole region. It particularly favors the connection from the north, Montenegro, and the Dinaric coast toward the south with the main centers of Albania (Tirana, Durrësi, Vlora, etc.) and further down to Greece.

To the south, the Karaburuni peninsula and the island of Sazani separate Albania by only 41 nautical miles from the coast of Italy and the Apennine Peninsula. Also, this border is a very important geopolitical line where three countries are divided and united: Albania, Italy, and Greece. It is from here that the Ionian Sea opens toward Corfu and further south. Therefore, political, military, and economic interests have historically been linked often to this area. This is also the reason that the maritime border between Albania and Greece has remained undefined until today and constitutes a major debate between the two countries (Doka 2015).

Besides its access to the sea, Albania also offers many natural resources, which, if used properly, increase its economic and geopolitical role in the region and beyond. Albania is a country rich in numerous mineral resources, such as chromium. The country is a very important exporter of chromium in the world, and the product has already crossed the

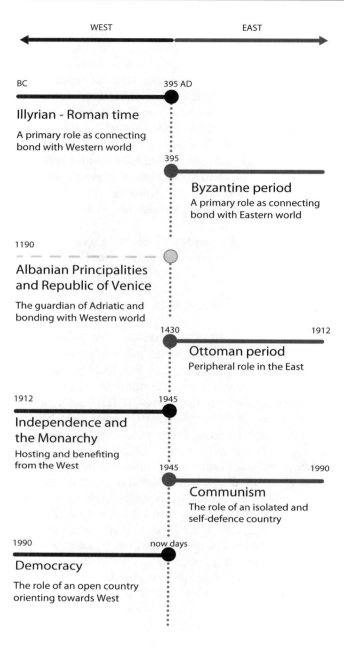

WEST   EAST

BC   395 AD

Illyrian - Roman time

A primary role as connecting
bond with Western world

395

Byzantine period

A primary role as connecting
bond with Eastern world

1190

**Albanian Principalities
and Republic of Venice**

The guardian of Adriatic and
bonding with Western world

1430   1912

Ottoman period

Peripheral role in the East

1912   1945

**Independence and
the Monarchy**

Hosting and benefiting
from the West

1945   1990

Communism

The role of an isolated and
self-defence country

1990   now days

Democracy

The role of an open country
orienting towards West

Fig. 2.4 The geopolitical role of Albania in different periods. (Source: Draft by Dhimitër Doka; Cartography: Ledjo Seferkolli)

borders of Europe, being exported in recent years mostly to China and India. Albania also exports its oil, bitumen, marble, and other products. Thanks to the favorable conditions of climate, land and water, Albania can produce and export more agricultural produce, as it is well known for it, but also export more for electricity, enhance tourism, and so on.

2.2.3 The Role of Cultural Factors

The role of the cultural factors is related to Albania's peripheral position and its territorial contact with two major world cultures, initially between the Roman and Byzantine Empires following the division of the Roman Empire in 395, and later, after the great division of 1054, between the Catholic and Orthodox religions. The effects of this division were manifested in several directions. While the north of Albania remained mostly under the influence of Roman culture and the Catholic religion, the south was dominated by the new Byzantine culture and the Orthodox religion. The border between them was the Shkumbini river valley, or the axis of the Egnatia Road. The influence can also be witnessed even today in the distribution of religious shrines, in the traditions and customs, the linguistic dialects, and so on (Peter Jordan 2003).

The penetration of Slavic tribes, starting from the sixth and the seventh centuries, would also affect the territory of present-day Albania with their cultural and linguistic elements. Although their culture failed to entrench itself and become dominant in these territories, its traces are still present today (Albania 2001).

The next great confrontation of cultures would occur at the beginning of the fifteenth century when Albanian territories were occupied by the Ottoman Empire for a very long time. The effects of this period were multifaceted and exert powerful influences even today.

All this confrontation of different cultures in the territory of a small country like Albania would be associated with different influences, among others in its geopolitical position and role. Among them, we can identify:

The existence of Different Languages and Dialects Although Albanian is a very early and special language with its roots in the Illyrian period, the confrontation of different cultures also meant the presence of other languages. Until 1908, Albanian had not managed to become the official language and enjoy its independent status; therefore, there was a mixture of languages in the country like Latin, Slavic, Greek, and Arabic. Under these conditions, it would take a lot of efforts and sacrifice for the Albanians to preserve their language and make Albanian their official language (Peter Jordan 2003).

The Diverse Religious Structure Although in the beginning Albanians were entirely Christian (Catholic and Orthodox) in religion, during the long Ottoman occupation with its mixture of other cultures, mass Islamization of the population would take place. At the end of the Second World

War, the religious structure of Albania was represented with 70% of people belonging to the Islamic faith, 20% Orthodox, and 10% Catholic (Albanien 2003). These figures have created to the outside world an image of Albania as a Muslim country, even though religion has not been an important factor for Albanians; therefore, that image is not in keeping with the reality. However, this has had its own impact from a geopolitical point of view.

The presence of Different Minorities Besides the dominant Albanian population, there are also different minorities in Albania, such as Greeks, Macedonians, Montenegrins, Roma, Aromanians, etc. A very large proportion of the Albanian population was left outside the current political borders of Albania due to various geopolitical factors. Thus, out of about 6 million Albanians living in the region, only about 58% of them live in Albania, 30% in the former Yugoslavia (primarily in Kosovo, but also in southern Serbia and Montenegro), 10% in North Macedonia, and 2% in Greece (not including the Albanian emigrants after 1990) (Peter Jordan 2003). This ethnic division and structure have historically been one of the main geopolitical problems of the region (*Peter Jordan: Albanien: Geographie – Historische Antropologie – Geschichte – Kultur. S.82. Wien 2003*)

The Economic Backwardness That Has Historically Accompanied the Country The confrontation of different powers and cultures and Albania's position in the peripheral part of different developments have caused the country to often face frequent conflicts, which have hindered the potential economic development. Therefore, agriculture and livestock have been mostly regarded as the only economic option for most Albanians. The communist regime that aimed at the country's industrialization did not achieve its goal. It was only after the opening of the country in 1990, especially in the last two decades, that there has been a change in the economic structure, a shift toward the tertiary sector with such dominant activities as trade, transport, tourism, and the like. This has provided "oxygen" and given an impetus to the economy, which is gradually being integrated into the world economy.

References

Albanien (2003) Geographie – Historische Antropologie – Geschichte – Kultur – Postkommunistische Transformation. In: Österreichisches Ost- und Südosteuropa Institut. Wien

Albania (2001) The State of the Nation 2001, ed. International Crisis Group (ICG), Tirana/Brussels

Bërxholi A, Doka Dh, Asche H (2003) Geographic Atlas of the Albanian Population

Doka Dh (2015) Gjeografia e Shqipërisë (Popullsia dhe Ekonomia). Tekst Universitar

Hatschikjan M, Troebst S (1999) Südosteuropa (Gesellschaft, Politik, Wirtschaft, Kultur). München

Herbert Louis (1925) Albanien – eine Landeskunde. Vornehmlich auf Grund eigener Reisen, Berlin

Johan Georg von Hahn (1854) Albanische Studien. Jena und Wien

Klaus-Detlev Grothusen (Hrg) (1993) Albanian. Südosteuropa-Handbuch, Band VII. München

Peter Jordan (2003) Albanien – Geographie – Historische Antropologie – Geschichte – Kultur – Postkommunistische Transformation. In: Österreichisches Ost- und Südosteuropa Institut, Wien, Faqe 77–95.

Ethnic Groups and Religions

3

Dhimitër Doka

Abstract

The population of Albania, in terms of its location and development as a transit country within the framework of the Balkan Peninsula, is distinguished for its ethnic variety and religious diversity.

In terms of its ethnic structure, the population of Albania is clearly distinguished for its dominant Albanian population over other ethnic minorities. In all the areas of the country, the ethnic Albanian population predominates. In some areas, mainly in the South and Southeast of the country, there is a significant presence of other ethnic minorities. The ethnic minorities present in Albania are the Greek, Macedonian, and Montenegrin ethnic minorities, but also some other ethnic groups, such as Roma, Aromanians, Egyptians, etc.

The religious structure of the Albanian population is also diverse and important to study. Various historical, political, and social factors have altered and enriched this structure over the centuries. There are currently five officially recognized religions in Albania: Muslims, Bektashis, Catholics, Orthodox, and Evangelicals.

This ethnic and religious diversity of Albania should be treated as an important asset for the country, which needs to be recognized, explained, and studied continuously. The most important feature of all time has to do with the fact that different ethnic and religious groups coexist in harmony with each other even within very narrow geographical areas, and Albania can be taken as a very good example in this regard.

Keywords

Ethnic minorities · Ethnic structure · Religions · Religious structure

3.1 Ethnic Composition of Albanian Population

The history of the presence of ethnic minorities in Albania dates back in early times. However, it was the establishment of the political borders of the Albanian state in 1913 that determined the presence of ethnic minorities within its border (Elsie 2001).

Although the country has had a difficult and complicated path through its history, it is nevertheless a fact that Albanians have survived, and today, they constitute the overwhelming majority of over 90% of the entire population of the country. The other part is made up of ethnic minorities: the Greeks, Macedonians, Montenegrins, Roma, Aromanians, etc. (Berxholi et al. 2003).

In the meantime, the scientific treatment of the ethnic structure of the population of Albania has been both a difficult and a sensitive issue. This has happened because, on the one hand, the treatment of the ethnic composition of the population needs to be as factual and realistic as possible, and on the other hand, it has often been endangered by extremist or nationalist attitudes. The initial difficulty arises with the proper understanding of the concepts of ethnicity, nationality, and citizenship, as well as their misuse in the treatment of certain ethnic minorities (Klaus-Detlev 1993).

Such sensitive issues can be further aggravated in the Balkans, where various positions in the understanding and treatment of ethnic minorities abound, due to numerous past wars, border demarcation disputes, frequent dominant nationalist sentiments, and other factors.

As a Balkan country, Albania understands that it cannot escape such problems. Thus, in different periods of time, it has not been easy to properly address the issue of the ethnic composition structure of the population although the country has a clear dominance of the Albanian population and an early recognition of its ethnic minorities (Doka 2013).

This intricate situation is also reflected, among other fields, in the statistical database that exists about the eth-

Table 3.1 The ethnic structure of Albania, 1912–1989

Years	1912	1923	1930	1945	1960	1979	1989
Total population	759,300	803,959	968,092	1,109,043	1,623,128	2,590,260	3,182,417
Albanians	–	–	–	1,075,467	1,578,558	2535 573	3,117,601
Greeks	–	–	–	26,538	37,282	49,307	58,758
Macedonians	–	–	–	2572	3234	4097	4697
Montenegrins	–	–	–	998	1423	66	100
Aromanians	–	–	38,000	–	–	–	782
Roma	–	–	30,000	–	–	–	–
Others		–	–	3468	1356	1271	1245

Source: Central State Archives

Table 3.2 Ethnicity of Albania population, 2011

Nr.	Ethnicity	Number	In %
1.	Albanians	2.312.356	82.58
2.	Greeks	24.243	0.87
3.	Macedonians	5.512	0.20
4.	Montenegrins	366	0.01
5.	Aromanians	8.266	0.30
6.	Roma	8.301	0.30
7.	Egyptians	3.368	0.12
8.	Others	2.644	0.09
9.	Refused to answer	390.938	13.96
10.	Not valid	44.144	1.58
11.	**Total population**	2.831.741	100.00

Source: INSTAT. Results of the 2011 Census

nic composition of the population. Thus, it is very difficult to obtain permanent data on the statistical performance of the ethnic structure. The Central State Archives, one of the few sources available, provides the statistical data on the ethnic composition of the population of Albania for the period from 1912 to 1989 (Heller et al. 2007). The following table summarized the ethnic composition of Albania (Table 3.1).

The table above clearly shows that the data on the ethnic structure of the population in Albania, for the entire period from 1912 to 1989, have been scarce and unsystematic. The difficulties in reflecting these statistical data are related to the various criteria used in their collection (ethnicity, language, religion, etc.) but also to the recognition or nonrecognition of certain ethnic minorities during certain periods. As the data in the table indicate, for some periods, the registration of the ethnic structure has been complete, whereas for some other periods, it has been limited (Berxholi et al. 2003).

The democratization of the country (Doka 2013) after 1990 and the institutionalization of the human rights and freedoms created the appropriate conditions for the different ethnic groups to organize themselves and demand their rights but also for all the scholars who wanted to deal with the topic of the ethnic structure of the population.

However, addressing this issue has proved difficult even during this democratic period. The difficulty has mainly arisen in collecting statistical data, as the treatment of this issue has often been influenced by two contradictory attitudes. On the one hand, there are the representatives of various ethnic minorities who seek to inflate their presence in as large a quantity as possible, and on the other hand, there are the so-called Albanian "nationalists," who try to deflate or minimize the existence of different ethnic minorities in Albania (Doka 2013).

These conflicting interests have influenced the proper fulfillment of the population census based on their ethnicity. There have been two general censuses conducted after 1990. In the first one (in 2001), it was not possible to include questions about ethnicity. In the second one (in 2011), although a decade later, questions about ethnicity and religion were included. However, the way the census was prepared and conducted resulted in shortcomings and inaccuracies (see Table 3.2).

The data in the table show that along with the general decline in the population of Albania, there has been a decline in the number of almost all ethnic minorities compared to the data of 1989. This decline is related to the same factors that have influenced the general decline in the population of Albania, such as mass emigration abroad and the significant decrease in the natural population growth, which has particularly affected certain ethnic minorities (see Fig. 3.1).

One of the important achievements of the last census in terms of ethnicity is related to the expansion of the ethnic structure which included such minorities as Aromanians, Roma, Egyptians, etc., who were previously considered only as cultural groups (Doka 2013).

Because the questions on ethnicity and religion were not mandatory, but left at the respondent's free will, about 14% of the population opted not to answer such questions, thus creating a great deal of uncertainty about the real situation of the ethnic structure of population of Albania. Under these conditions, although a good effort was made with the census in 2011, the composition of the ethnic structure of the popu-

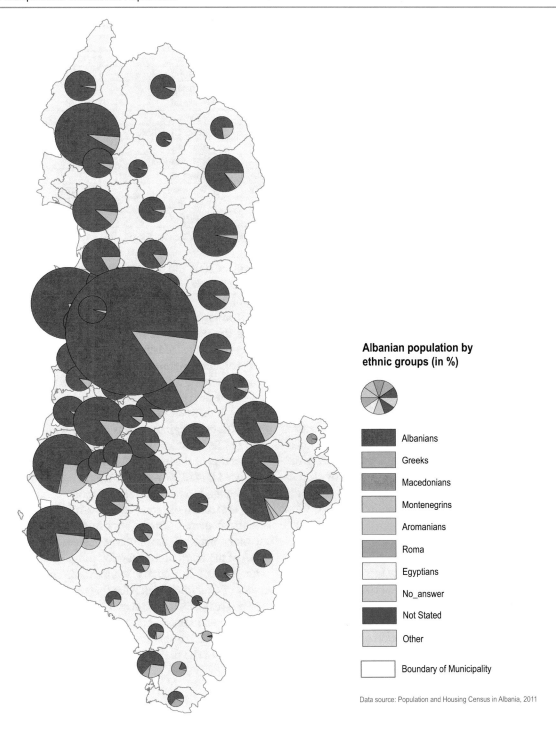

Albanian population by ethnic groups (in %)

- Albanians
- Greeks
- Macedonians
- Montenegrins
- Aromanians
- Roma
- Egyptians
- No_answer
- Not Stated
- Other

Boundary of Municipality

Data source: Population and Housing Census in Albania, 2011

Fig. 3.1 Ethnic structure

lation of Albania remains an important current issue that requires special attention and further study (INSTAT 2011).

3.2 Religions

At its roots, the Albanian population is Christian, but the long Ottoman occupation of the country caused the Islamic faith to spread widely throughout the country. From the beginning in the fifteenth century onward, many Albanians were converted mainly to Islam, either under duress, by their own free will, or by the need to survive.

Simultaneously with the conversion process, the expansion and enrichment of the religious structure of the population took place. This has led to a restructuring of other religious faiths within the two major religions, Christianity and Islam, such as Catholics and Orthodox within the Christian faith, or Sunni Muslims and Bektashi within the Islamic faith, etc. (Elsie 2001).

All these religious groups have lived in very good harmony among them for centuries, and this is the main feature of the religious structure of the population of Albania.

Statistical data on the religious structure of the population of Albania, like the data on the ethnic composition, are incomplete and inaccurate. The situation with these data has changed over time depending on various political and socioeconomic factors.

Consequently, for the early periods and up to 1960s, there are more consistent and complete data about the religious composition of our population. With the banning of all religious beliefs and the proclamation of Albania as the only atheist country in the world during the period 1967–1990, a large gap was created in the statistical data on the religious structure (Berxholi et al. 2003).

Since 1990, when the freedom of religion was reintroduced, the opportunity was created not only for the free exercise of different religious beliefs in Albania but also the right conditions for collecting data about the religious composition and the scientific study of religious structure.

However, it has been difficult to include questions on the religious structure in the recent general censuses. Thus, the

Table 3.3 The religious structure of Albania 2011

Nr.	Religious	Number	In %
1.	Muslim	1,587,608	56,70
2.	Catholic	280,921	10,02
3.	Orthodox	188,992	6,75
4.	Bektashi	58,628	2,09
5.	Protestants	3797	0,14
6.	Other Christians	1919	0,07
7.	Atheists	69,995	2,50
8.	Not identified	153,630	5,49
9.	Others	602	0,02
10.	Not answered	386,024	13,79
11.	Not valid	68,022	2,43
12.	***Total population***	**2,831,741**	**100.00**

Source: INSTAT. Census Results 2011

census in 2001 did not include any questions about religious affiliation, whereas in the census in 2011, although questions on this issue were included, the results revealed numerous shortcomings (see Table 3.3).

As the data of the above table reveal, the considerable presence of the Muslim faith in Albania is clear, but not insignificant in number are also the Christian faiths (Catholic and Orthodox), while about 14% of the population opted not to answer the question about their religious affiliation, thus making it difficult to properly calculate the accurate weight that different religious beliefs occupy within the entire population of the country (INSTAT 2011).

In terms of the territorial distribution of these religions, it is noticed that in the northern region and in the central part of the country, the Muslim faith prevails.

Apart from most areas in the North, the Bektashi religion, a world-centered Muslim sect in Albania, is represented in all other parts of the country. Likewise, except for some areas of the North, the Orthodox religion is represented in all other parts of the country, while Northern Albania is also the center of Catholicism in Albania (see Fig. 3.2).

The spatial distribution of religious beliefs is mostly a historical phenomenon, but nevertheless, it has been heavily influenced in the last two decades by the massive internal migration of the population.

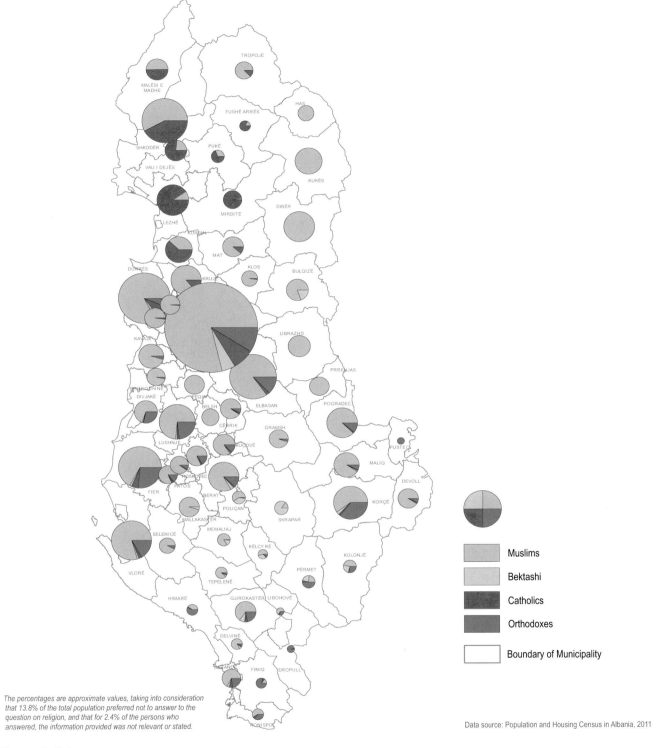

The percentages are approximate values, taking into consideration that 13.8% of the total population preferred not to answer to the question on religion, and that for 2.4% of the persons who answered, the information provided was not relevant or stated.

Data source: Population and Housing Census in Albania, 2011

Fig. 3.2 Religions structure

References

Berxholi A, Doka Dh, Asche H (2003) Demographic atlas of Albania. Tirana

Doka Dh (2013) Die ethnischen Minderheiten in Albanien im Licht der Volkszählung. In: Albanische Hefte 4/2013

Elsie R (2001) A dictionary of Albanian religion, mythology and folk culture. Hurst, London

Heller W, Becker J, Belina B, Lindner W (2007) Ethnizität in der Globalisierung. Südosteuropa-Studien 74 München

INSTAT (2011) Rregjistrimi i Popoullsisë dhe Banesave

Klaus-Detlev G (Hrg) (1993) Albanian. Südosteuropa-Handbuch. Band VII. München

Geology

4

Perikli Qiriazi

Abstract

The Albanian territory, although of a limited size, is distinguished for its complex and diverse geological building. This territory is a special node and key to the interpretation of several features of the geological building of the Balkan region and the Alpine-Mediterranean belt (Qendra e Studimeve Gjeografike 1990).

The territory of the country consists mainly of sedimentary (terrigenous, carbonate) and magmatic formations, but there are also metamorphic formations. Their age ranges from the Paleozoic to the Quaternary. The Albanian ophiolites are distinguished for their wide development. The country's geological structures are different and complicated by numerous tectonic faults (Aliaj 2012).

During the Alpine evolution, the nappe structures developed in the interior of the country while thrusts in its outer part. During the Plio-Quaternary neotectonic stage, the inland ancient structure is dislocated by normal faulting which created a horst-graben morphostructure. In the outer zones, the geological structure was partly subjected to pre-Pliocene compressional movements and partly to post-Pliocene compressional one. During the Plio-Quaternary, it was subjected to strong uplifting movements. They conditioned the formation of the mainly hilly-mountainous relief and its connection with the current geological structure (Aliaj 2012).

The geological evolution of the country continues even today. The proof for this is shown by the uplifting movements in most of the territory and the lowering ones in some grabens (Korça, Përrenjasi, Delvina, Ohrid, etc.) that are associated with a pronounced seismic activity.

With this diversity of geological formations and structures of Albanian territory are linked the numerous and diverse mineral resources. There are 61 types of known minerals, most of them of industrial importance: like chromium, copper, iron-nickel, coal, bitumen, oil, natural gas, etc.

Keywords

The Albanides fold · the Alpine-Mediterranean orogenic belt · The Dinaro-Albano-Hellenic segment · The internal zones · External Albanides · The Neotectonic Pliocene-Quaternary development · Seismically active

4.1 Geological Position of the Albanian Territory

The Albanides fold and thrust belt is a part of the system of the Alpine-Mediterranean orogenic belt. The Albanides orogenic belt is the result of the movements of large Eurasian and African plates, as well as of small plates separated from them (Aliaj 2012).

The Dinaro-Albano-Hellenic segment lies on the southern branch of this large orogenic belt (Giese and Reuter 1978). This segment follows the contour of the eastern edge of the Adria microplate. Tectonic transport in the Dinarides and Albanides was directed toward the Adriatic microplate. The tectonic boundaries of the Adriatic plate are seismically active (Anderson and Jackson 1987).

The Shkodra-Peja transversal divides the geological structures of the country into two parts: the northern, which follows into the Dinarides, and the southern, which follows into the Albanides and the Hellenides (Aliaj 2012).

4.2 Lithological and Structural Features of the Tectono-stratigrafic Units of Albanides

The internal and external tectonic zones distinguished in the Albanide are the following: the *internal zones* comprising the Korabi, Mirdita, and Gashi tectonic ones which lie to the east and the north of the country. They are distinguished from magmatic activity and their earlier folding movements to the end of the Jurassic – the beginning of the Cretaceous. They are consisted mainly of Paleozoic and Mesozoic magmatic, carbonate, and terrigenous complexes (Sh. Aliaj 2012) (Figs. 4.1, 4.2 and 4.3).

External Albanides include the Vermoshi, Albanian Alps, Krasta, Cukali, Kruja, Ionian, and Sazan tectonic zones which lie in the northern, western, and southern part of the country. They were folded from the Late Eocene to the end of the Miocene and consist of mainly Mesozoic and Cenozoic formations of the carbonate and terrigenous (molasse and flysch) formations (Instituti i Studimeve dhe i Projektimeve të Gjeologjisë 2002).

4.3 Magmatism in Albania

Magmatic rocks are widespread and are mainly found in inland tectonic zones. They have mainly of ultrabasic, basic, and medium-acid compositions, intrusive and effusive ones, which are related with magmatic activity developed in different geological periods, according to which they are divided into Paleozoic, Triassic, and Jurassic ophiolitic magmatism and the new, post-Jurassic magmatism. The most developed and most widespread is Jurassic ophiolitic magmatism. This magmatism has played a major role in the paleogeographical development of the Albanides and has great mineral potential: chromium mineralization, olivine, magnesite, chrysotile-asbestos nickel sulfide, etc. With the new magmatism (after the Jurassic), with a small extent and represented by intrusive acidic rocks (granitic, granodiorite), is associated the mineralization of realgar, auripigment, mercury, etc. (Instituti i Studimeve dhe i Projektimeve të Gjeologjisë 2002).

4.4 The Neotectonics Structure of Albania

In the Mediterranean region, the neotectonic Pliocene-Quaternary development was associated with strong and progressive uplifting movemens. During the Pliocene, within inland parts of the country are created the horst-graben structures, where Pliocene lakes were formed (Sulstarova and Aliaj 1990). The Middle and Upper Miocene molasse basins are found at altitudes around 1800 m east of the country (the Mali i Bardhe Mountain into the Korabi area). This testifies for the large rises that occurred during the Pliocene and Quaternary, from which were formed the horst structures and the high mountains. The neotectonic structure of Albania is divided into four large units (Aliaj 2012).

– *The inner unit (territory)* includes the inner sectors of the Alpine ranges, east of the Kruja tectonic zone. During the Pliocene-Quaternary, it was involved by extensional tectonics, which led to the formation of the highly developed horst-graben structure. The Shkodër-Pejë and Elbasan-Dibër main transverse faults cross the structure of Albanides. Large mountain blocks and ranges were formed in the horst areas, while large pits and valleys were formed in the grabens (Aliaj 2012).

Most of the new grabens in this unit were submerged in the Pliocene or up to the Early Pleistocene, while they were raised during the Middle Holocene-Pleistocene. Therefore, in the pits formed in these grabens, the erosion of river flows is activated, terrace levels are formed, and their post-lake bottom is broken. Such are the pit of Kolonja, Devolli, Peshkopia, Kukësi, Kruma, Tropoja, etc. Meanwhile, the grabens of Korça, Përrenjas, Ohrid, and Shkodra continued to sink in the quaternary, including currently. This was reflected in the flat morphology of the pits, formed in these grabens and in the absence of river terraces. The formation of large lakes relates to the quaternary diving movements of graben structures: Ohrid, Prespa, and Shkodra (Fig. 4.4).

– *The outer unit (territory)* includes the outer sectors of the Alpine ranges of the Kruja, Ionian, and Sazan tectonic zones, up to the Adriatic and Ionian seas. The structure was inherited from the pre-Pliocene one, strongly affected by pre-Pliocene compressional movements and that of Pliocene-Quaternary. These movements deformed the pre-Pliocene structure, creating the folds, reverse faults, and thrusts and backthrusts. At the same time, during this time, this structure underwent to the great and progressive uplifting (Aliaj 2012).

The structure of this area consists mainly of anticline ranges (expressed in relief with high mountain ranges), intertwined with syncline ranges, where valleys are formed. The small areas are affected by the Pliocene-Quaternary extensional tectonics, expressed with graben structures (Butrinti, Elbasani, Dukati, etc.) (Aliaj 2012). The compressional and uplifting movements continue even up to nowadays. This is evidenced by the considerable heights above sea level (up to over 2480 m) and river terraces in the valleys of the Shushica, Vjosa, Osumi rivers, etc. (Qiriazi 2019).

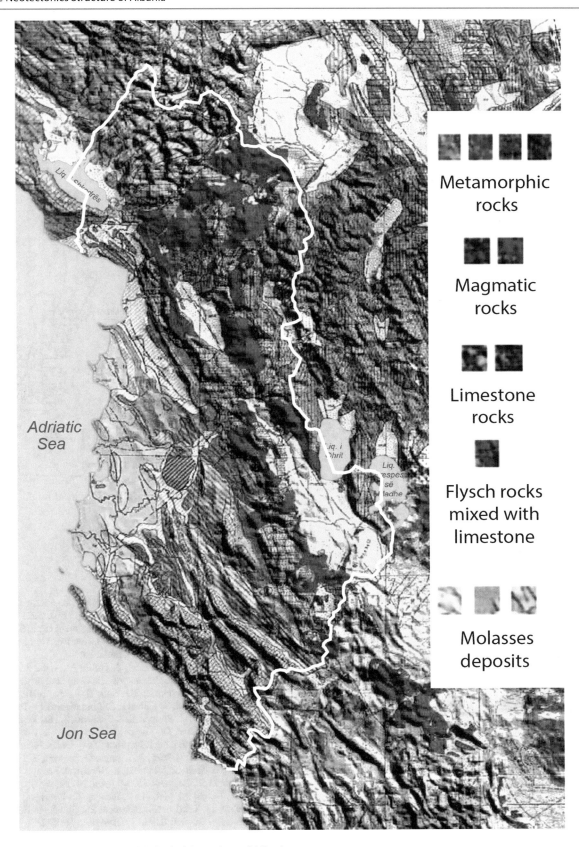

Fig. 4.1 Schematic map of the main lithological formations of Albania
Source: IDEART, Gjeografia 11/2019, p. 44

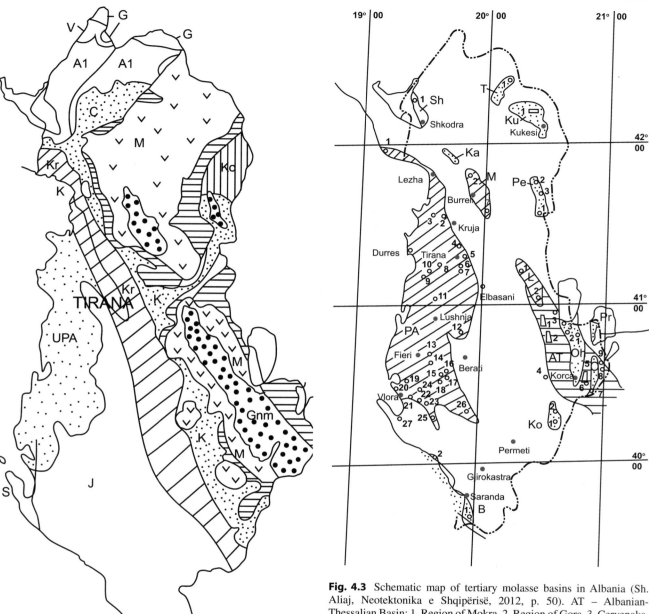

Fig. 4.2 Tectonic zones of Albania (according to Vranaj 2000). Ko – Korabi, M – Mirdita, G – Gashi, A – Albanian Alps (A1 – Malësia e Madhe Subzone, A2 – Valbona Subzone), V – Vermoshi, C – Cukali, Kr – Krasta, K – Kruja, J – Ionic, S – Sazani. Und – Intermountain Plain, UPA – Peri-Adriatic Lowland. Source: Mediaprint. Gjeografia fizike e Shqipërisë, 2019, p. 24

Fig. 4.3 Schematic map of tertiary molasse basins in Albania (Sh. Aliaj, Neotektonika e Shqipërisë, 2012, p. 50). AT – Albanian-Thessalian Basin: 1. Region of Mokra, 2. Region of Gora, 3. Çervenaka, 4. Voskopoja, 5. Mountains of Morava (Drenova, Drenica, Plasa); 6. Bozdoveci; 7. Mirasi; 8. Bilishti; 9. Bitincka; L – Librazhdi Basin: 1. Librazhd, 2. Golik; M. – the Mati Basin: 1. Klosi, 2. Rrësheni. Lake Basins of Pliocene and Plio-Quaternary: Ko-Kolonja Basin 1. Erseka, 2. Bezhani; D – Devolli Basin and Oh – Ohrid Basin: 1. Alarupi, 2. Çërrava, 3. Pogradeci; Pr – Prespa Basin; Pe – Peshkopia Basin: 1. Shupenza, 2. Kastrioti, 3. Peshkopia Ku-Kukes Basin: 1. Region of Has; T – Tropoja Basin 1. Tropoja. Pliocene marine Basins: Sh – Shkodra Basin: 1. Kopliku; Ka – the Kashnjet Basin. PA – Peri-Adriatic Basin: 1. Ulcinj, 2. Thumanë-Mamuras; 3. Ishmi; 4. Zall Heri; 5. Brari; 6. Mulleti; 7. Petrela; 8. Ndroqi; 9. Kavaja; 10. Helmësi; 11. Rrogozhina; 12. Kuçova; 13. Marinza 14. Patosi; 15. Ballshi; 16. Lavdani; 17. Usoja; 18. Hekali 19. Panaja; 20. Zvërneci; 21. Kanina; 22. Kropishti; 23. Gerdec. 24. Selenica; 25. Shushica; 26. Memaliaj; 27. Orikumi. B – Butrinti Basin: 1. Himara; 2. Butrinti

The western flanks of the anticline folds contact with the syncline folds through the reverse faults-thrusts of long extent which condition the asymmetry, which is also reflected in the expressed morphological asymmetry (Bega 1995). To the north of the large Vlora-Tepelena flexure, ionic anticline strings are submerged under the deposits of the Pre-Adriatic

Lowland. This is expressed by the transition from mountainous relief to hilly and plain relief (Qiriazi 2019).

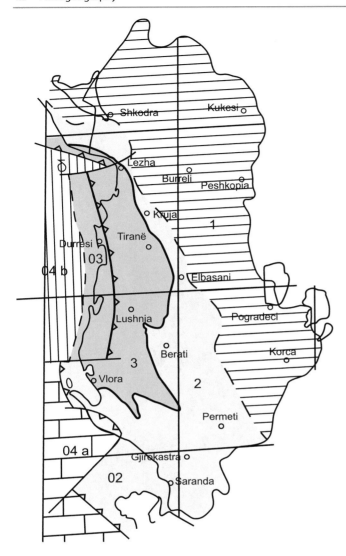

Fig. 4.4 Map of neotectonic zonation of Albania (Sh. Aliaj, Neotektonika e Shqipërisë, 2012, p. 101). Internal unit affected by extensional tectonics. 2. External unit affected by pre-Pliocene compressional movements (02 marine sectors). 3. Pre-Adriatic Lowland affected by post-Pliocene compressional movements (03 maritime sectors). 4. Platforms in the Adriatic and Ionian seas (04a Apulian platform, 04 b Albanian Basin). Transverse faults, which cut and displace the forehead of the Albanian orogen: α, the island of Othon-Dhërmi; β, the Gjiri i Ariut-Dukat; γ, north of Sazan Island; δ, Drini Bay-Lezha

deformations continue up to nowadays. In syncline structures, plains are formed, while in anticline structures, hills and hilly ranges. The thrusts and backthrusts are still active today, evidenced by the powerful earthquakes generated by them (Aliaj 2000).

– *The platform in the Adriatic and Ionian seas* includes the sectors in these seas, captured by vertical diving movements in the Pliocene-Quaternary, but also before. This platform is undeformed or poorly deformed with normal detachment and, in some cases, with wrinkles (Aliaj 2012).

4.5 Paleogeography

Today's landscape of the country is the result of long and complex geological development. Geological evolution dates back to the Paleozoic, about 500 million years ago (Aliaj 2012). The Korabi area was originated by Hercinian tectogenesis at the Middle-Upper Devonian boundary. But the main features of the country landscape were formed during the alpine cycle, which starts from the end of the Paleozoic to beginning of the Mesozoic and continues until the Pleistocene and to the present day.

This evolution is related to the subduction-collision mechanism, with which are associated the strong tangential movements, that created the folding, thrusting, and nappe setting of tectonic zones on top of each other. During the last period (Plio-Quaternary neotectonic period), in the country's hinterland, the earliest structure was fragmented by normal faulting, forming the horst-graben structure, while on its outer coastal part, thrusts and anticline structures developed, where hilly ridges and syncline structures, where, in turn, valleys and plains were formed (Aliaj 2012) (Figs. 4.6, 4.7, 4.8 and 4.9).

Related to this period is the formation of the present-day Mediterranean Sea. The Adriatic and Ionian Sea basins gradually took on the present appearance during the Pliocene (Ionian Sea) and in the Quaternary (Adriatic Sea), when their basins sank and the present configuration of their coast was outlined, the island of Sazani was separated from Karaburuni, and the Strait of Otranto was created, connecting the Adriatic with the Ionian.

The new Pliocene-Quaternary movements of today played the main role in the formation of the neotectonic structure and, in connection with it, also of the current relief, mainly the hilly-mountainous one. This relief was formed through the rising phases, interrupted by the flattening phase, until it took on its present appearance. In the Pliocene-Quaternary period, the present hydrographic network was formed gradually (Prifti 1984, 1996a, b). In the Quaternary, there were several glaciers, which modeled the heights of the mountains with glacial shapes.

The unit is crossed by large transverse dislocations (Borsh-Kardhiq, Vlora-Tepelena, Vermik-Bënçë, etc.), also currently active. The earthquakes generated during them testify this (Aliaj 2012) (Fig. 4.5).

– *The Pre-Adriatic Lateral Lowland Unit* lies between the Shkodra-Peja traverse fault and the Vlora-Tepelena flexure; it includes the hilly and plain territories of the molassic basin on the shores of the Adriatic Sea. Founded in the Middle Miocene and strongly affected by post-Pliocene compressional movements, it is deformed by folds, reverse faults, thrusts, and backthrusts. Compressional

Fig. 4.5 Neotectonic map of Albania (Sh. Aliaj, Neotektonika e Shqipërisë, 2012, p. 106)

I. Regions subjected to uplifting during the neotectonic stage: 1. regions subjected to uplifting since the Middle Miocene but more powerful since the Pliocene, with isolines of neotectonic deformation in m. 2. Molasse basins of the Lower Oligocene-Miocene (om1), subjected to uplifting since the Middle Miocene, with isolines of neotectonic deformation in m.

II. Regions subjected to immersion during the neotectonic stage: 3 Miocene molasse basins (from the Middle Miocene), subjected to post-Miocene uplifting (m); Mio-Pliocene molasse basins (from the Middle Miocene), subject to uplifting from the beginning of the Quaternary (m_2pq_1); 4 Pliocene marine or lake basins (p), subjected to Quaternary uplifting; Lower Pliocene-Pleistocene lake basins (pq_1), subjected to uplifting from the Middle Pliocene; Plocene-Quaternary basins (pq); 5. Quaternary basins (q); 6. isolines in m of Pliocene floor in some basins.

III. Active structural elements during the neotectonic stage and other elements: 7 the nappe boundaries deformed by normal faulting during the neotectonic stage; 8. reverse-thrust faults; 9. normal faults; 10. flexures; 11. faults and flexures determined by seismic data; 12. strike slips; 13. active diapiric domes; 14 inactive thrusts

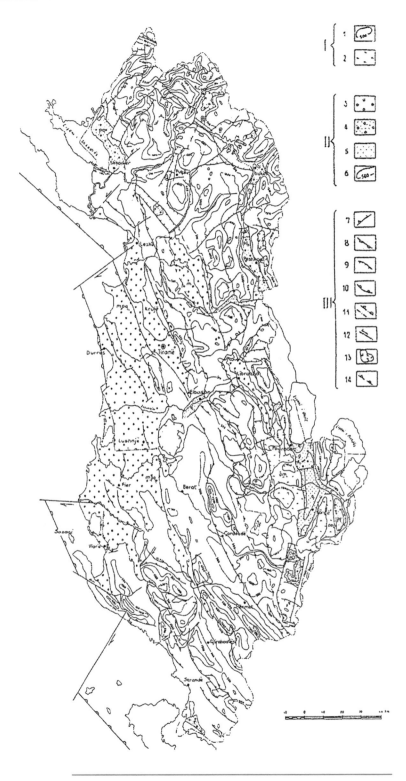

The evolution of geological terrain continues even today. The proofs for this are upward movements in most of the territory and downward movements in some grabens, and especially the marked seismic activity.

4.6 Seismicity of Albania

Albania is part of the Alpine-Mediterranean seismic belt, which lies at the contact between the lithospheric plates of Africa and Eurasia. In this belt, the Aegean and the surround-

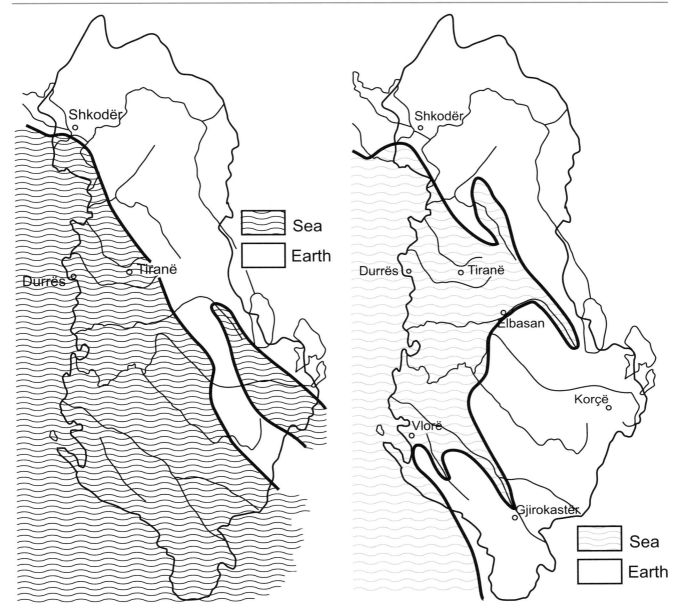

Fig. 4.6 Paleogeographical scheme in the Middle Oligocene (Aliaj and Melo 1987)
Source: Mediaprint P. Qiriazi Gjeografia fizike e Shqipërisë, 2019, p. 43

Fig. 4.7 Paleogeographical scheme in the Tortonian (Aliaj and Melo 1987)
Source: Mediaprint P. Qiriazi Gjeografia fizike e Shqipërisë, 2019, p. 43

ing territories (Greece, Albania, Montenegro, Kosovo, Northern Macedonia, Southern Bulgaria, and Western Turkey) are the most seismically active areas. The seismicity of Albania is mainly related to the contact of Adria with the Albanian Orogen, where deformations are constantly accumulated and active longitudinal and transverse tectonic detachments are set in motion, which, by interrupting this contact, enter the interior of the country and the Balkans. Seismic energy accumulated from these tectonic deformations is released along them (Aliaj 2012).

Albania is one of the most seismically active countries in Europe. But in its orogeny, there are three longitudinal areas (Ionian-Adriatic area, Drini (Peshkopia-Korca), and Shkodra-Mati-Librazhdi area) and three transverse areas of main seismic active disconnections (Shkodra-Peja area, Lushnja-Elbasani-Dibra Vlora, Tepelena). Most earthquakes are generated in these areas, especially strong ones (Muço 1995, 1999) (Fig. 4.10).

Historical records show that from third to second centuries B.C. and up until 1900, the country has been hit by 55

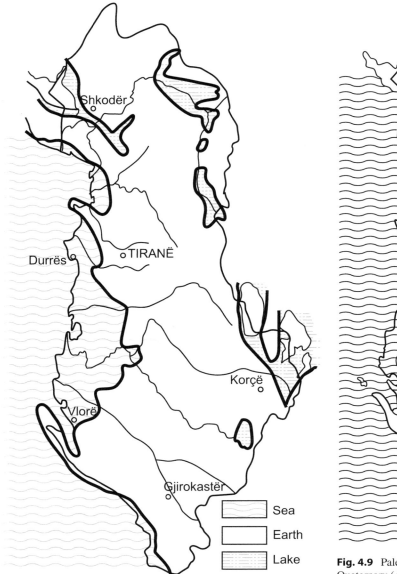

Fig. 4.8 Paleogeographical scheme in the early Pliocene (Aliaj and Melo 1987)
Source: Mediaprint P. Qiriazi Gjeografia fizike e Shqipërisë, 2019, p. 43

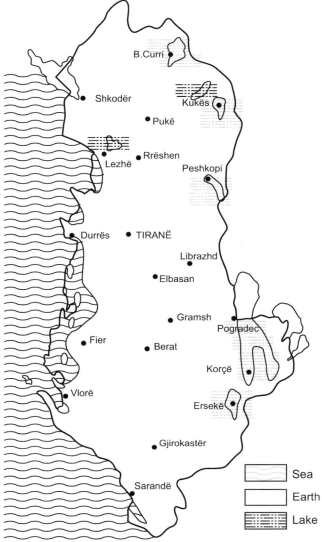

Fig. 4.9 Paleogeographical scheme of Albania at the beginning of the Quaternary (A. Papa 1990)
Source: Mediaprint P. Qiriazi Gjeografia fizike e Shqipërisë, 2019, p. 46

powerful earthquakes (magnitude 6.4) with intensities up to VIII–IX magnitude. These earthquakes destroyed ancient cities (Durrësi, Butrinti, Apolonia, etc.) and medieval ones (Durrësi, Vlora, Berati, Elbasani, Gjirokastra, Shkodra, etc.) (Aliaj 2012).

Over the past century, the country has been hit by several strong earthquakes, which have caused extensive damage and casualties to the people. Among them are the earthquake in Shkodra (1.06.1905); the earthquake in Lake Ohrid (18.02.1911), which claimed human lives and caused economic damage but also raised the level of the lake about 50 cm; the earthquake in Tepelena (26. 11. 1920), which devastated almost the entire city; and the earthquake in the Albanian-Montenegrin border area (15.04.1979) (Sulstarova and Koçiaj 1980), which caused casualties, thousands of destroyed houses, landslides, pseudovolcanic features, and rock falls (Kabo and Qiriazi 1985).

In 2019, the country was hit by three earthquakes: the earthquake on June 1, 2019, in the area of Korca; the earthquake in September 21, 2019, with epicenter on the coast, near Cape Rodon with magnitude 5.8 Richter; and the earthquake of November 26, 2019, the largest in the last 40 years (magnitude 6.3 Richter and intensity up to IX front). This earthquake was considered catastrophic, with the epicenter 38 km depth, 22 km, from Durrës, 30 km from

Fig. 4.10 Map of seismoactive zones in Albania with maximum expected magnitude (Sh. Aliaj, Neotektonika e Shqipërisë, 2012, p. 170). 1. Ionian-Adriatic, 2. Shkodër-Mat-Bilisht, 3. Peshkopia-Korça, 4 Shkodra-Peja, 5. Lushnja-Elbasani-Dibra, 6. Vlora-Tepelena

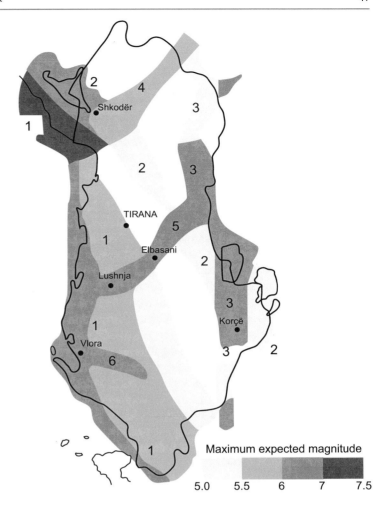

Tirana. It had severe natural consequences, causing an increase of the territory of Durrës area about 10 cm, and it also claimed human lives. It caused 51 victims and 2000 injured and affected over 200,000 people, while 10,000 inhabitants were left homeless. The housing damage was estimated at around 1 billion Euros (KMRSH 2020). Immediately after this great earthquake, great support and help came to Albania.

4.7   The Emergence of Man and His Role in the Environment

In Albania, the earliest traces of man belong to the Middle Paleolithic, from 100.000 to 30.000 years ago, as evidenced by the stone and bone tools found in Xara, Kryegjati, and Gajtan in Shkodra (Akademia e Shkencave e Shqipërisë 2002).

During the last Riss glacial geological period, the climate underwent major changes. At altitudes over 1000 m above sea level, ice appeared in the Albanian territory. This very cold climate was accompanied by a change in the flora and fauna of the country. Besides the plants and animals that could withstand the cold, animals typical of the cold climate appeared (like mammoth, hairy rhinoceros, bison, reindeer, bear, cave hyena, wild goat, etc.) (Akademia e Shkencave e Shqipërisë 2002).

In the struggle to cope with the very harsh conditions of the climate for survival, the man of this time took important steps toward progress: He took refuge in caves, dressed in animal skins, and learned to light the fire by himself, which he then used to protect himself from the cold and wild beasts and specially to improve the structure of food, etc. These eventually separated him from the animal world.

The people of the Middle Paleolithic lived on ready-made products, which they found in nature and especially through hunting. The hunting of large game (animals) could only be done collectively. This helped strengthen the internal ties between the members of the group and, together with the blood ties, organized the primitive community.

During the late Paleolithic (from 30.000 to 10.000 years ago), the population increased further, which extended over most of the country. The main means of food provision continued to be animal hunting and the collection of finished products in nature. During this time, the long and very complicated process of anthropogenesis ended (Akademia e Shkencave e Shqipërisë 2002).

Improving stone and bone tools increased the possibility of providing more food in bulk. This also brought about changes in the organization of the primitive community. More stable groups of people are now formed united by the common form of production, by blood ties, and by common origin. It was a gradual shift to matriarchy. This is because women play a key role in the social organization and social life of this gender community. The main form of family in this gender community was group marriage, and the origin of children was determined only through the mother lineage.

In the Mesolithic era (from 10.000 to 7.000 years B.C.), the climate gradually warmed up and took on the Mediterranean features of today. As a result, the fauna and flora changed: Pleistocene animals (mammoth, hairy rhinoceros, etc.) disappeared. The geographical distribution of plants and animals also changed. In this era, the Mesolithic man began to move from hunting and gathering ready-made products in nature to the beginnings of primitive agriculture and livestock. This marked the great step forward, which will make man an important factor in the environment, with his great changes. The evolution of human society continues with increasing intensity in subsequent periods, when the human impact on the environment increases (Akademia e Shkencave e Shqipërisë 2002).

4.8 Mineral Resources

Due to the marked lithological and structural diversity, geological position, and complex evolution, the country has numerous and diverse mineral resources: 61 types of minerals, most of them of industrial importance (Shërbimi gjeologjik Shqiptar 1999).

Useful metallic minerals: *Chromium* with wide distribution, especially in the eastern belt of ultrabasic massifs of the Mirdita tectonic zone (Bulqiza, Batra, etc.), with reserves over 57 million tons, continues to be extracted only in some of the many mines of the last century; *copper ores*, with reserves of over 69 million tons, are mainly associated with Jurassic ophiolitic volcanism, containing up to 1–3% pure copper; *iron-nickel and nickel-silicate ores*, with cobalt content and reserves of over 300 million tons (Aliaj 2016); *bauxite ores* among limestones; *manganese ores* mainly in Paleozoic effusive-sedimentary; *minerals of rutile, zircon, chromite grenades,* etc. in beaches; *polymetals* (lead, zinc, copper, etc.) in the Korabi and Gashi area; *rare metals*: vanadium, gold, platinum, mercury, titanium, uranium, etc., in limited appearance.

Useful nonmetallic minerals have magmatic and sedimentary origins, such as asbestos, talc, volcanic glass, albitophytes, quartz, magnesite, arsenic-realgar, fluorite, kaolin, sulfur, phosphorite, rock salt and gypsum, clay, sand, limestone and dolomite, travertines, trepels, etc.

Fuels (caustobiolite) are numerous: layered coals in molas of different ages, with calorific value up to 27.3 MJ/kg or 6500 kcal/kg and reserves over 800 million tons; *peat* in the former swamps (Maliqi, Tërbufi, Kakariqi, etc.); and *asphalt* (natural bitumen) and *bituminous sands*; *oil*, with reserves of over 437 million tons, is found especially in sandstones and carbonates, where, recently, large reserves have been discovered; *natural gas* as an accompanying oil and in the state of gas caps, its reserves are projected to reach 13.76 billion N m^3 (Aliaj 2016).

The level of economic development of the country does not correspond to the wealth and great variety of natural resources. By 1990, copper and chromium alone had given the country $2.28 billion, while the mining and oil industries provided about $450 million a year (Aliaj 2016). The degree of development of the country has been influenced by many other factors, for instance, political, human, technical, financial, organizational, and other factors, which have not been positive.

References

Akademia e Shkencave e Shqipërisë. Historia e Popullit Shqiptar. Shtëpia botuese "Toena", 2002

Aliaj Sh (2000) Active fault zones in Albania. XVII Gen. Ass. Of ESC, Book of Abstracts and Papers, 1 p. Lisbon, Portugal 10–13 September 2000

Aliaj Sh (2012) Neotektonika e Shqipërisë

Aliaj Sh (2016) Përparësitë konkurruese të Shqipërisë drejt BE-së

Aliaj I (2016) Përparësitë konkuruese të Shqipërisë drej Komunitetit Europian

Anderson H, Jackson J (1987) Active tectonics of the Adriatic Region (1987). Gjeophys. J.R, Astr. Soc. W, 91

Bega Z (1995) Sistemi mbihipës dhe kundërhipës në zonat e jashtme të Albanideve

Giese P, Reuter K (1978) Crustal and structural features of the margins of the Adria Microplate (1978). In: Apennines, Hellenides. Stuttgard (Cituar nga Sh. Aliaj, 2012)

Instituti i Studimeve dhe i Projektimeve të Gjeologjisë, Harta Gjeologjike e Shqipërisë (2002)

Kabo M, Qiriazi P. Pasoja gjeografike të termetit të 15 prillit, 1979, Studime Gjeografike 1/1985.

KMRSH (Këshilli i Ministrave të Republikës së Shqipërisë) (2020) Shqipëria vlerësimi i nevojave pas fatkeqësisë së tërmetit të 26 nëntorit 2019

Muço B (1995) The seasonality of Abanian earthquakes and crosscorrelation with rainfall. Phys Earth Planet Int 88(3–4):285–291. (Cituar nga Sh. Aliaj)

Muço B (1999) Monitoring of indivend seismicity in Albania, v. engineering Geology and Environment. eds. Marinos, P. et al. 1999

Papa A (1990) Zhvillimi paleogjeografik i Shqipërisë në pliocen dhe kuaternar (1990). Studime Gjeografike, 4/1990. Botim i Akademisë së Shkencave të Shqipërisë

Prifti K (1984) Kuaternari dhe veçoritë gjeomorfologjike në luginat e rrjedhjes së mesme të lumenjve Vjosë., Osum e Devoll

Prifti K. Probleme të formimit të tarracave lumore në vendin tonë (1996a). Studime Gjeografike, nr. 6. Botim i Akademisë së Shkencave të Shqipërisë

Prifti K. Neotektonika e gjeomorfologjia e luginës së Vjosë, Osumit, Devollit, Shkumbinit dhe Erzenit (1996b), Stdudime Gjeografike nr. 7. Botim i Akademisë së Shkencave të Shqipërisë

Qendra e Studimeve Gjeografke, Gjeografia fizike e Shqipërisë, vol. I, 1990

Qiriazi P. Gjeografia fizike e Shqipërisë., Shtëpia Botuese Mediaprint, 2019

Shërbimi Gjeologjik Shqipëtar. Harta metalogjenike e Shqipërisë, 1999

Sulstarova S., Koçiaj. The Dibra (Albania) earthquake of November 1967. Tectonophysics, vol. 67, nr 3–4/1980

Sulstarova S., Aliaj, Konvergjenca e sotme e orogjenit të albanideve me pllakën Adriatike, kufiri dhe funksionimi i saj. Buletini i Shkencave Gjeologjike, 2/1990

The Relief

5

Perikli Qiriazi

Abstract

Due to the action of structural and lithological factors, differentiating tectonic movements and different climatic conditions, the relief has undergone a very complex morphotectonic and morphoclimatic evolution, and during this process, it has been flattened (peneplanation) and again renewed; climate that favors erosion (rhexistasy) is combined with climate that favors rock alienation (biostasis); the hydrographic network was formed, which has been reorganized several times. These elements conditioned the mainly hilly-mountainous character of the relief with marked morphological and morphogenetic diversity (the type of structural, karstic, river, glacial, erosive, coastal relief), making the relief the basic component of the geographical landscape. The country is facing the process of desertification of landscapes and the crisis of beaches.

Keywords

Morphotectonic and morphoclimatic evolution ·
Pretortonian peneplain · Differential neotectonic
movements · Horstanticlinear structure · Structural relief
· The karst relief · Desertification of landscapes · Crisis
of beaches

5.1 Morphotectonic and Morphoclimatic Evolution of Relief

There are few studies on morphotectonic and morphoclimatic evolution of relief. These studies help determine the cycles and stages of morphotectonic evolution of the relief, of several morphosculptural stages, and, in general, the age of the country's relief (Qiriazi and Kristo 1987).

The Albanian relief has been flattened (peneplenized) and has been renewed again. Each stage of relief development begins with elevation, i.e., with the emergence of the terrains above water, due to the tecto-orogenic compressive phases, during which the relief is formed. Then, they are followed by morphogenic phases, during which the relief is flattened, which is then accompanied by a crust alienation. Flattening occurred when differentiating tectonic or neotectonic movements had relative calm (Aliaj and Melo 1987; Melo 1996).

After the flattening, strong tectonic uplifts resume, which mark the beginning of a new tecto-orogenic phase, during which the amplitude of differentiating tectonic movements increases to the maximum. This leads to the restoration of flattened relief, which enlivens the erosion, which breaks the morphology of peneplain and, in many cases, destroys it (the remains of the old peneplain are preserved only in buried form under the deposits of newer marine transgressions), while the remains of the new peneplain are also stored in watersheds in the form of flat surfaces, which are not related to the geological structure (Pumo 1990; Qiriazi 2019).

Based on the deposits and mineralization of iron-nickel, kaolin bauxites, and some residues in the morphology of the current relief, some flattening surfaces have been determined, from which are distinguished: flattening surfaces of pre-Cretaceous and Cretaceous age, buried and evidenced through mineralization crust of iron-nickel and bauxite. But the largest and most well argued is the paratortonian flattening surface, which marks the beginning of a new morphostructural evolution. At the end of the Miocene, the tectonic deformations of the interrupted structures by the peneplain in question resume. This marks the beginning of the formation of today's relief on this flat surface. These deformations, which continue even today, tore the peneplain. Its remains are preserved in the form of flat ridges at different altitudes, from 400–600 m to 1800–2000 m (Korabi), and covered by tortonian molas (Pre-Adriatic Lowlands) (Aliaj and Melo 1987; Aliaj 2012).

As it was said, in the tecto-orogenic phases of the neotectonic period, the upward movements of the structures of the overwater territories at that time prevailed. The amplitude of these rises has steadily increased from 0.1 mm/year (Middle and Late Miocene) to 0.3–0.5 mm/year (Late Pliocene-Quaternary). As a result, the height of the relief increased from 200–300 m (beginning of the tortonian) to about 1000 m (end of the Miocene). In continuation and especially during the second half of the Pliocene, the structures of the continental territories were engulfed by strong upward movements, which led to the increase of relief heights up to 1500–2000 m, at the end of the Pliocene, and up to over 2000–2500 m in the Quaternary. The last phase of uplifts continues even now (Aliaj and Melo 1987).

During this time, because of the interplay of tectonic uplift movements with their calming period, there have been several successive cycles of erosion and deposition along the river valleys. This combination brought the formation of several terraced levels in the valleys of the rivers Vjosa, Devolli, Osumi, Mati, etc. Climate change has also affected the formation of terraces (Melo 1961, 1964; Prifti 1984, 1996a, b).

Even in the Quaternary, there were sectors that were submerged, where large graben pits were formed, which were filled with water, thus creating lakes. In the graben not compensated by the deposits, the lakes continue to exist today (Shkodra, Ohrid, Prespa, Butrinti), while in the other graben compensated by the deposits (Korça, Bilishti, Përrenjasi, Kolonja, Kukësi, Iballa, Tropoja, etc.), the lakes were dried up. In most of them (Prespa, Kolonja, Tropoja, Kukësi, etc.), the downward movements were replaced by the upward ones. Consequently, the relief of the holes formed in these grabens was fragmented by the hydrographic network. In the other grabens (Korça, Përrenjasi, Delvina), the downward movements continued during the quaternary and continue even currently. Therefore, the holes formed in these grabens have a completely flat bottoms, thus taking the form of plains (Qiriazi 1985; Aliaj 2012).

Throughout this, climates that favor erosion (rhexistasy) are intertwined with climates that favor rock alienation (biostasis). The existence of laterites and faunal and morphological remains indicate that the climate during this morphotectonic evolution has changed from tropical to glacial cold. Each has conditioned a special modeling of relief: remnants of tropical karst, glacial forms, etc. (Qiriazi 2019).

The Mediterranean climate eventually stabilized as such after the Pleistocene. This is considered one of the most "aggressive" climates, associated with the reverse temperature rise with the most rainfall, often with great intensity, falling during the cold period of the year, when most of the vegetation is leafless, etc. Large modeling of the post-Pleistocene relief, in the conditions of the Mediterranean climate, brought about numerous changes in its forms (Qiriazi and Sala 1998).

5.1.1 Formation of Hydrographic Network and River Valleys

Under the influence of structural and lithological factors, various differentiating movements, but also the climatic conditions of the period from the end of the Miocene during the Pliocene and Quaternary, the hydrographic network, and its valleys, were formed. During the Miocene, massive water flows were dominant. This network began to form after the formation of the pretortonian peneplain, when the upward movements were activated. The earliest (probably the end of the Miocene) should be the Vjosa valley (the upper sector) and the Shushica valley (up to the lower sector). From the end of the Miocene to the beginning of the Pliocene, the Tortonian lakes of Mati and Shkumbini were dried up, and the valleys of Mati and the Upper Shkumbini were formed. But still in the Pliocene, water flows traverse shallow river valleys. These flows run mainly into inland Pliocene lake basins (Aliaj 2012).

During the Plio-Quaternary, because of large tectonic subsidence, the redistribution of the hydrographic network occurred inland. The largest river in the country, the Drini, has the youngest (quaternary) valley. This is evidenced by the placement of river terraces in the extension of Peshkopi on the lake deposits of the Lower Pleio-Pleistocene (Melo 1964).

The strong upward movements of the Quaternary gave rise to intense erosion at depth, which led to the continuous deepening of the valleys. At this time, especially from the end of the Quaternary, the lower sections of the rivers were formed, across the Western Lowlands, which gradually rose above the water. This is evidenced by the immediate turns of the main rivers from the southeast-northwest direction to the west and the formation of transverse valleys, in the form of narrow gorges, between their upper and lower sectors. In this way, the hydrographic network and its valleys gradually took on their present appearance (Prifti 1984).

The process of relief evolution continues even now: The differential neotectonic movements continue with great amplitude, which bring about not only morphological changes, but they also condition to a large extent the character and external modeling of the relief, in which the lithological composition and the Mediterranean climate play a major role.

The current modeling of the relief is the result of the continuation of the action of some morphogenetic processes of the previous periods of its evolution and the activation of some new processes, related to the current physical-geographical conditions. Based on the current dominant processes of relief modeling, in Albania, the following are distinguished: (1) *Mountainous areas*, with high density and depth of relief fragmentation, very steep slopes, and heavy rainfall (up to 2000–2500 mm/year), are modeled mainly from river-torrential and cryonival processes, slides, and slow movements; (2) *pits*, the structural basis of which is in the current elevations (the Great Prespa, Kolonja, Mati, Tropoja, etc.), are modeled mainly by deep erosion of running water and surface rinsing; (3) *plains*, the structural basis of which is currently decreasing (Korça, Përrenjas, Delvinë, etc.), are modeled mainly by accumulation; (4) *hills*, especially built of terrigenous, are distinguished by the predominance of river-torrential processes, slides, rinsing surface, etc; (5) *Western Lowlands* is where the accumulation processes of the rivers that flow through it dominate. Accumulation prevails on the low coast, abrasion on the high elevation, while karst predominates on the territories built of soluble formations (Qendra e Studimeve Gjeografike 1990; Qiriazi 2019).

5.2 Morphological Features of the Relief (Morphography and Morphometry)

Due to the uneven geological structure and complex morphotectonic and morphoclimatic evolution, the relief is distinguished by a great variety of shapes. As a result of the evolution of the terrain and relief from east to west, in this direction, the transition from the mountains to the hills and the plains is made, and the contrasts and variety of its forms are reduced.

Hills and mountains sometimes take the form of ridges, sometimes clusters or highlands, and sometimes the form of isolated hills and mountains or the high mountain blocks. Along with the predominant northwest-southeast direction of the hilly and mountainous ranges and many valleys, inland and in the north of the country, there is a deviation from this direction characteristic for the western part of the Balkans. Among the most important deviations are distinguished: (1) *arched shape* (Korabi range), which reflects the arched shape of the horstanticlinear structure, where this string is formed; (2) *radial distribution of mountain ridges and valleys* (especially in the Albanian Alps), associated with differentiating neotectonic uplift movements, with greater amplitude in the center of the horstic structure, accompanied in the surroundings with large subsidence of graben pit; and (3) *the north-south direction of the mountain ranges* that meet inland and are more related to the new Plio-Quaternary detachment tectonics, with a predominant north-south direction (Qendra e Studimeve Gjeografike 1990; Qiriazi 2019).

The valleys are distinguished for their large width, up to the shape of pits in sectors developed in the former lake pits (Mati, the extensions of the Black Drini, etc.). Trough-shaped glacial valleys are also found (Valbona, Boga, etc.). The valleys are generally asymmetric, especially the subsequent and contact ones. The larger, more developed valleys are on the right bank. This is related to the greater amplitude of the upward movements of the left-wing structures, which has shifted the river to this part, and to the greater develop-

ment of the right tributaries, in the estuaries of which larger deposition cones have been formed, which have pushed the main river from the left (Qendra e Studimeve Gjeografike 1990; Qiriazi 2019).

Fields have different shapes and sizes. To the west, in a north-south direction, lies the largest lowland in the entire western part of the Balkans, thus constituting a specific morphological feature. Inland, the plains are smaller and are associated with Plio-Quaternary tectonic subsidence or more developed river terraces. Flat-bottom plains meet here with pits with fragmented relief.

– *Hypsometry.* The most characteristic hypsometric feature of the relief is the distinct hilly-mountainous character. Elevations range from cryptodepressions (Tërbufi – 8 m) to 2751 m (Korabi). The large hypsometric amplitude (above 2750 m) testifies of large differences in the intensity and direction of neotectonic movements of geological structures, their geological construction, as well as relief modeling processes. This large amplitude has conditioned the vertical landing of climatic, hydrographic, soil, and vegetation features and consequently of relief modeling.

According to the data of the hypsometric map and the hypsographic curve of the country, it turns out that 23.4% of the territory is located at an altitude of up to 200 m above sea level, 48.1% extends from 200 to 1000 m and 28.5% over 1000 m, while altitudes above 200 m occupy 76.6% of the territory. Most of the mountains are up to 2000 m high, and only 1% of the country territory is above this altitude. The average height of the country is 708 m, twice the average height of the European continent (Qendra e Studimeve Gjeografike 1990; Qiriazi 2019).

– *The density of relief fragmentation is distinguished by* the predominance of medium values (2–3 km/km2) and large values (3–5 km/km^2), which is associated with the large spread of soft formations, with the predominance of upward movements in most of the territory, with the Mediterranean climate, and with human influence. Values of this density range from 0.1 km/km^2 to over 8–10 km/km^2. Territories with high density and very high relief fragmentation also had the highest rate of erroneous intervention in the environment (especially on vegetation and soils) by the rather dense population prior to 1990 (Qiriazi 2019).

– *The depth of relief fragmentation is distinguished by* the predominance of average values (100–300 m/km^2), but its values range from about 0 m/km^2 to over 600–700 m/km^2. This is related to the large extent of the mountainous relief, highly fragmented by the hydrographic network, which, not infrequently, intersects the mountain ranges, with the spread of strong formations; relates to the large amplitude of neotectonic movements of their structures, etc. (Qiriazi 2019).

Among other factors, the predominance of medium and large values of density and depth of relief fragmentation and the rather differentiated distribution of their values, conditioning the great potential energy of landscape degradation processes.

5.3   Morphogenetic Features

Although the relief is relatively young, its evolution and morphogenetic features vary from area to area. The types of structural, lithological, river, glacial, erosive, and coastal relief can be distinguished.

5.3.1   Structural Relief

This is related to the lithological composition and to the wrinkling and detaching structure (scaly and block). In the

neotectonic period, tectonic movements had a regime in oppression, dominant in the outer region, and in retreat, dominant in the inner territory of the country. These neotectonic movements, even of today, reorganized the structures formed during the tectogenic phases.

This reorganization was more powerful in the interior, where the wrinkled geological structures were restructured and took on a predominantly horst-graben block shape, with scaling to the west and a mainly north-south direction. For these reasons, today's relief in this area is more related to the detached structure than to the wrinkling one (Fig. 5.1).

In the outer area, monoclinal type over-sliding and overlying structures were formed. The new structure of this area was developed as a legacy of the older structures. Therefore, its current relief is more related to the wrinkling structure than to the detached one, while the morphology of the relief is almost completely consistent with the structure (Fig. 5.2).

Among the forms of structural relief modeled in detached structures are distinguished: (1) monoclinic ridges, formed especially in scaly and covering structures, mainly limestone; (2) contact and tectonic depressions, associated with

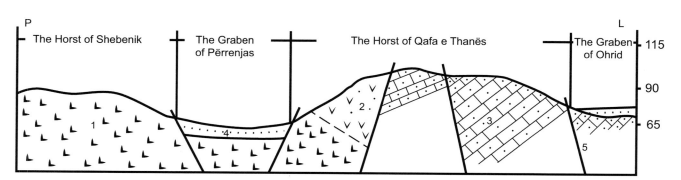

Fig. 5.1 Geomorphological profile in the horst-graben structure of the Southeast Pits (Geology Sh. Aliaj 1985, Source: Mediaprint P. Qiriazi Gjeografia fizike e Shqipërisë, 2019, p. 70). 1. Ultrabasic. 2. Volcanoes.

3. Triassic-Jurassic limestones. 4. Quaternary deposits. 5. Tectonic detachments

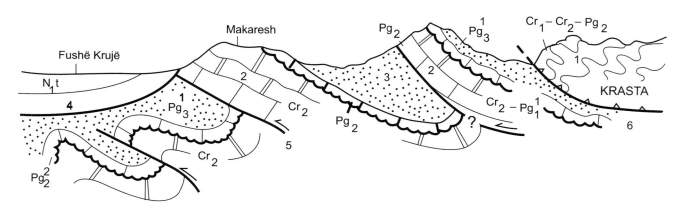

Fig. 5.2 Geological profile in the outer area, Kruja and Krasta area (V. Melo, Sh. Aliaj, 1987. Source: Mediaprint P. Qiriazi Gjeografia fizike e Shqipërisë 2019, p. 71). 1 Old flysch; 2 Limestones, 3. New flysch; 4 Molasse deposits, 5. Tectonic detachments; 6. Owerthrust

lithological diversity and overlying detachment structures, which bring rocks in contact with dissimilar strengths, as well as large tectonic subsidence, where new (Plio-Quaternary) graben are distinguished, as the most large tectonic system with complete structural reconstruction of the Ohrid tectonic system, where the Plio-Quaternary graben forms a large chain with north-south direction (Kolonja, Korça, Ohrid, the Black Drini, Kruma), the graben system, Prespa-Devoll, and in the west the graben of Përrenjasi; (3) structural thresholds, which are related to neotectonic differentiating movements and separate graben pits from each other; (4) horstic mountain and hilly ridges formed mainly in the horses and semi-horses of the inner neotectonic territory; (5) structural scales, related to the sliding of gravitational blocks, caused by normal tectonic fractures, but also meet in monoclinal structures, especially molas; (6) tectonic, tectonic-lithological, and lithological contact valleys, which meet in grabens and tectonic faults; (7) fracture junctions and fracture lines, along the supernatant and normal tectonic faults, which are connected to the monocline ridges and accompany the graben pits (Qendra e Studimeve Gjeografike 1990; Qiriazi 2019).

Among the forms of structural relief modeled in wrinkled structures are distinguished: (1) *quests* formed in monocline structures, especially in neogenic molas basins, where there is a combination of layers with different strengths; (2) anticline mountainous and hilly valleys, formed in anticline structures, with greater extension in the outer tectonic area; (3) syncline mountain and hilly ranges, which express discrepancies between morphology and syncline structure, where they are formed, related to the composition of syncline nuclei from erosion-resistant formations, while block tectonics gave the form of horst-synclinal, with which the morphology is related there in the form of plateaus and highlands; syncline valleys, expressing complete correspondence between relief morphology and syncline structure; (4) anticline valleys, with limited extent, express discrepancies between relief morphology and anticline structure; (5) syncline holes formed mainly behind the overhang front of the Mirdita tectonic zone (northwestern sector); (6) syncline plains, extending almost entirely in the Western Lowlands, where anticline hilly ranges intersect with syncline plains; (7) structural surfaces formed mainly in the backs of quests, monocline ridges, and Cretaceous limestone cover; (8) numerous penetrating gorges of still controversial origin but possessed by the thought of their antecedent genesis. There are attempts to distinguish inherited and antecedent transverse gorges, epigenetic, rare, and small and lake crossing (the Black Drini gorges) (Qendra e Studimeve Gjeografike 1990; Qiriazi 2019).

5.3.2 Karst Relief and Ecosystems

Soluble formations (limestone, evaporite, etc.) make up about 23% of the territory (Fig. 5.13). This factor, among others, has conditioned the large spread of karst, which accounts for the karstic landscapes and ecosystems (Fig. 5.3) (Kristo and Krutaj 1987; Krutaj 1994; Qiriazi et al. 1999).

The karst is of the complete type, with all its forms, ranging from the initial ones (limestone pavements) to the karst fields to the surface and underground forms; it is of the Mediterranean type, stripped of insulating materials and of polygenetic and polycyclic type, mainly driven by tectonics. It developed in two main phases: the first one begins with the emergence of the territory over water and continues until the Pliocene epoch, while the second one from the Pliocene epoch to the present day (Qiriazi 2019) (Photos 5.1 and 5.2).

The shapes created in the first phase constitute the paleokarst, as evidenced by the mineralization of bauxite in the old karstic voids (the Alps, the Dajti, Mali i Thatë, etc.), terra rossa (red soil) and karst caves with several floors, evidenced hills, etc. The main morphology of present-day karst is related to the second phase, when the karst underwent great development, due to the large tectonic uplift movements and the transition from the hot and dry climate of the Messinian to the wet and cold climate of the Pliocene and especially after the lowermost Quaternary (Kristo and Krutaj 1987; Krutaj 1994).

The elevation of the carbonate blocks conditioned the penetration of rainwater into the karst fissures. As a result, small and dense karstic surface forms were preserved, such as those in "Mali me Gropa" (the Pitfall Mountain), the Oroshi Mountains, etc. Under these conditions, it is the underground karst that is developed the most, and following the continuous decrease of the basic level, it descends in the depth from one floor to another. The surface karst took precedence in the limestone territories of low altitude above sea level. This karst constantly expands the elementary karstic forms into the karst fields (Fig. 5.4).

The same phenomenon is observed in the territories composed by evaporites at different altitudes above sea level. In the evaporites P-T1 of Korab, at an altitude of over 1000 m above sea level, there are developed karsts, where funnels and wells predominate, while in Dumre, at an altitude of about 300 m, large karst forms predominate. Karst lakes speak of karst maturation, which is reactivated on hill ridges, less often in pits and at the bottom of lakes, where suddenly wells are formed. Karst caves are preserved in the new (Messinian) evaporites of the Mëngajve massif (in Kavaja), consisting of large, pressure-resistant crystals (Fig. 5.5).

The most suitable karst conditions are found in the interior and eastern part, as compared to the western and especially southwestern part, with greater and longer drought. The most developed karst forms are at the altitudes of 600–2000 m, where heavy rainfall falls, especially in the form of

Fig. 5.3 Location of karst
rocks and of the main karst
springs of Albania Karst
springs of Albania and their
management. Acta
Geographica Silesiana, Wno,
Sosnowiec, 2019, p. 3
(electronic version)

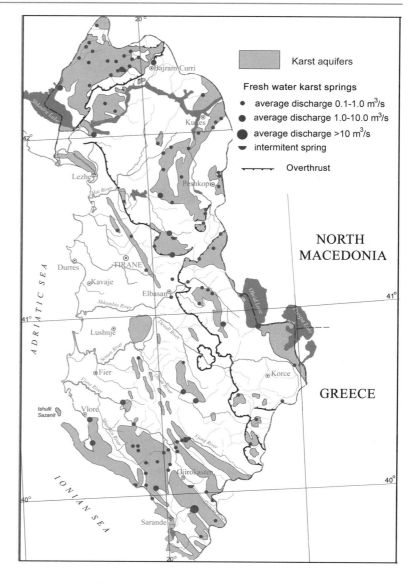

Photo 5.1 Lapies karstiques
on Mount Pashtrik (Qiriazi
2015)

Photo 5.2 Karst landscape in mountain with holes (Mali me Gropa), (Plaku 215)

Fig. 5.4 Diagramblock on the mountain with holes (Mali me Gropa) (Prifti 1985). Source: Mediaprint P. Qiriazi Gjeografia fizike e Shqipërisë 2019, p. 400

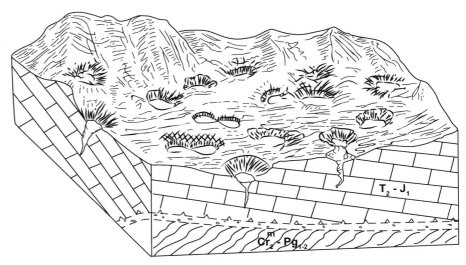

snow, and the temperatures are low. Above altitudes of 2000 m, the period of frost lasts longer. Below altitudes of 600 m, the thickness of insoluble deposits increases, to which the covered karst is connected. At altitudes of 1400 m, there are nivokarst and glaciokarst forms (Qiriazi 2019).

Among the forms of surface karst relief are distinguished: (1) limestone pavements and funnels, with density up to about 50 funnels/km², formed by solution (shallow and regular) and by the collapse of underground karstic cavities (deep, irregular); (2) doline and uvala, which are larger along large tectonic faults; (3) karst plains (polies) that are complex and larger forms (up to tens or more km²), formed in graben or syncline pits, pits, or by the union of several uvalas; (4) canyons formed by the collapse of the gap ceiling and by the erosive and karstic activity of the rivers; (5) death or (steephead) blind valleys, which lose flow in ponors, etc. (Kristo and Krutaj 1987).

Among the forms of underground karst relief, karst caves with large dimensions are distinguished, especially when they are formed along tectonic faults or at their intersection. Exploration and study of caves began after 1990, with the help of foreign speleological groups, especially Italian ones. So far, the largest known caves are found in the Albanian Alps: the Black Cave (or the Qireci Cave), 6.3 km long, the largest in the country; the Puci Cave, 5 km long, with five gallery floors; the Shtara Cave, about 5 km long, with several galleries, stunning beauty of concretions and a large colony of bats; the Ice Cave, with ice, maybe fossil, since the last ice age, etc. Interesting caves have also been discovered in the gypsum massif of Mëngaj (in Kavaja), etc. (Fig. 5.6).

The most studied one is the Black Cave (Pëllumbasi Cave, 360 m long), near Tirana. Here are found the skeletons of cave bears (Ursus spelaeus), which lived from 10,000 to 400,000 years ago, but also traces of human culture of the

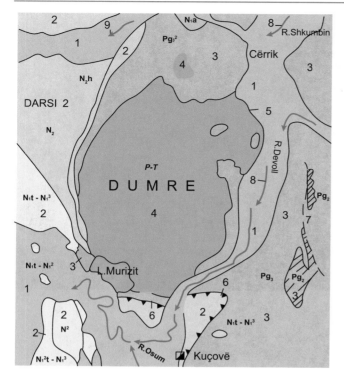

Fig. 5.5 Diapir of Dumre (Aliaj and Melo 1987, Source: Mediaprint P. Qiriazi Gjeografia fizike e Shqipërisë, 2019, p. 550). 1. Quaternary deposits. 2. Molasse. 3. Flysch. 4. Evaporite. 5. Tectonic boundary. 6. Overthrust. 7. Assumed tectonic detachment. 8 and 9. River

period from the Neolithic to the early Middle Ages. These elements make the cave rare in Europe and in the world. The project for its conversion into a geological-geographical, paleontological, and archeological tourist museum has been prepared.

The Faentino speleological group (Italy) discovered the Steam Well (in Gramshi), the deepest known in the country (probably 500 m deep). There are also other wells: Jack Mark's Well (in the Albanian Alps), 234 m deep; the Kakruka Well (in Tomorri), about 230 m, etc. (Figs. 5.7, 5.8 and 5.9; Photos 5.3 and 5.4)

5.3.2.1 Karst Ecosystems

They are related to the spread, development, and dynamism of the karst. They consist of two floors (surface and underground), which are connected and communicate through water, air, and materials brought by them. The physical, biological, and human components of karst ecosystems have distinct characteristics (Qiriazi et al. 1999).

The relief is distinguished for a variety of closed forms, without superficial flow, often quite deep and with great contrasts, with flaps, gaps, funnels, wells, sharp ridges that make movement difficult, etc.

Surface hydrography is poor, with temporary flows, running or going dry along the bed, swelling suddenly, or drying

instantly. Although in the catchment area of (the Dry Stream) the Përroi i Thatë falls the largest amount in the country (over 3100 mm/year), this stream is dry even in the wet period of the year. It has water only during long and intense rains, when it even floods suddenly (November 2010, etc.). There is great water wealth inside the karst massifs, and it emerges in the form of large springs in the massif surroundings, where karst rocks contact impermeable rocks (Mountain with Pits, etc.).

In karst ecosystem soils, pedogenetic processes operate in conditions of good aeration and scarcity or lack of water in hot and dry summers, while in the wet period of the year, great erosion degrades them. Therefore, they are often shallow and incomplete. Non-aerated soils have good physical, chemical, and biological qualities.

Vegetation in karst ecosystems with low erosion has very good development, up to forest formations, while in karst regions with high erosion activity and poor soils, vegetation is xerophilous and underdeveloped. But the steep limestone slopes and karst forms provide habitats for many special plant species, endemic and subendemic ones, such as *Wulfenia baldaccii*, *Carex markgrafi*, *Crepis albanicum*, etc. At the same time, these karst forms provide ecological niches – habitats and reproduction grounds – for many species of animals, especially birds that nest on steep limestone slopes (*Tichodroma muraria*, *Sitta neumayer*, etc.). Some animals use both floors of karst ecosystems (bats, bears, moths, reptiles, etc.). In the extreme living conditions of the caves, there are species of endemic fauna, still little studied.

Some caves have been inhabited by humans since prehistory, like the Xarra (Paleolithic), the Treni Cave (the Little Prespa), the Neziri Cave (Eneolithic), Zeza (Black) Cave (Neolithic-Early Middle Ages), etc. (Gjipali et al. 2015). The features of karst ecosystems condition their special human use and management: Often, the distribution of inhabited centers is related to karst water resources and the best soils, which are found in pits and karst fields. They have a special structure and architecture: solid limestone constructions, etc. Scarce agricultural land, in the form of small, scattered plots, is protected by a system of terraces with dry stone walls (without mortar), characteristic of Mediterranean karst territories. In the conditions of the climatic drought from the karst, drought-resistant plants (olives, vineyards, etc.) have been cultivated since antiquity, and a special infrastructure has been built for water supply (aquifers and canals protected from water infiltration in the karst cracks).

Special landscapes, attractive and rich in natural monuments; special offer for the construction of the traditional, attractive infrastructure of settlements and tourist centers; large groundwater reserves, and the like make for significant values of karst ecosystems. Meanwhile, their problems are numerous: The complex circulation of karstic waters makes it difficult to find sources of pollution of these waters, the

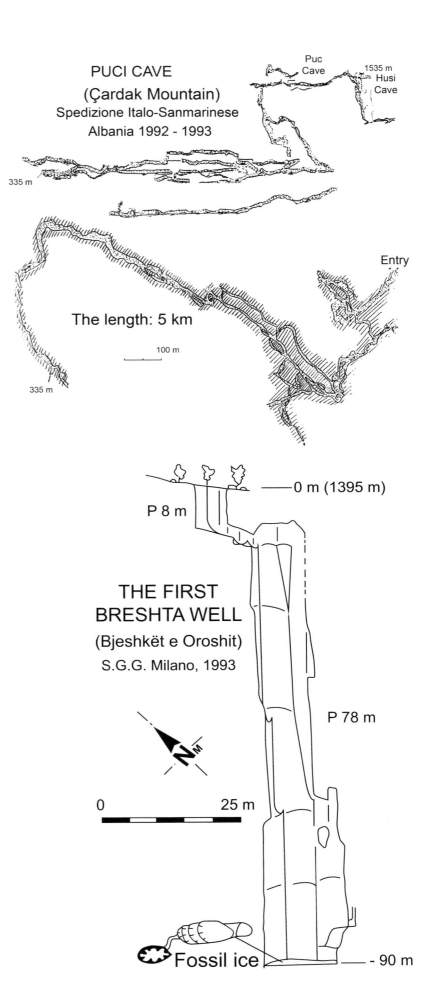

Fig. 5.6 Layout of the Puci Cave (Spedizione Italo-Sanmarinese, 1993, Source: Mediaprint P. Qiriazi Gjeografia fizike e Shqipërisë 2019, p. 85)

Fig. 5.7 Layout of the First Breshta Well (Milan 1993, Source: Mediaprint P. Qiriazi Gjeografia fizike e Shqipërisë 2019, p. 86

Fig. 5.8 Layout of the
Pirrogoshit Cave, Skrapar
(G. S. Puglia Grotte e Dauno,
1996, Source: Mediaprint
P. Qiriazi Gjeografia fizike e
Shqipërisë 2019, p. 86)

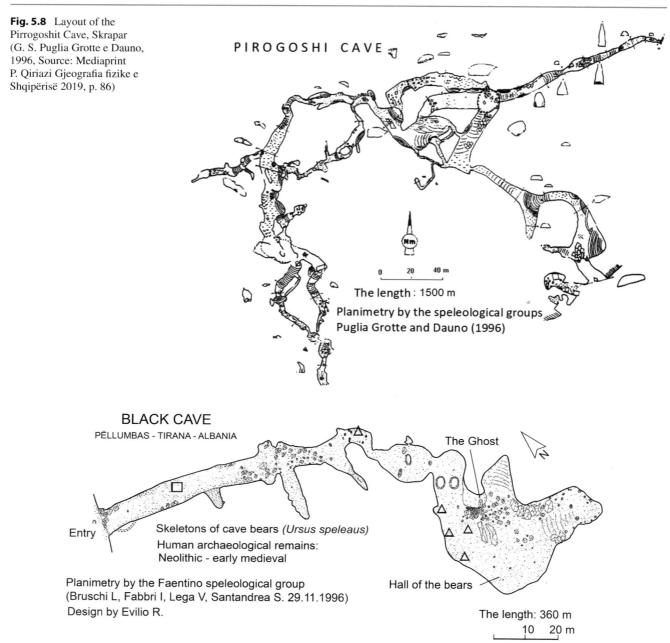

Fig. 5.9 Layout of the Black Cave (Pëllumbasi) (Faentino 1995, Source: Mediaprint P. Qiriazi Gjeografia fizike e Shqipërisë 2019, p. 87)

difficulties for the construction of infrastructure, especially the necessary irrigation, etc.

5.3.3   River Relief

At the country's latitude and climatic conditions, running water is the main modeling factor of the relief. Even in the past, except for the short glacial periods, river activity has been a major factor in modeling relief. The character and intensity of this activity were conditioned by (i) differentiat-

ing neotectonic and modern movements; (ii) tectonic faults; (iii) the formation of the Adriatic and Ionian Sea basin during the Plio-Quaternary, when the eastern territory returned to the mainland; (iv) lithological composition; (v) Mediterranean climatic conditions (Qiriazi 2019).

The rapid evolution of the dense river network was accompanied by the disruption of the internal hydrographic network and the formation of the external network and formation of penetrating valleys and gorges, with the formation of river captures, which brought about the distribution and

Photo 5.3 Farriti Cave
(Cmeta 2013)

redistribution of the hydrographic network and with the formation of river relief elements.

Among the forms of river relief are distinguished:

- River valleys differ from their origin, stage of evolution, their formation along or across geological structures, with lithological construction and unequal regime of neotectonic movements, etc. Related to this is the polygenetic character of the valleys and their sectors (Photos 5.5 and 5.6).

- River terrace levels indicate several cycles of erosion. In some valleys (Vjosa, Osumi, Devolli, Shkumbini, Mati, Black Drini), four to five levels of river terraces have been distinguished, mainly erosive accumulating, with different relative heights. The greatest development is concentrated in the middle sector of the river valleys (Figs. 5.10, 5.11 and 5.12) (Carcaillet et al. 2009).

Cultivated lands and inhabited centers are located on the river terraces. They also have great importance in the search for useful minerals (magnesites, etc.), building materials, groundwater, etc. Therefore, the river valleys are among the most populous relief forms in the country.

- River sandstones (rivers) have started to form since the Holocene. They are widely developed in plain regions and valley expansions.
- Meanders are quite developed. There are two types of meanders: the ones in the valley of most rivers in mountainous and hilly areas and the meanders of the alluvial plain, in the Western Lowlands. Their very rapid evolution is expressed in their migration to the west and especially in the interruption of two neighboring meanders.
- Deposition cones were formed especially after the last glaciation, when they were eroded and transported and solid materials were deposited in river estuaries, and at

the foot of mountains, in the vicinity of pits and plains. Many of them, joined together, form glacis of different types (slope glacis, stream glacis, and cone glacis). In some cases, they have taken the form of an intermountain piedmont (Cologne pit).

Stabilized cones offer surfaces of small inclined sloping, fast water drainage, good and ventilated soils, water sources, etc. Therefore, they are used for agricultural land, but also for residential centers. However, in unusual meteorological cases, there have been torrential floods, in the form of mudslides and rock leaks with great destructive force causing economic damage or even casualties.

5.3.4 Glacial Relief

Glacial relief has limited extent. It was modeled during the Quaternary, when, at altitudes above 1200 m, mountain and valley ice appeared. Its traces belong to the last two ice periods, Riss and Würm. During the Vyrmian period, in the Balkans, the largest were the Lim Glacier, which began in the Vermosh valley, and the Valbona and Boga Glacier (Almagia 1914; Palmentola and Baboçi 1994).

The Albanian Alps are distinguished for their large distribution of glacial forms, due to their northernmost geographical position and greater elevation of the relief, which condition the coldest climate and the most abundant rainfall. When moving toward the south of the country, glacial forms shrink and become rarer and smaller. Their southernmost border is marked in Nemërçka and Çika, while the westernmost in the Scanderbeg, Çika, and Tomorri mountains. In the north, glacial accumulation forms descend up to about 300 m above sea level, while in the south up to 1200 m. These are related to the climate change and the most numerous erosive or karstic valleys and pits in the Alps (Qiriazi 2019).

Photo 5.4   Haxhi Aliu Cave (Plaku 2019)

Photo 5.5   Shoshan Canyon
(Qiriazi 2014)

Photo 5.6 Osumi Canyon (Qiriazi 2012)

The most preserved glacial forms are in magmatic territories, especially ultrabasic erosion-resistant and insoluble (Gashi, Martaneshi, Lura, Shebeniku, Valamare, etc.), while in terrigenous territories, these forms are very little preserved, usually on the northern and northeastern slopes in altitudes above 1700–1800 m. In karstified limestone territories, glacial forms are almost altered.

Among the most widespread glacial forms created by erosion are cirques, troughs (lugjet), valleys, and glacial ridges, while from glacial accumulation, moraines and deposition cones (Photos 5.7, 5.8 and 5.9).

At altitudes between 1690 and 2200 m, in the northeast of the Albanian Alps, 16 glacial forms of the type "Rock glaciers" were identified, with a length of 100 m to 1 km. They are found on the northern and northeastern slopes and with a slope of 35% to 50%, while the longest rock glaciers are found on slopes of 20% to 40%. The conditions for their formation must have been created approximately 17000 and 14000 years ago (Palmentola and Baboçi 1994).

5.3.5 Erosive Relief and Degradation and Desertification of Landscapes

The large extent of this relief is related to the conditions suitable for intense erosion (Mediterranean climate with a lot of rainfall, mainly hilly-mountainous relief, large extent of terrigenous, active tectonic detachment, etc.). Erosion was intensified by the millennial indiscriminate human activity in the environment (overuse of vegetation and lands, opening of new lands, etc.). Erosive processes and forms are more prevalent in hilly and mountainous regions (Qiriazi and Sala 1999).

There are (1) river-torrential forms, which are related to the erosion and accumulation of rivers and especially streams, and (2) rockfalls, which are found mainly in mountainous areas, high abrasive coasts, etc. The most widespread landslides are in mountainous areas, especially in the alpine ridges, where they create the so-called rocky seas, which cover the vegetation and the lands, degrading and desertify-

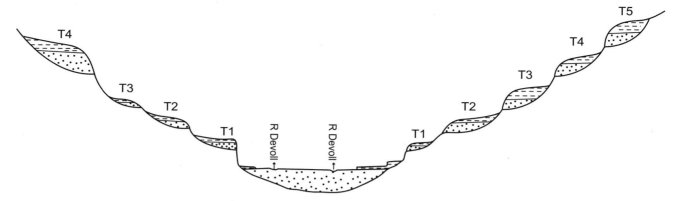

Fig. 5.10 Schematic profile of the Devolli river terraces in the sector of Kokël-Gostimë (Prifti K, 1996, Source: Mediaprint P. Qiriazi Gjeografia fizike e Shqipërisë 2019, p. 91)

Fig. 5.11 Schematic profile of the Osum River terraces, in the sector of Miçan-Berat (K. Prifti, 1996, Source: Mediaprint P. Qiriazi Gjeografia fizike e Shqipërisë 2019, p. 91)

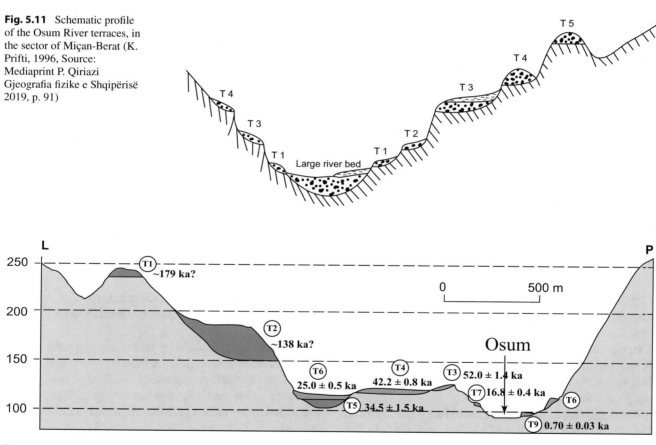

Fig. 5.12 Cross section in lower stream of Osumi river that illustrates 9 tarrace levels with their age in ka (Carcaillet et al. 2009, cited by Sh. Aliaj, Neotektonika e Shqipërisë 2012, p. 96.). T1, T2... terrace levels, – age in their thousand years

ing the landscapes; (3) landslides, most frequent in hilly areas, occur in cases of intense and prolonged rains, causing economic damage; (4) mud currents found in terrigenous often take on very large dimensions and turn into quarry-mud streams, causing severe damage; (5) flows occur during intense rains and when combined with snowmelt (Qiriazi 1994, 2004, 2019).

5.3.5.1 Landscape Degradation and Desertification

The world is facing rapid rates of desertification of lands and landscapes, with major economic, social, and environmental consequences. Therefore, this grave problem has been put in the center of attention of scholars, politics, and national and world organizations. The UN adopted the World Convention

Photo 5.7 Valbona glacial valley (Plaku 2019)

against Desertification (1994) to provide legal support against desertification. Albania is a member and participant in the action plan "The Decade of Deserts and the Fight against Desertification, 2010–2020," which is being implemented by the UNCCD, in cooperation with PNUE, UNDP, and FIDA.

This convention emphasizes that desertification and drought are a global problem because they affect all regions of the world. But this problem is greater in the Southern Region of the World, especially in its arid and semiarid countries, but it is also found in the Northern Region of the World, in the form of wet desert areas, which tend to expand and merge (Melo et al. 1999; Qiriazi 2019).

In the northern region, the Mediterranean basin is particularly vulnerable to desertification because there are optimal conditions for landscape desertification: major seasonal climate changes, variable and intense rains, and droughts; rugged relief with steep slopes; poor and fragile soil and vegetation; ancient and ever-growing population; and intensive development of political events and processes, including frequent wars accompanied by great environmental damage, application of wrong policies, which have accelerated the desertification of landscapes, etc. Even now, the Mediterranean is distinguished for its overuse of agricultural land, irrigation without criteria

and overgrazing of vegetation, rapid urban growth, industry, tourism, pollution of the coastal belt, etc. (Qiriazi 2019).

The IPCC report states that, due to global warming, the temperature rise will be greater in Europe, while the amount of precipitation will be further reduced in most of the Mediterranean. This will increase the risk of drought and desertification, which will increase hunger and socioeconomic instability. Therefore, developing Mediterranean countries will have more intensive landscape desertification (CNULCD 2007).

The Anti-Desertification Convention states that degradation and desertification are, first and foremost, social, economic, political, and then natural processes, which means that human causes take precedence over natural causes (CNULCD 2007).

In Albania, the phenomena of desertification of the northern region and the Northern Mediterranean are present in the areas of wet deserts, which tend to expand continuously (Photos 5.10, 5.11, 5.12, and 5.13).

The causes of landscape degradation and desertification in Albania are related to intensive erosion, many times greater than in some European countries. According to the empirical calculations of the values of the modulus of the solid flow and the average annual erosive layer, which give an idea of the degree of desertification of landscapes, it

Photo 5.8   Glacial cirques on Mount Nemercka (Qiriazi 2010)

Photo 5.9   Glacial moraine
in the Albanian Alps (Qiriazi
2000)

Photo 5.10 Desertified landscapes in the Danglli highlands (Qiriazi 2000)

Photo 5.11 Deserted landscapes on the coast (Qiriazi 2000)

Photo 5.12   The big slide of Bago (Qiriazi 2010)

Photo 5.13   Destruction of the road by landslide (Qiriazi 2021)

Fig. 5.13 Erosion intensity map in Albania (Sala and Qiriazi 1994, Source: Mediaprint P. Qiriazi Gjeografia fizike e Shqipërisë 2019, p. 103). 1 No erosion or very weak, flat surface. 2. Vegetated surfaces. 3. Small erosion, up to 300 t/km²/year and corrosive layer 0.11 mm/year. 4. Average erosion, 300–1200 t/km²/year and 0.11–0.45 mm/year. 5. Major erosion 1200–6000 t/km²/year and 0.45–2.26 mm/year. 6. Very large erosion, 6000–18000 t/km²/year and 2.26–6.7 mm/year

results that (1) territories with very high values of these parameters (6000–18000 t/km²/year and 2.26–6.7 mm/year) occupy 10% of the land (some hilly areas with very large extent of deserted landscapes) and (2) territories with large values of these parameters (1200–6000 t/km²/year and 0.45–2.26 mm/year) occupy 14% of the country and are distinguished by a large degree of degradation and desertification of landscapes (Fig. 5.13) (Sala and Qiriazi 1994).

The measured values of the average solid flow modulus for the site reach 1489 tons/km²/year and the average erosion layer about 0.6 mm/year. This intensive erosion is related to

the Mediterranean position of Albania and the specific natural and human conditions of the country.

The high intensity of erosion is related to several natural causes: (1) the predominance of tectonic uplift movements and easily eroded rocks, which occupy 58% of the territory, while rocks with medium strength occupy 28% and strong rocks only 14% of it; (2) mainly hilly-mountainous character of the relief and the predominance of the slope with a dominant slope over 25–30%, with large hypsometric amplitude, over 2700 m and medium and large degree of density and depth of cleavage; (3) Mediterranean climate with irregular rainfall regime and the amount and intensity of them (over 20–100 mm in 24 hours, meet in over 45 days a year), which fall after the dry and hot summer; climate aggressiveness determined by the average annual amount of precipitation, which takes small values (20% of the country), medium (42%) and large (38%); (4) predominance of lands not resistant to erosion (58.5% of the country), while lands with medium resistance (32.9% of it) and large (8.6% of the territory); and (5) limited protective role from the erosion played by natural vegetation, which covers about half of the territory of the country, but in cold and wet weather, most of this vegetation is leafless (Qiriazi 1994, 2019).

As can be seen, natural factors are optimal for the intensive development of landscape degradation and desertification. This would require that human activity on the natural environment be extremely careful. But human activity overall has been brutal and with great environmental consequences.

Among the main socioeconomic and political causes of the great intensity of erosion are distinguished: (1) ancient habitation, since the Middle Paleolithic, and constantly growing; (2) limited relief offer for arable land; (3) intensive developments of political and social events, frequent wars, and long occupations by foreigners; (4) high concentration of population in highly rugged hilly and mountainous areas with limited supply of livelihood resources (natural resources only) and isolated position; (5) application of economic and ideological-political principle in the humanization of the environment; and (6) mainly agricultural and livestock activ-

ity, pressure from foreign companies to intensify the use of natural resources and discharge their environmental problems in the Albanian environment, etc. Forest fires have had great consequences on vegetation, along with indiscriminate exploitation, and overgrazing (Qiriazi 1994, 2019).

Strong stress was inflicted on natural vegetation during the communist period, when the principle of "self-reliance" (relying entirely on our own capacities) was embraced. This led to the implementation of policies which had great consequences for the environment: The population of all economic units, in the plains, on the hills, and in the mountains, had to "provide their own food" in the country at all costs. Therefore, without being accompanied by anti-erosion measures, new lands were opened on the slopes of hills and mountains: Over 280,000 ha of forests and shrubs on these slopes, often with a steep slope, were turned into arable land, 70% of which are located on the slope. Forest damage continued due to overexploitation beyond their regenerative rate and vegetation, in general, through overgrazing by livestock, to which the area of natural pasture was constantly narrowed. Pollution from industrial acid rain or from the oil industry was also a problem (Photos 5.14, 5.15 and 5.16) (Qiriazi 1994, 2019).

These problems were exacerbated by the high rate of population growth (in 50 years, i.e., by 1989, the population had tripled) and by the ban on free movement of the population (in 1960s). As a result, population pressure on the environment, land, and vegetation increased in hilly and mountainous areas. This pressure was also intensified by the dictated

Photo 5.14 Desertification of landscapes from acid rain (P. Qiriazi 1993)

Photo 5.15 View of the landscape after about 30 years (Qiriazi 2021)

Photo 5.16 Desertification of landscapes from mining industrial activity (Qiriazi 2008)

structure of agricultural crops (over 83% was occupied by field crops); the backward technology of tillage and irrigation; the creation of large plots by the collectivization of agriculture, which increased their watersheds; overuse of chemical fertilizers; state ownership of land, etc. These factors disturbed the morphobioclimatic balance of the slopes and stimulated the processes of landscape desertification (Table 5.1).

During the communist period, there were some positive factors: new afforestation, construction of dams and mountain systems, functioning of regulatory plans and green belts of inhabited centers, etc. However, their effect on reducing the pace of degradation processes and desertification of landscapes was very small.

During the political change and its aftermath, after 1990, several positive factors have been operating, which have

Table 5.1 Changes in arable land, forests, pastures, structure of agricultural crops (per thousand ha), and the population number per thousand inhabitants (INSTAT and Ministry of Agriculture)

Years	1938	1950	1960	1970	1980	1982	1990	2012	2016
Total area	292	391	457	599	702	714	704	696	696
Crops	–	374	417	512	585	590	579	410	418
Orchards*	–	3,2	15	30	55,7	59,6	60	8992	10575
Vineyards	–	2,68	12	17,7	20,3	20,3	20	9.348	10011
Olive groves*	–	11,4	17	36	43,4	44,5	45	4829	6643
Forests	–	1356	1282	1293	1233	1024	1045	1.041	1052
Pastures	–	816	777	631	416	443	403	505	478
Population	1003	1215	1607	2135	2670	3138	3255	2900	2876

Source: Mediaprint P. Qiriazi Gjeografia fizike e Shqipërisë 2019, p. 106)
Note: there are no data; * in thousands

Photo 5.17 Abandonment of arable land after 1990 caused intensive development of erosion (Qiriazi 2015)

slowed down the desertification of landscapes: (1) providing legal and institutional support for environmental protection; (2) international, national, and local cooperation to protect against desertification and implementation of new concepts of environmental assessment and protection; (3) land privatization, which increased the owner's interest in protecting it from land degradation and division into smaller plots, which reduced their catchment area; (4) population migration from mountainous and hilly areas with a double effect (on the one hand, it relieved the environment from the great pressure, while on the other hand, it brought about the abandonment of arable lands, which are degrading); and (5) positive trend toward the increase of fodder, orchards, olive groves, vineyards, forestry, ecotourism, livestock, etc.; the cessation of the opening of new lands on the slopes of mountains and hills; improvements in tillage and irrigation technology, etc. (Photos 5.17, 5.18, 5.19, and 5.20) (Xinxo 1986; Qiriazi 1994, 2019).

Meanwhile, during this period, a number of negative factors have also been operating which create conditions for the acceleration of landscape desertification: (1) the risk of using economic criteria in humanization of the environment and foreign pressure for intensive use of natural resources; (2) damage to forests and vegetation at faster rates, by illegal logging and overexploitation and by environmental pollution, more frequent fires, and more intensive overgrazing by the largest livestock; (3) cessation of afforestation and anti-erosion measures and land use with chaotic constructions; and (4) abandonment of arable land on the slopes of mountains and hills, which are turning into desertified areas. As

Photo 5.18 The abandonment of terraces after 1990 caused the desertification of the landscape (Qiriazi 2015)

Photo 5.19 Massive
deforestation (Qiriazi 2010)

Photo 5.20 Damage to
forests by fires (Qiriazi 2007)

wrong as it was to open these new lands in the communist period, it is just as wrong to abandon them after this period (Photos 5.21, 5.22, and 5.23).

The trend toward increasing landscape desertification continues. Therefore, by cooperating in the programs of the Convention Against Desertification, the best economic alternatives to eradicate poverty are being identified. The problem of land, forest, and pasture ownership is being addressed with the aim of at solving it. Progress has been made also on sustainable management of natural resources, improving the legal and institutional framework, building the appropriate capacities, and applying technical and biological protection measures: anti-erosion systems, green infrastructures, etc.

5.3.6 Type of Coastal Relief

The country has a long coastline, 427 km, which gives the coast great natural, economic, social, historical, cultural, and spiritual values for Albanians. There are both low or accumulating seashores but also high or abrasive ones.

5.3.6.1 Low or Accumulation Coast

It lies mainly on the Adriatic coast. Its modeling by marine and river deposition activity is influenced by neotectonic movements and longitudinal and transverse tectonic faults, which its two largest bays (Drini and Durrësi) relate to. As a result of the complex evolution, with the sea repeatedly advancing and receding, many forms of

relief have been created on the accumulating coast: underwater mounds and ridges, arrows and coastal cordons, large beaches, deltas, lagoons, dunes, and, with limited extent, forms of abrasive coastal relief on the shores of the capes (Qiriazi 2019).

It is distinguished for its great dynamism, which is related to the progress of the land toward the sea in some sectors and vice versa. It used to be the Dalmatian-type coast, with bays and islands, but because of a series of processes, it changed to the low-type coast. The coastal plain was formed along the coast.

Dynamism is expressed by the instability of the relief forms and in the pendular and continuous oscillation of the river estuaries. The river Seman is distinguished in this respect. These oscillations are associated with large solid river flows, shallowness of the sea, the upward tectonic movements, and the small slope of the field (Photos 5.24, 5.25, and 5.26) (Qiriazi and Nikolli 1999).

As a result of coastal processes, the land has continuously advanced toward the sea. At the beginning of the Holocene (about 10,000 years ago), the coastline stretched about 6–8 km further east in the Drini Bay sector and 5–6 km east of the shores of Greater Myzeqe and the present bay of Lalzi (Qiriazi 2019).

During the most recent historical period, the pace of coastline progress has changed. Map measurements show that during the period 1916–1978, the Western Lowlands increased toward the sea by 3530 ha, i.e., on average about 60 ha/year. At the same time, about 1200 ha of it have been

Photo 5.21 Desertification of landscapes on Dajti mountain from stone quarries (Qiriazi 2015)

Photo 5.22 Desertification of landscapes from road construction (Qiriazi 2010)

Photo 5.23   Desertification of landscapes from the construction of hydropower (Qiriazi 2010)

Photo 5.24   Accumulation coast of Velipoja (Plaku 2019)

Photo 5.25 Buna River Delta (Plaku 2019)

Photo 5.26 Shëngjin beach (Plaku 2019)

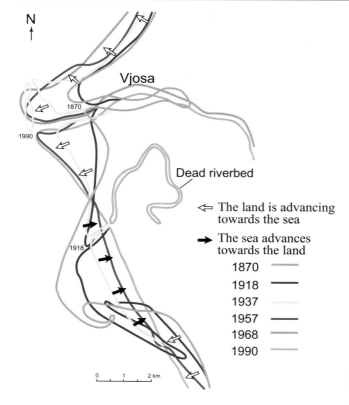

Fig. 5.14 The evolution of the Vjosa delta in the period 1870–1990 (Fauché 2001, Source: Mediaprint P. Qiriazi Gjeografia fizike e Shqipërisë 2019, p. 114)

eroded, on average about 20 ha/year, so the ratio between accumulation and erosion has been 3:1. Subsequently, this ratio dropped to 1:2.4. This was accompanied by the beach crisis (see below) (Boçi 1981, 2002).

Land progress toward the sea is expressed by (1) formation of river deltas (conditioned by large solid river inflows, exceeding three times the rinsing capacity of the sea; shallow depth of the sea near the shore (the isobath 100 m goes 50 km from the shore); small amplitude of ebb and high tides, about 30–40 cm; predominance of upward movements of the structural foundation); (2) the formation and evolution of beaches, arrows, coastal cordons, and lagoons in the sectors between two neighboring deltas; and (3) the formation of raffles (in the sector of the cape of Rodon and cape of Lagji cape and in the section of Narta-Vlora), which separated parts of the sea that evolved toward salt marshes and swamps, which were reclaimed in the 1970s (Fig. 5.14) (Kabo 1988; Fauché 2001; Ciavola 1999; Qiriazi 2019).

Being a few meters above the level of the plain, which was flooded by rivers, the remains of old cordons have, in some cases, been used as land for the settlement of inhabited centers (Patog, Fushë Kuqe, Gorre). Opposite the lagoons of Patoku and Karavasta, new cordons and lagoons

are being formed, while the old ones are evolving toward swamps. By reclaiming swamps and marshlands, man accelerated the land's advance toward the sea (Paskoff 1989; Nikolli 2010).

The large extent of wetlands (lagoons, swamps, marshlands, abandoned or active riverbeds) gives the coast a marsh-type character, with rich biodiversity.

Aeolian forms, mostly in the form of dunes with a height of 1–6 m, are found in some wide and open beaches (Povelce, Darzeza, etc.). Dunes have great ecological and scientific values. Therefore, some have been declared natural monuments.

The beaches are most extensive on the low coast, often up to several tens of kilometers long and up to several hundred meters wide. Beaches are postglacial formations. They are found at river estuaries, in bays between capes, and on the banks of coastal cordons. Beaches have tourist values, some even scientific and ecological values (Shehu 1996).

5.3.6.2 The Crisis of the Beaches

Albanian beaches have been affected by crisis in recent decades, as have other beaches around the world. Marine erosion has destroyed the beaches of Semani and Patoku and partly destroyed the beaches of Kuna, Tale, Rrushkulli, etc. The trend is toward increasing erosion rates, which threatens to destroy other beaches, causing both economic and ecological damage. Stable beach sectors or those in an equilibrium are few. There are cases of replacement of accumulation through erosion and vice versa related to the position of river estuaries (Photos 5.27, 5.28 and 5.29).

From field measurements, but also from maps and satellite photographs of different periods, as well as from the comparison of the current position of the bunkers built in the 1970s, it was found that in recent decades, many sectors of the accumulation shores were included in intense erosion, up to 35 m/year (Dyrmishi 2005). The causes are of general and local character, natural and anthropogenic. In the first category operate (Figs. 5.15 and 5.16):

- Reduction of solid river inflows for natural and human reasons. Today's beaches were formed after the last glaciation, when thick layers of easily eroded material were created. Therefore, the solid inflows of the full rivers were much greater than the rinsing capacity of the sea. But now, more erosion-resistant root rock has surfaced. As a result, rivers erode and deposit less than the rinsing capacity of the sea. This depositing decrease is also related to human activity, such as anti-erosion measures, afforestation, dams, artificial change of estuaries and riverbeds, etc. (Paskoff 1989).
- The continuous rise of the planetary ocean level, a result of the melting of polar and mountain ice due to global

Photo 5.27 The crisis of the beaches of Rrushkulli (Qiriazi 2008)

Photo 5.28 The crisis of the beaches of Seman (Qiriazi 2010)

warming. According to the report of the 51st Session of the Intergovernmental Panel on Climate Change (20-24 September 2019), "While sea level rise is currently rising more than twice as fast now as during the 20th century and accelerating, the report notes a projected rise by 30–60 cm more by 2100 even if emissions significantly decrease and temperature rise is limited to below 2°C. This figure will be much greater if emissions continue to rise unabated." This would be a catastrophe for humanity because islands, plains, and coastal cities will be submerged, and their ecosystems and biodiversity will disappear.

Photo 5.29 Chaotic
constructions on the coast
endangered by beach erosion
(Qiriazi 2018)

Fig. 5.15 Change of the
coast configuration in the
sector between the Shkumbin
and Vjosa estuaries (Ciavola
P. 1999, Source: Mediaprint
P. Qiriazi Gjeografia fizike e
Shqipërisë, 2019, p. 111)

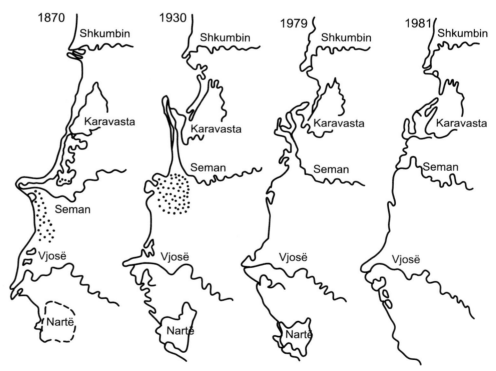

Fig. 5.16 Change of the coast configuration in the sector between the Mati and Ishmi estuaries (Ciavola 1999, Source: Mediaprint P. Qiriazi Gjeografia fizike e Shqipërisë, 2019, p. 111)

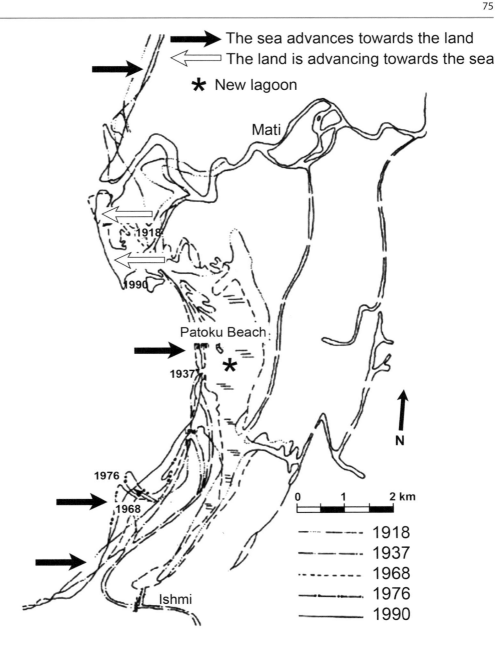

Local causes are natural (subsurface tectonic movements, change of river estuaries, etc.) and human, which operate in the catchment area, riverbed, estuary, beaches, and marine shallows. Human activity in the catchment area and the riverbeds have been as follows: (1) Anti-erosion works in the second half of the twentieth century (afforestation, mountain systems, etc.) and artificial reservoirs were created (about 700), where the solid depositing of water streams that flow into them is deposited (the construction of the large hydropower plants of the Drini river strongly influenced the rapid erosion of the Buna-Shengjin coast); (2) riverbeds and estu-aries were artificially diverted (the Drini in Lezha was diverted into the Buna River, which interrupted the solid deposit flows in its delta, the Ishmi estuary was diverted into the Patoku lagoon to fill it, which quickly destroyed the Patoku beach, etc.); (3) inert materials from river beds were used, especially after 1990, which, besides the erosion in the beaches, was associated with damage to the road infrastructure and erosion of arable land; (4) abandonment of downstream riverbed management increased floods and reduced their solid inflows into the sea. On the beaches and the marine shallows, (1) the taking of sand has intensified ero-

Photo 5.30 Protective
measures applied in Albania
proved ineffective (Qiriazi
2018)

Photo 5.31 Water pollution
conditioned by protective
measures (Qiriazi 2018)

sion, and (2) the numerous unlicensed constructions (the new port of Vlora jetty, about 1 km long, built in the 1990s) caused the rapid erosion of the beach and the artificial forest to its south) (Qiriazi 2019).

Protective measures applied in the world, and in Albania, such as covering the seashore with stone blocks or building parallel or vertical shore tiers, proved ineffective, transmitted marine erosion to neighboring sectors, denatured the beaches, and increased their degree of pollution. These affected the ecological and tourist values of the beaches (Photos 5.30 and 5.31).

As can be seen, the beach crisis is a serious and complex problem, which requires in-depth studies and concrete programs for protection measures.

5.3.6.3 High or Abrasive Ionian Coast

This type of coastline starts in the Uji i Ftohtë (Cold Water) in Vlora and continues intermittently to the Cape of Stili, in the bay of Ftelia. The coast offers different views: the rugged coastline, in some place with deep precipices and in some other places with secluded beaches, narrow valleys, gorges, and canyons, and mountains running parallel to the coastline, somewhere with cliffs and somewhere else with small fields. There are several tectonic bays in this part (the bay of Vlora, Himara, Portopalermo, Borshi, Saranda, etc.), some small islands, and some sectors with (predominantly) high but also low coast, the latter limited to the bays and stream estuaries.

The intensity of abrasion is affected by the great depth of the sea (isobath 100 m goes to about 50 from the shore), the large size of sea waves (up to or over 4 m high), and the not-so-large solid flow of the streams. The action of the waves is also related to the lithological composition and structural features of the shores. In the limestone sectors, the coast is rough, with ridges or jagged rock formations, intertwined with karst forms (above-water and underwater caves, canyons, etc.). In the terrigenous sectors, the coast is softer and with a slight slope (Sala et al. 2006).

Along the coast, there are anticline and syncline structures, complicated by Plio-Quaternary tectonics, which are associated with the formation of the graben structure (Vlora, Himara, Portopalermo, Saranda). When the rocky layers fall toward the ground, ridges are formed. When rock layers fall toward the sea, marine abrasion is low, and the coast is rocky, jagged, and full of very small beaches. On the lower coast, there are lagoons and several pebbly beaches of different sizes (Photos 5.32, 5.33, 5.34, 5.35 and 5.36).

All this variety of forms of coastal relief and nearby relief, together with the great climatic, floristic, and terrestrial riches, give great tourist values to the Albanian Riviera.

Photo 5.32 Borsh Bay, beach, and plain on the Ionian marine erosion coast (Qiriazi 2000)

Photo 5.33 Small bays with pebble beaches in high or abrasive Ionian coast (Plaku 2018)

Photo 5.34 Ksamil Islands on the Ionian marine erosion coast (Qiriazi 2000)

Photo 5.35 Indented coast
on the Ionian marine erosion
coast (Plaku 2018)

Photo 5.36 Cliff and entrance of Haxhi Aliu cave on the Ionian Sea erosion coast (Plaku 2018)

References

Aliaj S (2012) Neotektonika e Shqipërisë

Aliaj S, Melo V, Tipare të evolucionit morfotektonik të relievit tonë. Studime Gjeografike, nr.2/1987

Almagia R (1914) Tracce della glaciazione diluviale nelle montagne dell'Albania., Roma

Boçi S (1981) Studim topografik i vijës bregdetare të Adriatikut

Boçi S. Gjendja e studimeve bregdetare dhe detyra për të ardhmen. Studime gjeografike, nr.14/2002.

Carcaillet, J, Mugnier J, Koçi R. etj. Uplift and active tectonics of southern Albania inferred from incision of alluvial tarraces. Quaternary research, 71/2009, 465–476

Ciavola P. Relation between river dynamics and coastal changes in Albania: an assessment integrating satellite imagery with historical data (1999), Int. J. Remonte Sensing, vol. 20, No. 3

Cmeta A. Personal photo archive, 2000–2021

Cmeta A (2012) Tirana through lens. Albas Publishing House

CNULCD (2007) Convention des Nations Unies sur la lutte contre la désertifcation (Guide des négociations)

Dyrmishi Ç (2005) Studimi i hapësirës bregdetare të Shqipërisë. Botim i Shërbimit Gjeologjik Shqiptar

Fauché E. Evolucioni gjeomorfologjik holocenik dhe historik i deltave të Vjosë dhe të Semanit nëpërmjet imazheve satelitore. Studime Gjeografike, nr.13/2001

Gruppo speleologico Milan, 1993

Gruppo speleologico Faentino, 1995

G. S. Puglia Grotte e Dauno, 1996

Gjipali R, Ruka D. White etj. Shpella e Pëllumbasit. Iliria, XXXVII/275/2015

Kabo M, Veçori të bregdetit Shqiptar të Adriatikut. Studime Gjeografike, 3/1988

Kristo V, Krutaj F. etj. Visoret karstike të Shqipërisë, probleme të shfrytëzimit. Studime Gjeografike, 2/1987

Krutaj F (1994) Veçori të relievit karstik. Studime Gjeografike, nr.5/1994

Melo V Pasqyrimi i lëvizjeve neotektonike në tarracat e Shkumbinit (sektori Elbasan–Peqin) Bul USHT, nr.2/1961

Melo V. Tarracat e vjetra aluvio-proluviale të Drinit të Zi, historia neotektonike. Bul. USHT. Shk. Nat. nr.2/1964

Melo V, Qiriazi P Sala S (1999) Dezertifikimi i terreneve në zonën e Cërrujës ... Studime Gjeografike. nr.12/1999

Nikolli P. (2010) Evolucioni morfotektonik dhe gjeomorfik i sektorit bregdetar Shëngjin-Patok

Palmentola G, Baboçi etc. Shenjat e para të egzistencës së "Rock-Glaciers" në Alpet Shqiptare. Studime Gjeografike, nr.5/1994

Paskoff R. Les cotes d'Albanie. Aspects geomorphologiques. Bull Geog.2/1989, Paris

Plaku F. Personal photo archive, 2000–2021

Plaku F (2018) Homeland Albanian's hidden gems

Prifti K (1984) Kuaternari dhe veçoritë gjeomorfologjike në luginat e rrjedhjes së mesme të lumenjve Vjosë., Osum e Devoll

Prifti K (1985) Praktikum për gjeologjinë e përgjithshme

Prifti K., Probleme të formimit të tarracave lumore në vendin tonë (1996a). Studime Gjeografike, nr. 6. Botim i Akademisë së Shkencave të Shqipërisë

Prifti K Neotektonika e gjeomorfologjia e luginës së Vjosë, Osumit, Devollit, Shkumbinit dhe Erzenit (1996b), Stdudime Gjeografike nr. 7. Botim i Akademisë së Shkencave të Shqipërisë.

Pumo E, Lateritet në ndihmë të deshifrimit të paleogjeografisë. Studime Gjeografike. nr. 4/1990

Qendra e Studimeve Gjeografike, Gjeografia fizike e Shqipërisë, Pjesa e parë, 1990

Qiriazi P. Personal photo archive, 1980–2021

Qiriazi P (1985) Morfologjia dhe Morfogjeneza e Gropave Juglindore dhe e maleve përreth tyre

Qiriazi P (1994) Degradimi i terreneve ne vendin tonë e problemet sociale. Sudime Gjeografike, nr. 5/1994

Qiriazi P (2004) Probleme të degradimit e të shkretirizimit të peizazheve.... Revista Ekonomia e Biznesi, nr. 1(14)

Qiriazi P (2019) Gjeografia fizike e Shqipërisë. Mediaprint

Qiriazi P, Kristo V etj. Tipare themelore gjeomorfologjike të Shqipërisë. Studime Gjeografike, nr. 2/1987

Qiriazi P, Nikolli N., etj. Evolucioni i peizazhit gjeografik në shekullin XX në zonën fushore midis deltës së Drinit të Lezhës dhe deltës së Matit. Gjeomorfologjia e Aplikuar, 1999

Qiriazi P, Sala S (1998) Les activitées humaines et la degradation des montagnes en Albanie, 1998, Observatoire de Montagne de Moussala OM2, Sofia

Qiriazi P, Sala S (1999) Les mouvements de Terrain en Albania et le Rôle de l'Homme. Transactions Japanese, Geomorphological Union, Vol. 20, nr 3/1999. Selected Papers on Landslinde, presented at ICG, Bologna

Qiriazi P, Sala S, Melo M, Bego F etj. Ekosistemet karstike të Shqipërisë, 1999

Sala S, Qiriazi P (1994). Harta e intensitetit të erozionit në Shqipëri

Sala S, Krutaj F, Meçaj N (2006) Gjeomorfologjia e bregdetit jonian

Shehu A (1996) Mbrojtja, administrimi dhe parashikimi i evolucionit të vijës bregore të Shqipërisë

Spedizione Italo-Sanmarinese, 1993

The report of the 51st Session of the Intergovernmental Panel on Climate Change (20–24 September 2019)

Xinxo M (1986) Erozioni i tokës nga uji dhe masat për luftimin e tij në Shqipëri

Climate

6

Perikli Qiriazi

Abstract

The climate of the country is formed by the action of radiative, planetary, and local factors. The climate is Mediterranean, especially in the western and southwestern parts. Winters are generally mild and wet, while summers are hot and dry. The climate is distinguished by the marked fluctuations of weather and the values of the climatic elements and its great variety. Drought is a summer phenomenon, but it is not excluded in the autumn and spring months. There are four climatic zones (lowland Mediterranean, hilly Mediterranean, pre-mountainous Mediterranean, and mountainous Mediterranean) and 13 sub-climatic zones. The country has great climatic resources (variety, thermal, pluviometric, large number of sunny days, especially in summer, etc.). However, there are also negative climatic phenomena: the reverse course between temperature and precipitation, its capricious character, etc. The country is also facing the consequences of global warming.

Keywords

Mediterranean climate · Cosmic or radiative · Planetary and local factors · Sunshine · Wind · Air temperature · Atmospheric precipitation · Climate assets and negative phenomena · Global warming of climate

6.1 General Features

The country has a Mediterranean climate, which is distinguished by (i) the marked seasonal character of the sunshine regime, the atmosphere circulation, and other climatic elements; (ii) winters that are generally mild and wet and summers that are hot and dry; (iii) unpredictable character, with large fluctuations of weather and values of climatic elements from their perennial average; and (iv) its great variety, so much so, that rarely can one

find another such small country with so many climatic differences between its provinces. These Mediterranean climate features are prominent in the coastal part, while inland, because of the diminishing influence of the sea, the climate displays some characteristic phenomena for the high mountain areas and the continental climate, which are secondary, because inland and in eastern areas, the Mediterranean climate features also predominate (Qiriazi 2019).

6.2 Climate Formation Factors

– *Cosmic or radiative factors.* The country is in the geographical area where the angle of incidence of sunlight reaches up to 730 52′. The astronomical length of the day (theoretical maximum of sunshine hours) varies from 14.9 h (June) to 9 h (December). The average annual amount of radiated energy is 2107.5 kwh/m² of horizontal surface area which is average on a planetary scale (Mandili 1986).

Due to the altitude above sea level, the slope, and exposure of the slopes and especially the overcast, the average annual amount of solar radiation energy varies from one area to another: It ranges from 1686.5 kwh/m² in Myzeqe (in Fieri), which is the area with the largest number of sunlight hours (2840 h per year), to 1205.6 kwh/m² in Kukësi. The main role in this distribution is played by the uneven overcast sky. The maximum of solar radiation is in July, which is related to the greater overcast in June than in July, while the minimum is in January. The direct impact of latitude on temperature values in winter is 19%, in autumn 12%, in spring 5%, and in summer 1.3% (Sanxhaku 1983; Qendra e Studimeve Gjeografke 1990).

– *Planetary factors.* Although the circulated energy is less than the radiated energy, again the circulation of air masses plays a significant role in the climate of the country. It conditions the periodic and nonperiodic changes of the weather and the irregular distribution of precipitation during the year. It affects the thermal regime, atmospheric pressure and wind, air humidity, in unusual meteorological events, etc.

The Mediterranean climate is formed mainly under the influence of marine air masses created in the North Atlantic and the Mediterranean Sea. In the interior and eastern regions of the country, there are also average continental air masses coming from Eastern Europe, the interior of the Balkans, and less often Central Asia. Air masses mainly come from medium and subtropical latitudes. In rare cases, there are also tropical air masses from North Africa and Asia Minor but also arctic air masses from large latitudes.

The territory of the country, as an integral part of the synoptic region of the Mediterranean and the Southeast Europe, is under the direct, sometimes indirect, influence of baric centers of atmospheric action, stable throughout the year (the Azores and Greenland anticyclone and the Iceland cyclone) and seasonal in nature. The Mediterranean cyclones (most notably the Genoa cyclone), the winter anticyclone of Mongolia (Siberia), and summer lows over the Sahara (where, in some cases, anticyclones also form) and over Asia Minor (Institut Hidrometeorologjik 1975).

The change of position and intensity of these baric centers, during the year but also from one year to the other, conditions the change of weather over the Northern Mediterranean, and therefore in Albania. The main role is played by the Azores anticyclone and the Icelandic cyclone. In Europe and especially in the Mediterranean belt, including Albania, there are (1) the type of western circulation (air masses move from west to east) which prevails in winter and summer; (2) eastern type (air masses move from east and northeast to west), appearing on average one day a month in autumn, winter, and spring; and (3) meridional (air masses move from north to south and vice versa), prevailing in spring, on average 11 days a month.

Between the Azores anticyclone and the Iceland cyclone, a polar or middle front is created which makes pendular movements from north to south and vice versa. This motion is conditioned by the pendular displacement of the dynamic Azores anticyclone. During the warm half of the year, especially in the summer, the Azores anticyclone gathers strength and shifts north. The polar front, pushed by it at this time, lies north of the parallel 44° north latitude (Gracianski 1971). Mediterranean Europe, including Albania, falls under the strong influence of the Azores anticyclone. As a result, the weather during the warm half of the year, especially in summer, is mostly clear and without precipitation. During this time, the subtropical and tropical air masses predominate (Qiriazi 2019).

The opposite happens during the cold half of the year and especially in winter. During this time, the Azores anticyclone weakens and retreats to the south, releasing the Mediterranean territories of Europe and Albania from its influence. The polar front descends to the south and is usually placed between the parallels 30–40° north latitude. The Atlantic cyclones developed on this front, moving more and more to the southeast, become the predominant elements of the weather in Mediterranean Europe, where they bring air masses from subtropical latitudes and especially average (polar) and prevailing cloudy and rainy weather.

This predominant, seasonal circulation of these baric systems and air masses associated with them conditions the Mediterranean rainfall regime, which is expressed by their concentration mainly during the cold half of the year (Qiriazi 2019).

In special synoptic situations, there are other forms of air mass circulation, which condition completely different features of the weather, compared to the prevailing weather. These forms, with minor coincidence, are related to the meridional and eastern circulation of air masses. During meridional circulation in winter, cold polar and arctic air masses penetrate the Mediterranean through corridors forming between high mountain systems. For Albania, the corridor between the Alps and the Dinaric Alps and the other corridor between the latter in the Carpathians are important. This cold air causes a marked decrease in temperature. In cases where the cold air of the cyclone, formed in the Gulf of Genoa and in the Adriatic, warm air moves over the Balkans from south to north, an unusual rise in temperature is observed.

During the winter, especially in January, the Mongolian anticyclone, which conditions the eastern circulation of air masses, gains strength and moves to the west. Therefore, during this time, the anticyclonic weather regime prevails. It is clear and cold, and it usually lasts several days.

There are also cases of extraordinary meteorological events, which have increased in the last decades. They are expressed with temperatures much lower or much higher than the perennial average and prolonged droughts or with extremely wet periods (Dautaj and Berisha 1974). These features underscore the capricious character of the climate and are especially related to the disturbances of the ordinary mechanism of circulation of the air masses, which are conditioned by the displacement beyond the average position of the Azores anticyclone and the Icelandic cyclone. In cases when their displacement beyond the average position is done toward the north, the anticyclonic regime and the penetration of air masses from tropical latitudes prevail. This brings hot and dry weather. When shifting beyond the average position is done toward the south, the cyclonic regime is dominant, which brings intense and prolonged rainfall. Significant temperature drops are also present. The normal weather conditions are restored when the main baric centers occupy an average position, bringing about the normal circulation of air masses.

The theory that links these events to global warming prevails. It is predicted that such weather disturbances will be more frequent. In Albania, these processes are intensified by the relief, which slows down the movement of cyclones due to its high degree of density and depth of fragmentation, and in some cases, the relief also intensifies the slope of the plane of the cyclonic front.

Some of these extraordinary events are the winter of 1962–1963 and the winter of 1985, with intense rainfall, heavy snowfall, and very low temperatures; very cold winters of 2004–2005 and dry winters of 2006–2007; winter of 2009–2010 and January 2011, with intense rainfall and flooding in the Shkodra area; prolonged drought, nearly 10 years (1980–1990); heavy rainfall and floods in February 2015; drought and high temperatures of the summer of 2017, etc.

– *Local factors.* The diverse character of the surface layer plays a major role in the climatic features of the country as a whole and especially of the specific physical-geographical units. Local factors include (1) relief, with considerable height, density, and great depth of its fragmentation and the dominant northwest-southeast direction of the main mountain ranges, the main local factor of great climatic diversity (Jaho 1982); (2) the influence of the seas expressed by the change of temperature and its amplitude; overcast and air humidity; in the local atmospheric circulation, etc.; (3) soils and vegetation condition special microclimatic features; (4) anthropogenic transformations of the environment (drying of swamps, creation of large artificial aquifers, afforestation and deforestation, construction of urban and industrial areas, etc.) that have all been accompanied by microclimatic changes (Qendra e Studimeve Gjeografke 1990; Qiriazi 2019).

6.3 Climatic Elements

– *Sunshine.* The country is distinguished for its large number of sunshine hours, which range from about 2000 h (in Kukësi) to 2841 h/year (in Fieri). This change is mainly related to the greater cloudiness in Kukësi, which is surrounded by mountains and the open horizon in Fieri. Due to more cloudiness in June, the highest value of the actual extension of sunshine is recorded in July and August, while the lowest value is met in December, which is associated with the smallest astronomical extension of the day and the largest cloudiness in this month. Most sunny days are in summer. The opposite happens in winter (Table 6.1) (Instituti Hidrometeorologjik 1980b).

– *Wind.* The average calm amounts to 40–50% of cases. The largest occurrence of calm is found in some areas protected by high relief forms, while the smallest occurs in the coastal area, which is associated especially with local air circulation, and is in front of narrow gorges.

Albania, like other Mediterranean countries, has an unstable regime of wind direction and speed. This is related to migratory cyclones and anticyclones, which pass over or near the country, with unequal intensity, as well as the impact of seas and lakes, etc. However, it is noticed that in every season, especially in the west of the country, certain winds prevail. The impact of the relief in the interior of the country upsets the laws of air movement (in Kukësi, the northern and northeastern winds prevail throughout the year) (Table 6.2).

Average annual wind speeds range from 1.5 m/s to 5.9 m/s. The highest average speeds are observed in the winter and spring seasons (up to 5.4 m/s) while the lowest in summer and especially in July to August (up to 4 m/s), which are related to the predominance of the anticyclonic regime (Table 6.3) (Instituti Hidrometeorologjik 1980d).

Maximum wind speeds are very high (above 30–40 m/sec) and are related to the passage of atmospheric fronts. They occur mainly in winter, and they cause economic damage. The coastal strip, the inland and eastern regions are distinguished for high-speed winds.

The main local winds are as follows: (1) The *murlani* is a characteristic winter wind, cold and strong, more frequent inland and in the east of the country; (2) the *shiroku* blows mainly in winter and autumn; the cyclonic type is warm and loaded with dust from the Sahara. While passing over the Mediterranean Sea, it absorbs moisture, becomes wet, and brings precipitation, and not infrequently storms also; (3) the *juga* (the south wind) is a warm wind, more predominant on the coast, and depending on the synoptic situation, it may be wet (especially in winter and autumn) or dry (late spring and summer). The periodic winds are sea breezes and land gusts, Föhns winds, and mountain and valley winds (Naçi 1983, 1984, 1986).

– *Air temperature.* The average annual and January temperature isotherms, reduced at sea level, have a northwest-southeast direction, thus coinciding with the direction of the mountain ranges. The map of isotherms of average annual temperatures, reduced at sea level, resembles the hypsometric map, which reveals the significant role of relief, especially of its height in their geographical distribution. Temperatures drop from south to north, from west to east, and especially in the southwest-northeast direction. Factual annual average temperatures range from 7 °C (in Vermoshi) to 17–18 °C (in the Riviera), January temperature range from around -3 °C (in Vermoshi) to around 10 °C (Himara),

Table 6.1 Multiannual average of actual length of sunshine per hour (period 1956–1980), Instituti Hidrometeorologjik, Manuali i Diellëzmit, 1980b

Stacion	I	II	III	IV	V	VI	VII	VIII	IX	X	XI	XII	Annual
Shkodër	116	117	167	189	248	292	342	316	246	195	110	105	2443
Tiranë	125	124	163	191	256	297	350	328	257	207	124	108	2530
Elbasan	129	130	161	186	242	277	328	314	243	207	131	112	2460
Fier	136	144	188	228	292	339	391	362	268	228	145	120	2841
Vlorë	131	138	179	220	281	324	370	344	270	218	140	119	2734
Xarrë	133	136	177	210	283	320	359	337	262	209	146	124	2696
Kukësi	61	87	128	162	220	262	312	287	206	149	75	51	2000
Peshkopi	90	110	150	177	229	268	322	301	232	184	102	80	2245
Korçë	102	122	155	187	239	279	332	307	230	185	122	95	2355

Source: Mediaprint P. Qiriazi Gjeografia fizike e Shqipërisë 2019, p. 137

Table 6.2 Annual coincidence in percentage of wind directions and calm Instituti Hidrometeorologjik. Klima e Shqipërisë (Manuali i erës), 1980d

Stacion	C	N	NE	E	SE	S	SW	W	NW
Shkodër	60.7	0.6	2.3	10.5	7.1	5.8	4.3	5.2	3.4
Tiranë	46.9	3.8	2.6	3.1	14.4	4.3	6.6	3.7	14.5
Kuçovë	48.1	2.3	1.4	2	24.4	2.3	1.8	6.7	10.9
Vlorë	43.3	2.9	5.6	15.2	3.6	7.2	5.2	6.4	10.6
Xarrë	0.0	43.2	8.2	8.2	16.9	11.9	2.1	1.5	7.9
Gjirokastër	46.1	2.6	0.4	1.3	14.9	9.1	2.7	7.6	15.3
Peshkopi	54	1.4	3.7	6.9	10.1	3.2	5.3	7.8	7.6
Korçë	54.3	10.3	4.4	6.3	5.2	1.6	13.4	2.3	2.2
Ersekë	48.8	9.7	3.1	2.6	3.2	13	7.9	4.3	7.5

Source: Mediaprint P. Qiriazi Gjeografia fizike e Shqipërisë, 2019, p.139

Table 6.3 Average wind speed in m/s, by directions (Instituti Hidrometeorologjik. Klima e Shqipërisë, Manuali i erës, 1980d)

Stacion	N	NE	E	SE	S	SW	W	NW
Shkodër	1.5	4	4.7	4.4	4.8	3.7	3.3	3.4
Tiranë	2.2	2	1.4	2.5	2.5	2.8	2.5	3
Kuçovë	1.9	1.8	2	3.4	3	2.9	3	2.7
Vlorë	3.3	2.9	3	3.9	5.9	5.3	3.8	4.8
Xarrë	4.1	3.7	4.3	5	4.7	3.3	2.6	5.4
Gjirokastër	4.1	1	2.1	3.4	3.5	3.2	3.8	4.6
Peshkopi	2.8	2.7	2.9	3.8	2.8	3	3.1	4.1
Korçë	3.4	2.8	4.4	4.8	2.8	5	3.2	2.7
Ersekë	3.4	2.2	3.3	4	5.3	3.5	2.6	2.9

Source: Mediaprint P. Qiriazi Gjeografia fizike e Shqipërisë, 2019, p.140

and July temperatures range from 15.9 °C (in Vermoshi) to 25.8 °C (in Saranda) (Table 6.4) (Instituti Hidrometeorologjik 1980e).

By comparing the spring average temperature with the autumn ones, everywhere in Albania, spring is cooler than autumn. The difference between them decreases from the coast to the inland, and especially from the southwest to the northeast (Saranda 4 °C, Kukësi 2 °C). This highlights the presence of a maritime climate throughout the country and its weakening effects in the east.

The average temperature amplitude increases from the coast (14–15 °C) inland (around 22 °C) and decreases in the vertical direction. This change is related to the impact of sea air masses in the west and southwest and of the continental air masses inland (Instituti Hidrometeorologjik 1980e).

The highest absolute maximum temperatures are recorded in the interior of the Western Lowlands, especially in the Myzeqe (in Kuçova 43,9 °C, the highest in the country, 18.07.1973), and in the southern part of the country. They are mainly related to the penetration of hot tropical and Mediterranean air masses, which create a state of sultriness, which impairs normal human respiration and damages agricultural crops. This condition usually lasts for a few days. These maximums in the interior of the country have lower values: Vermosh 33.4 °C and Kukësi 39.5 °C. They are typical in the period July 15 to August 15. Often, the January maximum absolutes are greater than the annual averages; the maximum absolutes of March are higher than the July average (Table 6.5) (Jaho 1981; Boriçi and Heba 1981; Mici 1988).

The most typical absolute minimums are inland, especially in the closed plains of Sheqerasi −26.9 °C (January 1963) and Vermoshi -27 °C (8.01.2017), which are the minimum absolute lowest of the country. As a result of very low temperature values, several lakes have frozen: Lake Fierza (January 1985 and 2017), Lake Ohrid and Lake Shkodra, Lagoons of Narta, Karavasta and Kune Vaini, and the Drini of Lezha (January 2017). Lower minimums are found in January, December, and February, and often, the absolute

Table 6.4 Perennial average monthly and annual temperatures for some stations in ° C

Stacion	I	II	III	IV	V	VI	VII	VIII	IX	X	XI	XII	Annual
Tiranë	6,7	7,9	9,9	13,3	17,7	21,6	23,8	23,8	20,6	16,1	11,8	8,2	15,1
Shkodër	5	6,5	9,5	13,5	18	22	24,6	24,7	20,9	15,7	10,9	6,9	14,8
Durrës	8,2	9,2	10,9	14	18,1	21,8	23,8	23,8	21,1	17,3	13,4	10	16
Vlorë	9.2	10	11.4	14.4	18.3	22	24.1	24.2	21.6	17.9	14.1	10.8	16.5
Sarandë	10,3	10,9	12,2	15,3	19,6	22,7	25,1	25,8	23,3	19	15	11,6	17,6
Xarrë	9.1	9.9	11.4	14	18.3	22	24.2	24.6	21.8	18	14	10.8	16.5
Fier	7.2	8.3	10.1	13.3	17.4	21.3	23.1	23	20.3	16.3	12.2	8.7	15.1
Burrel	3.8	5.4	8.1	12	16.4	20	22.6	22.8	19.1	14.1	9.6	5.7	13.3
Korçë	0,5	2,2	4,9	9,2	13,9	17,6	20	20,2	16,5	11,3	6,8	2,5	10,5
Kukësi	0,5	3,1	6,4	11,4	16,2	20	22	22	17,8	12,2	7,6	2,8	11,8
Peshkopi	0.1	2.1	5.4	10.2	15	18.8	21	21.1	17.3	11.8	6.8	2.1	11
Tropojë	0.4	3	6.6	11.2	15.7	19.4	21.2	21.1	17.6	12.3	7.7	2.8	11.6

Source: Mediaprint P. Qiriazi Gjeografia fizike e Shqipërisë, 2019, p.143

Table 6.5 Absolute temperature maximums for some stations at ° C (Instituti Hidrometeorologjik, Manuali i Temperaturave, 1980)

Stacion	I	II	III	IV	V	VI	VII	VIII	IX	X	XI	XII	Annual	Date
Tiranë	20.6	27.7	29.6	28.1	35.8	37.9	41.5	40.3	37	31.4	26.9	22.5	41.5	30.08.1954
Shkodër	18.2	21.7	26.1	27.5	34.5	35.8	39.8	39.4	36.4	29.9	24.5	21.7	39.8	28.08.1962
Vlorë	22.6	29.6	30.6	29.3	37.2	36	39.5	38.6	35.9	31.9	27.7	24.5	39.5	26.08.1965
Xarrë	19.3	26.6	26	28	31.7	34.9	38.6	36.4	33.4	31	25.1	24.6	38.6	17.08.1974
Korçë	17.4	21.8	26.3	26.7	31.6	34.3	37	36.5	33.1	27.6	22.1	16.1	37	26.07.1965
Kukësi	18.1	23.1	28.9	29.8	35	37	39	39.5	36.6	29.5	25.4	21.6	39.5	13.08.1957
Peshkopi	16.5	21.9	27.6	27.6	32.4	34.6	38.3	37.3	33.6	28.4	23	19.6	38.3	26.07.1965
Tropojë	17.2	21.6	26	28.2	33.8	34.2	36.4	35.6	32.4	30	22.8	18.5	36.4	18.08.1973

Source: Mediaprint P. Qiriazi Gjeografia fizike e Shqipërisë, 2019, p. 145)

Photo 6.1 Shkodra lake freezes, January 10, 2017 (Qiriazi 2017)

minimums of July are lower than the annual averages (Photo 6.1 and Table 6.6) (Jaho 1981; Mici 1988).

The amplitudes between the extreme temperatures are also very large for different stations (69.7 °C) and for the same station (up to about 63 °C, Sheqerasi). The smaller are in the coastal regions, especially in the southwestern part (in Himara 38.2 °C), and quite large (above 12 °C) is the fluctuation of the average monthly temperature for different years, especially for the winter months (Instituti Hidrometeorologjik 1980e).

The greatest extension of the period with high temperature, above 30 °C, is observed in the south and inside the plain area (in Gjirokastra, 374 h per year, etc.). Icy (frosty) days vary from about 2 days in warmer regions to over 131 days (in Voskopoja). The northeastern and eastern region is distinguished for its uninterrupted maximum extension of the negative temperature period (Peshkopia 444 h, etc.). Peshkopia is distinguished for the largest number of hours with negative temperature during the year (1004 h) and Vlora for the smallest (25 h) (Mici 1988).

These extreme temperatures and their recurrent frequency indicate the marked thermal uncertainty of the climate, which is expressed in large fluctuations of temperature values from one year to the next. The escalation of these phenomena in recent decades also testifies to the impact of global warming.

– *Atmospheric precipitation.* The average perennial annual rainfall reaches 1480 mm. This rainfall value ranks Albania among the richest countries in rainfall in the Mediterranean. It is especially related to the predominance of the mainly hilly-mountainous, highly rugged relief, which slows down the movement of cyclones, but it is also related to the predominant direction of the mountain ranges northwest-southeast, where the rain-laden winds collide, blowing usually in a southwest direction.

Despite the limited size of the country, the geographical distribution of rainfall is highly irregular. The average annual amount varies from about 620 mm (in the Korça plain) to about 3100 mm (Boga, in the Alps) (Instituti Hidrometeorologjik 1980). There are regions with high rainfall (the Albanian Alps and the Kurveleshi plateau) and regions with scarcer rainfall (the western and eastern regions). By analogy with neighboring countries, it is thought that precipitation increases to altitudes of 1500–2000 m, where the amount of precipitation should be reduced. The very irregular geographical distribution of the annual rainfall is conditioned by the morphological features of the relief (Fig. 6.1).

Everywhere in the country, the rainfall has a disorderly Mediterranean regime. It is mainly concentrated in the cold half of the year, during which fall from 60% (in east and northeast) to about 80% (in the coast, especially in the Riviera) of annual rainfall. The wettest season is winter and partly autumn. The percentage of its precipitation fluctuates from 44.1% (in Himara) to about 30–31% (in Kukësi, Korça). The driest season is summer. But summer precipitation as a percentage of the annual and as an absolute value decreases from the interior and east (10–15%) onto the coast, especially on the Riviera (3–4%) (Table 6.7).

The wettest month is November or December, when 19% (in the Riviera) to 13% (in the northeast) of annual rainfall falls. The driest month is July or August. The disproportion between the driest and the wettest month varies from about 33 times (Piqerasi) to 2.5 times (Kukësi) (Instituti Hidrometeorologjik 1980c).

Differences between the western region, especially between the Riviera and the interior eastern and northeastern regions, show that firstly, the climate is Mediterranean everywhere and, secondly, that inland, its Mediterranean character weakens, and the influence of the continental climate appears, although it does not become dominant.

The irregular character of the precipitation regime is also expressed in its high-intensity values, which distinguished regions with the most annual rainfall. Precipitation intensity values range from 420.4 mm/24 h (Boga, in 15.12.1963), which is the absolute country maximum, to 73.9 mm/24 h (Sheqeras, in 15.11.1962) (Table 6.8) (Instituti Hidrometeorologjik 1980c).

Table 6.6   Absolute temperature minimums for some stations at ° C (Instituti Hidrometeorologjik, Manuali i Temperaturave, 1980e)

Stacion	I	II	III	IV	V	VI	VII	VIII	IX	X	XI	XII	Annual	Date
Shkodër	−13	−12.4	−4.6	0.1	3.6	9.1	11	11.4	7.2	−0.1	−4.5	−6.6	−13	26.1.1954
Vlorë	−7	−4.8	−3.1	0.4	4.6	10.1	11.4	13	6.6	3	−0.8	−4	−7	14.1.1968
Xarrë	−5.4	−3.7	−0.6	5.4	7	10.2	12.8	14.8	10.2	4.1	0.8	0.3	−5.4	14.1.1968
Tiranë	−10.4	−7.6	−5.3	−0.7	1.8	5.6	9.4	10	3.8	−1.3	−6.1	−6.9	−10.4	15.1.968
Korçë	−20.9	−17.3	−16.5	−10.5	0	2.6	4.9	6.6	−0.5	−7.4	−9.9	−19	−20.9	28.1.1963
Kukësi	−21	−17.9	−13	−2.8	0.3	4	6.9	6.4	0	−3	−16	−16.4	−21	27.1.1963
Peshkopi	−20.8	−18.2	−12.9	−4.6	−0.4	1.6	6.5	5	−1.2	−4.3	−12.9	−15.4	−20.8	27.1.1954
Tropojë	−17.7	−13.4	−11.6	−1.9	1	3.2	6.6	5.6	1	−3.1	−10.2	−18.4	−18.4	1.12.1973

Source: Mediaprint P. Qiriazi Gjeografia fizike e Shqipërisë, 2019, p. 146)

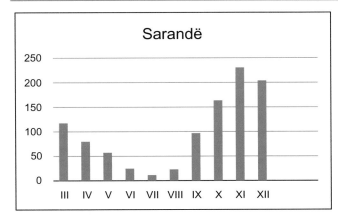

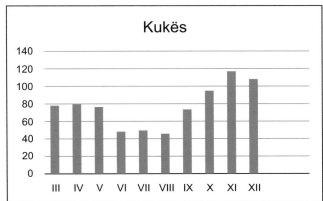

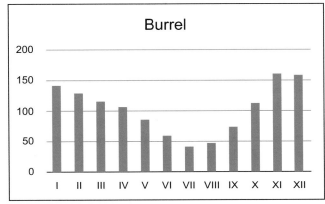

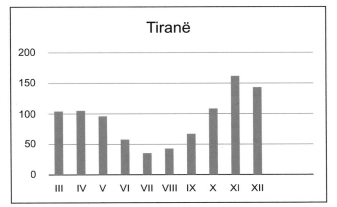

Fig. 6.1 Graphs on precipitation regime, P. Qiriazi, 2019, Source: Mediaprint P. Qiriazi Gjeografia fizike e Shqipërisë, 2019, p. 150)

Table 6.7 Multiyear average rainfall in mm for some stations (1951–1980), (Instituti Hidrometeorologjik. Manuali i Reshjeve, 1980b

Stacioni	I	II	III	IV	V	VI	VII	VIII	IX	X	XI	XII	Annual
Shkodër	243	200	180	174	126	67	42	70	179	230	274	280	2065
Durrës (Plazh)	143	132	115	105	104	67	42	49	78	116	174	148	1273
Tiranë	136	130	106	108	98	60	37	45	69	110	164	146	1209
Burrel	142	130	116	107	86	60	42	47	73	113	161	159	1241
Kukësi	92	81	78	79	77	48	49	46	74	95	118	109	946
Peshkopi	120	108	96	79	70	41	31	36	58	85	130	135	989
Korçë	78	73	59	60	74	42	32	31	48	85	109	98	789
Përmet	164	151	102	89	68	41	26	29	60	138	221	218	1307
Sarandë	188	151	118	81	57	25	11	23	97	164	232	204	1351
Bogë	319	280	257	236	197	148	87	125	158	234	631	421	3093

Source: Mediaprint P. Qiriazi Gjeografia fizike e Shqipërisë, 2019, p. 150)

Table 6.8 Daily maximum precipitation in mm for some stations (1951–1980), (Instituti Hidrometeorologjik. Manuali i Reshjeve, 1980c)

Stacioni	I	II	III	IV	V	VI	VII	VIII	IX	X	XI	XII	Annual	Date
Shkodër	130	116	189	202	131	102	70	206	291	183	186	135	291	26.11.1952
Vlorë	98	56	37	81	48	37	78	72	67	98	103	71	103	17.11.1979
Sarandë	123	59	66	48	50	136	31	38	207	164	112	89	207	28.9.1955
Tiranë	79	102	67	62	91	101	59	71	71	85	150	112	150	14.10.1951
Korçë	50	41	32	52	50	60	49	45	63	55	77	66	77	15.11.1962
Kukësi	83	49	66	41	53	54	74	38	100	61	76	58	100	24.9.1968
Peshkopi	87	100	59	73	52	36	34	61	50	75	83	58	100	2.2.1957
B. Curri	95	86	110	80	71	47	51	74	180	171	109	148	180	16.9.1972

Source: Mediaprint P. Qiriazi Gjeografia fizike e Shqipërisë, 2019, p. 151)

The highly irregular character of the precipitation regime is also expressed in the marked differences between the amount of precipitation that fell for the same period but in different years. The examples are numerous: The year 1955 had the most precipitation in Selca, where 5352 mm of rainfall fell, while the year 1958 in Boga, with 5238 mm. The least rainfall for Sheqerasi (in Korça plain) was in 1961, with 423 mm. Dry summers have been frequent: The summer of 1954 had only 0.1 mm of rainfall (Tirana), 3.6 mm (Himara), etc. It is noteworthy the extremely low rainfall in the decade 1980–1990, when all years were dry, with few exceptions. Dry winters and autumns were also present. But in certain years, there has been much more rainfall than their perennial average, and there have even been years of wet summers. These constitute unusual meteorological events, which are also accompanied with considerable economic damage (Instituti Hidrometeorologjik 1985f).

Most of the precipitation falls in the form of rain. Snow falls in the high, inland, and eastern regions of the country, while for the low regions, especially in the Riviera, snow is extremely rare. The number of snowy days and the duration of the snow cover vary from lowland regions (several centimeters to one day per year) to the high mountainous regions (over 150 cm and to about 100 days per year). Numerous snowfalls fell during the extraordinary events of the winter of 1962–1963 and especially in the winter of 1985, when snow fell all over the country, while its thickness reached over 2–3 m in the northern and eastern part of the country. At the time, the roads and inhabited centers were blocked and avalanches took place, causing economic damage and human casualties. There were also some special cases of snowfall: in April 2014 and 2020 in Korça, Dibra, Kukësi, etc., while on August 19, 1968, most of the Alps were covered with a layer of snow 30 cm thick, etc. (Instituti Hidrometeorologjik 1980e).

Storms, which are numerous in all seasons of the year, are often accompanied by hail. There are more hail days in the winter months, in the second half of autumn, and in the first half of spring. There were also special cases of hailstorms: In Elbasan, on 25.06.1962, hail stones fell for 20 minutes, weighing 80–120 grams/grain, causing damage to agricultural crops; the hail stones on August 5, 2015 in Gramshi and Memaliaj were so big that they broke car windows; the big hail in Shkodra formed a layer of about 30–40 cm (May 14, 2021).

– *Drought.* The small amount of rainfall during the warm half of the year, especially during the summer, combined with its high temperatures, create drought that coincides with the period when the plants and everything else require maximum water. The definition of drought in Albania is based on several indicators: the total of decades with rainfall and their deviation from normal, the duration of decades without rainfall, indicators of drought severity, xerothermic and hydrothermal coefficient, etc. (Jaho and Mici 1984).

Drought is a summer phenomenon, but it is not excluded in the autumn and spring months. Changes from one year to the next are mainly related to drought duration and intensity; dry and isolated decades appear, but successive dry decades with three to nine such cases are of special importance, especially in the period July and August and then in June and May and September and October. The southern and southwestern regions are the driest, while in the interior, especially the high mountain regions, the drought occurs rarely. In recent years, there have been significant droughts, such as droughts of almost 10 years (1980–1990) and summer drought of 2017, 2019, etc.

Drought brings about catastrophic consequences for the economy, especially for agriculture and energy. Over 1.800 km of irrigation canals were opened, and about 700 reservoirs and pumping stations were built to reduce the damaging effects of droughts. They are used to irrigate over half of the arable land. The rehabilitation of the country's irrigation system has already begun.

6.4   Climate Assets and Negative Phenomena

– *The variety of climatic features and conditions* allows for the growth of a large array of natural and cultivated plants, from plants that require plenty of heat and light (like citrus, cotton, etc.) to plants with more limited heat and light requirements (plums, chestnuts, rye, sugar beet, and the like). At the same time, this diversity enables a wide variety of tourist activities (Qiriazi 2019).

– *Large thermal assets.* In the Western Lowlands and in the lower southern parts, January average temperature does not fall below 5 °C, except for unusual meteorological cases. So plants with biological zero of 5 °C or above do not interrupt their vegetative cycle, or they interrupt it in special cases and for a short period. In the interior and eastern regions, the period with biological zero of 5 °C decreases gradually with the increasing altitude and eastward movement. The vegetative period with biological zero of 10 °C lasts spans from about 100 days to 320 days. The annual minimum isotherm −10 °C goes further inland, leaving to the west the Riviera, the Western Lowlands, and the hilly regions and low mountains to its east. This territory is the area of olive culture (Instituti Hidrometeorologjik 1980e).

– *Sufficient light* for normal vegetation development. Sunlight lasts from about 2000 h/year (in the northern and northeastern parts) to over 2800 h/year (in the Fieri region). This factor creates suitable conditions to generate solar energy for economic needs.

- *The large amount of rainfall* (on average over 1480 mm/year) and the significant percentage of snowfall create large water reserves which are important for hydropower, for the supply of residential centers, and for agriculture, industry, and tourism.
- *The predominance of clear weather* and continuous sunshine during the summer creates very good conditions for the development of tourism on the coastal areas and inland (in lakeside areas). The mountainous climate is also very pleasant for tourism.

At the same time, the country's climate has a series of negative phenomena which are harmful to the economy: i) The reverse course between temperature and rainfall creates difficulties for the development of cultivated and natural vegetation, the development and intensification of agriculture necessarily requires artificial irrigation, and the maximum demand for water coincides with the minimum of rainfall, making it difficult to provide water for agriculture, industry, and residential centers, which has been solved by building reservoirs for irrigation and hydropower. ii) The bizarre character of the climate, much more so recently: severe and prolonged droughts (spanning several years), which also deplete the underground water reserves, creating serious problems for the economy; highly wet winters and years which cause great flooding of lands and inhabited centers; very high or low temperatures, etc.

6.5 Climate Regionalization

Climatic diversity makes it difficult to carry out climate regionalization. This is revealed, among others, in the four types and climatograms of some stations in the country. (1) The first type (Saranda, Vlora) has a more or less regular shape, with gradual narrowing on the lower part, characteristic for the most prominent Mediterranean climate; (2) the second type (Thethi) is distinguished for its extremely elongated and irregular shape, characteristic of high areas with Mediterranean alpine climate; (3) the third type (Peshkopia, Korça) is distinguished by the swollen shape of the upper part and narrow in the lower part, even the tendency of the intersection is noticeable, characteristic for the Mediterranean climate with significant influence of the continental climate; (4) the fourth type (Burreli) is with a transitional character between the coastal and inland type (Fig. 6.2) (Qiriazi 2019).

The most accepted opinion on the climatic regionalization of the country was given by a group of climatologists in the work *Climate of Albania* (1975a). For these authors, the main factor that causes climate change from one area to another is the relief with all its features, and the key feature out of them is the altitude above sea level. Besides relief, these authors also single out these features of the country: the limited size of the territory, the meridional extension in the form of a narrow strip on the Adriatic and Ionian coasts, its mountainous and rugged character, and the northwest-southeast direction of the main mountain ranges. To identify the climatic zones and subzones, the regime of some climatic elements is analyzed, with priority given to the thermal regime, the precipitation regime, and the humidity amplitude. Referring to them, there are four climatic zones and 13 subzones, and they are the following (Fig. 6.3) (Instituti Hidrometeorologjik 1975, 1985f):

- *The Mediterranean plain zone* includes the entire lower and coastal part, occupying approximately 18% of the country's territory. It is subject to the greatest influence of the sea. It has much more prominent Mediterranean features: milder and wetter winters and hotter and drier summers. Its extension from the north to the south causes noticeable changes, on the basis of which three subzones are distinguished: northern, central, and southern.
- *The hilly Mediterranean zone* lies to the east of the former. It includes the hilly area and the Drino and Shushica valleys. Its climate is cooler than in the first area. There are four subzones: northern, central, southeastern, and southwestern.
- *Mediterranean para-mountainous zone* includes the valleys of the Black Drini, the Drini, the Valbona rivers, the valley of the Upper Shkumbini, the Southeast Plains, the Dangëllia Highlands, and the Postenan Mountain and the highest parts of the Kruja-Dajti-Çermenika Range. The general valley and mountain character, with considerable height, affect its features. Northern and southern subzones are distinguished.
- *The Mediterranean mountainous zone* lies above the altitudes 1000–1300 m and includes all the high mountain ranges. There are four subzones here: northern, eastern, southeastern, and southern.

By distinguishing many climatic zones and subzones, this regionalization is closer to the climatic reality of the country, distinguished for its. As such, it also has practical significance.

6.6 Global Warming, Its Consequences in Albania

In recent decades, global warming has brought about major negative consequences on a local and planetary scale. Among these are the following:

- *The melting of polar and mountain ice*, which has caused the rise of the global ocean level at a rapid pace (see

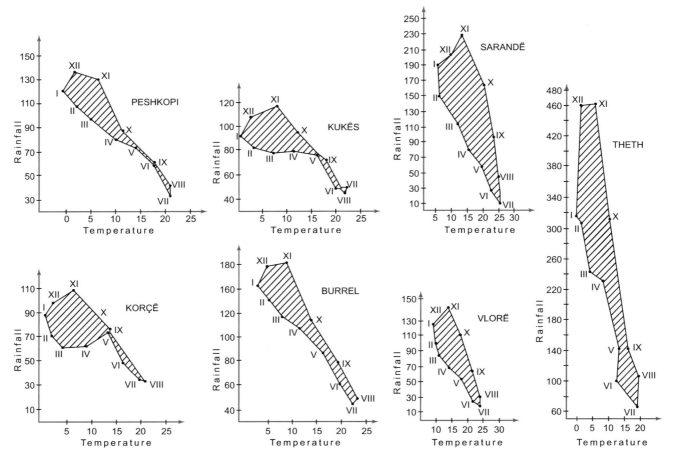

Fig. 6.2 Types of climatograms (Qiriazi 2019, Source: Mediaprint, Gjeografia fizike e Shqipërisë, 2019, p. 157)

coastal relief). This has deepened the beach crisis and threatens to inundate islands, coastal cities, and plains, among them the coastal plain of Albania and its cities.

– *Disruption of the natural balance* that leads to the increase, intensification, and frequency of extreme climatic events (much higher or lower temperatures, torrential rains, storms, severe and prolonged floods, or droughts). Average temperature values are increasing, and the average amount of precipitation is decreasing. It is expected that by 2050 or 2070 rainfall will be reduced by 15–20%, highlighting the water crisis in many regions of the world but also the expansion of deserts, deforestation, depletion of rivers, etc.

– *Great damage to biodiversity and ecosystems*. More than 21% of the world's plant species are in danger of extinction due to habitat loss (Royal Botanic Garden 2017). This will damage the food chain, which can reach the verge of collapse.

The ocean warming is associated with a decrease in the carbon absorption capacity, which increases the amount of carbon in the atmosphere, which then increases the greenhouse effect, thus further raising the temperature (Banka Botërore 2016).

It is predicted that climate change will deepen the water and food crisis, which will destabilize many countries, leading to huge waves of refugees. Climate projections show that from the southern latitudes to the northern latitudes, some diseases will spread (malaria, yellow fever, etc.) and waves of dangerous insects will invade, etc. (Banka Botërore 2016).

6.6.1 Negative Consequences of Global Warming in Albania

This is related to the country's location in the Mediterranean belt, considered as one of the regions most endangered by desertification, and other consequences of global warming, especially water reserves will be reduced, etc. It is also related to the social, historical, political, and economic conditions of the past and present.

Fig. 6.3 Climatic zones of Albania (Institut Hidrometeorologjik, Klima e Shqipërisë, 1975 (reworked), Source: IDEART Gjeografia 11, 2019, p. 52)

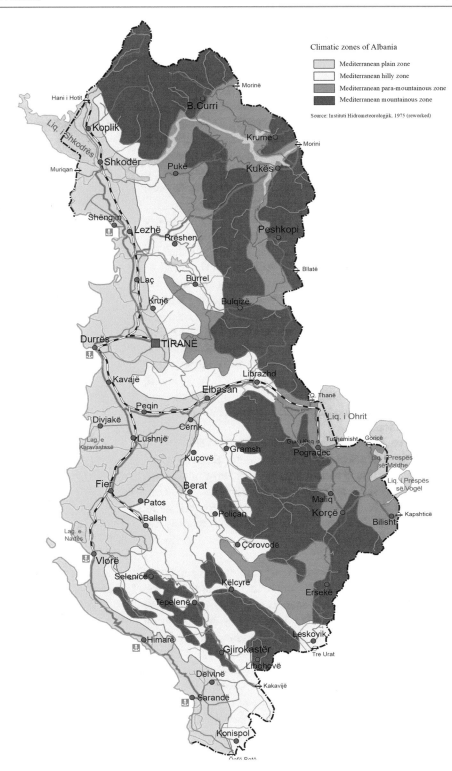

In the World Bank report on the consequences of global warming in Albania (2016), it is highlighted that in terms of global climate risks, Albania is positioned in the 58th place: hydrometeorological droughts, storms and torrential rains, floods, snowstorms, blockage by snow, avalanches, etc. This study talks about a reduction in the amount of average annual rainfall in Albania, about 50 mm by 2050, while in the Western Lowlands, the main agricultural region of the country, it is predicted that this amount will be reduced several times. At the same time, the report points out that the rains will be heavier in May to September, especially in the southern and northern mountains, causing floods in the spring, which will hinder the planting of seasonal crops, and during the summer, which will destroy agricultural produce.

The study also predicts that summer temperatures will rise by 4–5 °C in the northern mountains of Albania; grapes and olives will be more affected by the climate change. There will be a decline in agricultural production in all areas of the country, but especially in the plains. These will intensify the food and water crisis and the energy crisis, which relies almost entirely on hydropower resources, which are expected to decrease by 14% by 2050 due to climate change.

Albania has signed the UN Framework Convention on Climate Change (UNFCCC) and the Kyoto Protocol on Greenhouse Gas Reduction and the Global Climate Pact (COP21), Paris 1915, which aim to keep the rise in global temperatures below 1.5 °C in the twenty-first century.

The country has prepared and is implementing the strategy and programs for the implementation of measures to slow down global warming, as well as to adapt to the expected climate change. For this purpose, legislation on monitoring and regulation of sectors affected by climate change (energy, forests, etc.), the special law "On Climate Change," and the Council of Ministers Decision on monitoring and reporting of greenhouse gases and adaptation to these climate changes have been adopted.

Albania's National Targeted Contribution (NTC) stipulated in the Inter-Sectorial Environmental Strategy, as an integral part of the National Strategy for Development and Integration (NSDI), aims to reduce greenhouse gas emissions by 11.5%, compared to the forecast for the period 2016–2030, aiming to reach up to 2 tons/inhabitant by 2050 (Strategjia Kombëtare e Energjisë për periudhën 2015–2030).

In the context of EU membership, the implementation of the Energy Community Treaty and in line with the EU objectives 20-20-20, Albania will increase by 9% the energy efficiency in 2020, as compared to 2009, and by 38% the use of its alternative renewable energy sources (sun, wind, geothermal, etc.) (Agjensia Kombëtare e Mjedisit 2017).

Albania participates in the Global Network for "Adaptation to Climate Change," in the projects for "Clean Development Mechanism" and "Appropriate National Mitigation Measures." Albania is a member of the Adaptation and Green Fund for Climate, etc., and it has completed the national adaptation plan, which assesses the risks of climate change and adaptation scenarios for the coastal region, etc.

Global warming is the long-term problem that requires a solution but also the participation of all countries.

References

Agjensia Kombëtare e Mjedisit. Raport i Gjendjes në Mjedis, 2017

Banka Botërore (2016) Pasojat e ndryshimeve klimatke në mjedis dhe shëndet në Shqipëri

Boriçi M, Heba I, Shpërndarja e zgjatjes së temperaturave 30oC që ndikojnë në veprimtarinë e njeriut në mjedisin e hapur. Studime Meteorologjike e Hidrologjike, nr, 7/1981

Dautaj H, Berisha S. Shkaqet fziko-sinoptke të ngjarjeve të jashtëzakonshme të 30-31 dhjetor 1970 dhe 1 Janar 1971, Studime Meteorologjike dhe Hidrologjike, nr. 6/1974. Tiranë

Gracianski N (1971) Priroda Sredizemnomorja, Moskva

Institut Hidrometeorologjik, Klima e Shqipërise, 1975

Instituti Hidrometeorologjik. Klima e Shqipërise (Manuali agrometeorologjik), 1980a

Instituti Hidrometeorologjik. Klima e Shqipërise (Manuali i reshjeve), 1980b

Instituti Hidrometeorologjik. Klima e Shqipërise (Manuali i erës), 1980c

Instituti Hidrometeorologjik. Klima e Shqipërise (Manuali i temperaturës), 1980d

Instituti Hidrometeorologjik. Atlasi Klimatk i Shqipërise, 1985f

Jaho S. Temperaturat ekstreme të ajrit në Shqipëri, Studime Meteorologjike dhe Hidrologjike, nr. 7/1981

Jaho S. Përcaktmi i gradientëve vertkalë të temperaturës së ajrit, Studime Meteorologjike dhe Hidrolgjike, nr. 8/1982

Jaho S, Mici A, Thatësira në Shqipëri. Studime Hidrometeorologjike, nr. 10/1984

Mandili T. Vlerësimi i potencialit gjeografk mbi regjimin termik të territorit të Shqipërise. Studime Meteorologjike dhe Hidrologjike, 11/1986

Mici A. Shpërndarja gjeografke e temperaturave minimale dhe disa veçori të regjimit të ngricave në vendin tonë Studime Gjeografke, nr. 8/1988

Naçi R. Mbi disa karakteristka dhe veçori të regjimit të erërave në Shqipëri. Studime Meteorologjike dhe Hidrologjike, 9/1983

Naçi R. Brizat detare dhe koha e shfaqjes së tyre në vendin tonë, Studime Meteorologjike e Hidrologjike, nr.10/1984

Naçi R. Erërat e nxehta dhe të thata në Shqipëri. Studime Meteorologjike dhe Hidrologjike, nr. 11/1986

Qendra e Studimeve Gjeografke, Gjeografa fzike e Shqipërise, Pjesa e parë, 1990

Qiriazi P (2019) Gjeografa fzike e Shqipërise, Mediaprint

Royal Botanic Garden, Kew, State of the worlds plants. Report, 2017

Royal Botanic Gardens, Kew, State of the worlds plants Report, 2017

Sanxhaku M. Veçori të regjimit të rrezatmit diellor në Shqipëri. Studime Meteorologjike e Hidrologjike nr. 9/1983

Strategjia Kombëtare e Energjisë për periudhën 2015–2030

Waters

7

Perikli Qiriazi

Abstract

Albania is distinguished for its large water resources (total length of the hydrographic network reaches 49,027 km, its average density 1.4 km/km², its average river flow over 1308 m³/second, while the average annual volume of liquid flow is 41.25 km³ water per year), for large values of coefficient, modulus, and flow layer (an average 0.64, 30.2 l/sec/km² and 952 mm, respectively). The numerous rivers have irregular Mediterranean flow regime and cause floods and inundations with major economic consequences. They have large solid flow and medium mineralization. The country is rich in groundwater: Total usable reserves reach about 7 km³/year, in mineral and thermo-mineral springs and in glacial, karstic, saltwater lakes (lagoons), tectonic and tectonic-karstic. The country is washed by two Mediterranean seas, the Adriatic and the Ionian. Water is used in agriculture, especially for artificial irrigation, the hydropower industry, the supply of residential centers, etc.

As part of the Mediterranean region, (extremely sensitive to the effects of global warming and drying, and other global crises), the country faces some indicators of water crisis (lack of water supply, pollution, waste, incomplete base legal for water protection and management, etc.). Large water reserves calculated per unit area and per capita (among the highest in Europe) and recent measures create conditions to solve these problems and the consequences of this crisis.

Keywords

Morpho-hydrographic · Large water resources · Flow module · Mediterranean flow regime · Large solid river flows · Groundwater · Thermo-mineral waters · Water crisis

7.1 General Features

The territory of Albania is part of the Mediterranean watershed and has clear features of Mediterranean hydrography.

- Morpho-hydrographic: (i) large variety of water bodies, natural and artificial: (1) generally short rivers (the Drini being the longest river, 285 km) and with a small catchment area. However, the largest rivers of Albania (the Drini, Semani, and Vjosa) are the largest arteries in the Western Balkans; (2) the hydrographic network belongs mainly to the catchment area of the Adriatic and Ionian Seas and very little (about 200 km²) to the Black Sea and Aegean Seas; (3) considerable slope of riverbeds, varying from 0.1‰ (Buna) to 18‰ (Bistrica); (4) the great length of the river network: 49027 km within the borders of the country's territory and 60,323 km in the whole hydrographic basin; (5) high density of hydrographic network: average 1.4 km/km², the highest about 3 km/km², the lowest about 1 km/km² (Instituti Hidrometeorologjik 1985).
- Large water resources: the average annual inflow of rivers over 1308 m³/sec., while the average annual volume of liquid flow 41.25 km³ of water per year, which means 1,433,100 m³/km² of the country's area and over 15,000 m³/resident (Pano 2015).
- Very high flow coefficient: average value 0.64, conditioned by the concentration of precipitation mainly in the cold half of the year, the large extent of permeable formations, the predominance of slopes, even with a significant slope angle, the large extent of bare territories, the predominance of deciduous plants, and the average height of the basin area (Table 7.1) (Qiriazi 2019).

Table 7.1 Average annual parameters of the water flow of river network of Albania (N. Pano, Pasuritë ujore të Shqipërisë, 2015)

River	Surface of the Basin area km^2	Average height of the Basin area m	Flow m^3/sek	Flow module l/s/km^2	Water flow layer in mm per year	Water flow coefficient %
White Drin	4964	862	68.2	13.8	430	0.48
Black Drin	5885	1132	118	20	632	0.65
Drini spill in Buna	14,173	971	352	24.8	781	0.64
Buna at the lake exit	5187	770	320	61.7	1950	0.90
Buna and Drini	19,582	907	680	34.7	1090	0.50
Mat	2441	746	103	42.2	1331	0.80
Ishëm	673	357	20.9	31	977	0.68
Erzen	760	435	18.1	23.8	751	0.51
Shkumbin	2444	753	61.5	25.2	794	0.59
Devoll	3130	960	49.5	15.8	500	0.47
Osum	2073	828	32.5	15.7	494	0.45
Seman	5649	863	95.7	16.9	532	0.49
Drino	1324	746	42.5	32.1	1030	0.53
Vjosë	6706	855	195	29.1	917	0.61
Pavllë	373	521	7.2	19.3	608	0.39
Hydrographic basin	43,305	789	1308	30.2	952	0.64

Source: Mediaprint P. Qiriazi Gjeografia fizike e Shqipërisë (2019, p. 168)

- Very large flow module, on average 30.2 l/sec/km^2, which is conditioned by the factors of flow coefficient values and especially by the large amount of annual rainfall (Pano 2015).
- The flow layer has large values (average 952 mm) (Pano 2015), which is among the highest in Europe.

7.2 Flow Regime

As part of the Mediterranean watershed, the hydrographic network has a Mediterranean flow regime, which is distinguished for greater discharge during the cold and wet period (over 60–70%) and minimal one during the hot and dry period of the year. Surface water accounts for about 69% of river water, while groundwater accounts for 31%. The ratio between them varies from one area to another, from river to river, even in the same river. Water flow is distinguished by frequent and rapid increases and decreases of inflows (Table 7.2) (Pano 2015).

According to the seasons, the biggest flow is winter up to 53% (the Drino, the Pavlla). After winter comes spring and autumn, while summer is distinguished for very small water flow (Fig. 7.1).

In the Mediterranean type of flow, three subtypes are distinguished: (1) the pluvial regime, with more Mediterranean influence (the Vjosa, the Pavlla, the Erzeni rivers), is distinguished by the decrease of the water discharge from winter to summer and its increase from summer to winter; (2) the pluvio-nival regime, in which the increase of flow from the snow modifies the regime, which is expressed by the increase of discharge from January to March (the Shkumbini and the

Osumi, the Black Drini); and (3) the nivo-pluvial regime, in Alpine rivers (the Valbona, the Shala etc.), with maximum discharge in late spring-early summer, when the snow melts (Table 7.3) (Qiriazi 2019).

The rivers are distinguished for their torrential regime, i.e., for large fluctuations of the inflow, the immediate swelling during the long and intense rains and the rapid decrease of their levels. For the Drini river, the average monthly inflows vary by about seven times, while as far as the extreme values are concerned, it is up to 129 times (from 5180 m^3/second to about 40 m^3/second). This disproportion is more intense in rivers with mainly terrigenous catchment area and with higher impact of the Mediterranean climate (the Erzeni, the Semani, the Pavlla, etc.). The river inflows were at a minimal level especially in the last two decades of the twentieth century, when prolonged drought also affected groundwater reserves, greatly reducing groundwater discharge. This is proof of the significant uncertainty of river inflows, which has detrimental consequences for the economy, because the minimum water inflows coincide with the maximum water demands (Instituti Hidrometeorologjik 1987).

- River overflows and floods. These are characteristic of the lowland areas of the country during extraordinary meteorological events. Some of the most severe among them are the river overflows and major floods in the winter of 1962–1963, those of the late 1970 and early 1971, and the floods of the years 2009–2010, in 2011, in January 2016, in December 2017, and in March 2018. Besides the great economic damage, these floods have also caused social, environmental, and long-term ecological problems. These

Table 7.2 Flow distribution in % for the rivers and the hydrographic territory (Insituti Hidrometeorologjik, Hidrologjia e Shqipërisë, 1985)

River	Period					
	Water a little	Water a lot	Winter	Spring	Summer	Autumn
White Drin	92	8	34	40	12	14
Black Drin	94	6	31	39	15	15
Drini	90	10	35	35	13	17
Buna after joining Drin	91	9	37	33	13	17
Mat	93	7	42	31	7	20
Ishmi	94	6	43	32	8	17
Erzeni	92	8	45	31	8	16
Shkumbini	95	5	38	38	9	15
Devolli	95	5	39	39	7	15
Osumi	95	5	41	36	8	15
Semani	95	5	38	39	8	15
Drino	95	5	53	28	6	13
Vjosa	96	4	44	33	9	14
Bistrica	93	7	29	29	21	21
Pavlla	81	19	53	22	6	19
Other rivers	92	8	53	23	6	18
Hydrographic territory	92	8	40	33	11	16

Source: Mediaprint P. Qiriazi Gjeografia fizike e Shqipërisë (2019, p. 170)

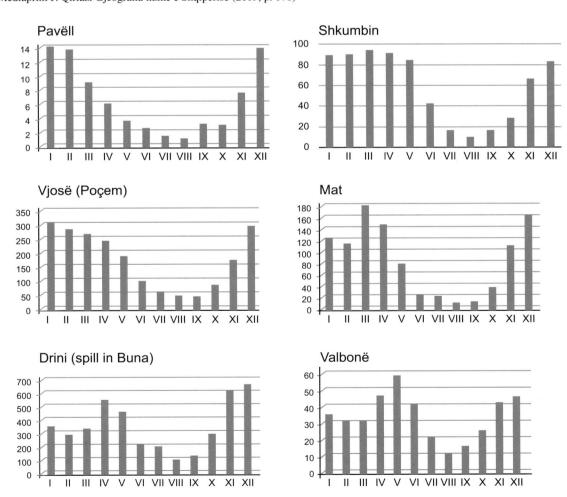

Fig. 7.1 Water flow regime of some rivers. (P. Qiriazi, 2019, Source: Mediaprint P. Qiriazi Gjeografia fizike e Shqipërisë, 2019, p. 171)

Table 7.3 Average monthly and annual discharge of some rivers in m³/sec (Instituti Hidrometeorologjik, Buletini Prurjet e ujit, Tiranë 1987)

River	I	II	III	IV	V	VI	VII	VIII	IX	X	XI	XII	Mes Vjet
Drini (Buna spill)	347	282	329	546	453	215	189	94	124	286	614	661	352
Buna (sea spill)	694	530	1452	1431	926	476	256	117	96	112	755	855	680
Mat	123	112	179	146	76	22	20	7	10	35	109	162	103
Erzeni	33	14	17	8,7	9,1	4,9	1,64	1,24	2,2	2,1	12,5	22,5	18.1
Shkumbini	89	90	93	91	84	41	16	9	16	28	66	83	61.5
Semani	143	156	161	143	116	52	19	12	22	41	100	117	95.7
Vjosa (Poçem)	305	283	261	240	183	94	60	43	41	80	170	290	167
Pavlla	14	13,6	9	6	3,5	2,4	1,4	1	3	3	7,3	13,7	7.2
White Drin	123	91	99	98	11	11,4	6,3	5,4	11	18	45	64	68.2
Valbona	34	31	31	45,6	57,8	40,5	20,4	11,1	14,9	25	41,4	45	33,5

Source: Mediaprint P. Qiriazi Gjeografia fizike e Shqipërisë (2019, p. 172)

Photo 7.1 Floods in the field of Shkodra

severe floods are related to natural and anthropogenic factors (Photo 7.1) (Qiriazi 2019).

The natural factors include (1) the climatic factors (the prolonged, intense, and frontal rainfall falling in most of the territory, especially when combined with the rapid snow melting or when the sea waves block river estuaries due to sea storms) and (2) the geomorphological factors (the large slope of the riverbeds and slopes of the catchment areas of the upper and middle river flows, which is immediately reduced in the sectors of their lower plain flow, where there is a large accumulation). Consequently, the riverbeds here are shallow, meandering, and, as such, unable to conduct large amounts of water during maximum discharges.

– Anthropogenic factors are related to human activity, which is divided into two periods: before and after 1990. In the decades 1950–1980, with the aim to discipline of river flows, many swamps were reclaimed, the riverbeds were deepened, lowland river sections were adjusted, and embankments and levees were built, which run along the downstream of the rivers, increasing the water flowing capacity of the riverbeds. Several riverbeds were diverted, and the Drini of Lezha was all redirected into the Buna and the Gjadri river into the Drini; the Kalasa and the

Bistrica rivers, which had flooded the Vurgu in Delvina, were diverted into Lake Butrinti, through the artificial canal of Çuka, thus flowing directly into the sea. The construction of lakes for hydropower plants on the Drini, the Mati, and the Bistrica rivers significantly affected the level of maximum inflows, thus reducing floods. At the same time, these interventions affected the erosion of the beaches, damaging biodiversity and landscapes (Hoxha 1999; Qiriazi 2019).

After 1990, major floods have increased. This is related to (1) the increase in the unusual meteorological events; (2) damage to vegetation, the abandonment of maintenance, damage to the drainage system (embankments, drainage canals, water-scooping plants, etc.), and damage to riverbeds from intensive exploitation of inert materials; (3) large movements of inhabitants toward the plain areas, accompanied by constructions following no standards or criteria and the wrong placement of dwellings and economic activity near the river, etc.

There are large floods in the Shkodra area, which are related to its geomorphological features and intense rainfall, also combined with the rapid melting of snow. However, the hydrographic node of the Lake-Drini-Buna has had a strong influence, acting in the unusual meteorological events, when the strong current of the Drini hinders the calm flow of the Buna River, thus affecting the Lake of Shkodra, making it without leakage system. Consequently, the lake floods large areas around it, including the city of Shkodra. This is not all. The Drini of Buna, especially when the Drini of Lezha was artificially diverted into Buna, it filled the Buna riverbed with alluvium, further reducing its capacity to discharge large amounts of water. The Buna River discharges about 1300–1400 m^3/second water, while its own and the Drini's inflows have reached about 6000–7000 m^3/second (Pano 2015).

To avoid or at least alleviate their negative consequences, floods should be treated as long-term issues, where sustainable solutions are envisaged and enforced, through implementing the "European Floods Directive," reorganizing flood monitoring and forecasting institutions, maintaining the drainage system and riverbeds, relocating the most endangered villages to safe terrain, banning the hewing of forests for a period of 10 years, and undertaking long-forgotten and much-needed anti-erosion measures.

7.3 Large Solids River Flow

The proof of this large solid river flows is (1) the perennial average reaches up to 1650 kg/second, which corresponds to a total volume of 53.2 million tons/year; (2) the average modulus of flows of suspended alluvium (calculated on the basis of its measured values) reaches up to 1489 tons/km^2/

year, but its values vary from 282 tons/km^2/year to 5575 tons/km^2/year (the Devolli); (3) the average annual erosion layer, according to the measured values, reaches 0.6 mm and the highest is 1.5 mm (the Erzeni, the Ishmi, etc.), while, based on empirical calculations, it reaches over 5–6 mm per year (the Pari, the Tomorrica streams) which have mainly a terrigenous catchment area (Pano 1984).

As it can be seen, the values of the solid flow parameters testify of an intense erosion, which degrades and desertifies lands and landscapes. The fight against this phenomenon must create a new agricultural, pasture, forest, and aesthetic balance, as a response to the new relationship between natural conditions, suitable for erosion, and the current and future perspectives of society's intervention in the geographical landscape.

7.4 Chemical Composition and Thermal Regime of River Waters

In general, waters of the river network have an average mineralization that ranges from 162 to 461 mg/liter (Puka 1987, 1994). They are included in the bicarbonates class (calcium group), which is related to the considerable extent of limestones, dolomites, and evaporites in their watersheds. Generally, the value of mineralization goes in the opposite direction to the flow of rivers.

The thermal regime of river waters is significantly related to the altitude above sea level. The highest temperatures are observed during July and August (13–25 °C) and the lowest during January and February (0–4 °C). River waters freeze only in exceptional meteorological cases and, for a short time, even in their downstream flows.

7.5 The Main Rivers

– *The Drini river* (285 km), together with the Buna River and Shkodra Lake, constitutes the largest hydrographic system in the country. In terms of catchment area (14,173 km^2) and water level, the Drin river is the largest river in the country and the largest one in the Western Balkans. The average height of its basin is 971 m. It is formed by the White Drini and the Black Drini.

The Black Drini (149 km) originates from Lake Ohrid. After traversing a narrow valley between mountain ranges, it enters Albania with a wide bed. Subsequently, the river valley and its bed widen and narrow several times until the river flows into Lake Fierza of the hydropower plant, where the White Drini also ends. The relief of its watershed is very rugged and consists of terrigenous, limestone, gypsum, and magmatic. The climate has a continental influence, with little

Fig. 7.2 Drainage of the rivers of Albania (Insituti Hidrometeorologjik, Hidrologjia e Shqipërisë, 1985, reworked, Source: IDEART Gjeografia 11, 2019, p. 56)

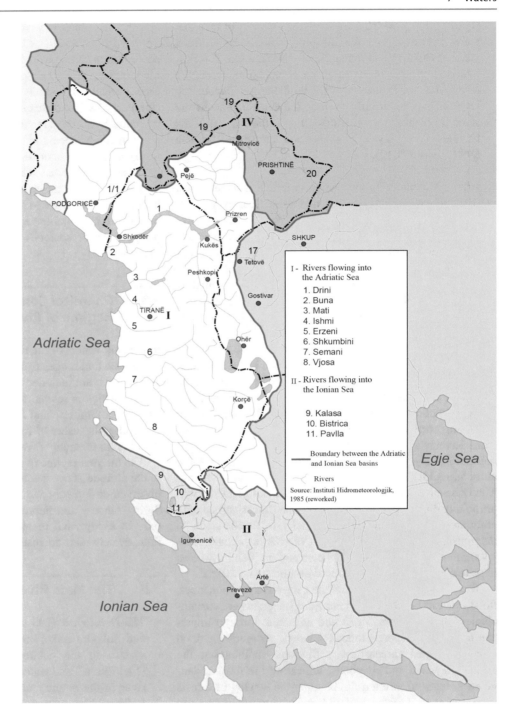

I - Rivers flowing into the Adriatic Sea

1. Drini
2. Buna
3. Mati
4. Ishmi
5. Erzeni
6. Shkumbini
7. Semani
8. Vjosa

II - Rivers flowing into the Ionian Sea

9. Kalasa
10. Bistrica
11. Pavlla

―――― Boundary between the Adriatic and Ionian Sea basins

〜 Rivers

Source: Instituti Hidrometeorologjik, 1985 (reworked)

rainfall, a lot of snow, and significant percentages of rainfall in summer. The plant cover is generally dense. The river's average annual flow reaches 118 m³/second, and the flow module is 20 l/sec/km² (Fig. 7.2) (Pano 2015).

The White Drini (136 km) originates at the foot of the mountain block of Rusulia (Kosovo). It crosses the Dukagjini valley, where many branches flow into the river and enters the territory of Albania in the Vërbnica Gorge and then flows into Lake Fierza. Its basin consists mainly of terrigenous (molas) and limestone, and it has slightly rugged and fragmented relief. The climate has marked continental influence: lower temperatures, less rainfall, and more snowfall. Its average annual flow reaches 68.2 m³/second, and the flow module 13.8 l/sec/ km² (Pano 2015).

The Drini river consists of hydropower lakes built on this river, up to Vau i Deja. Its catchment area consists mainly of limestone and magmatic, and it has a relief with sharp contrasts, very tall height and fragmentation, heavy rainfall and

Photo 7.2 Buna River, lake and the city of Shkodra (P. Qiriazi, 2020)

snow, and dense vegetation. From Vau i Deja to Buna, the riverbed extends up to 2–3 km in the plain relief, consisting of Quaternary molas. At this part, the Drini has constantly changed its bed several times. The division of the Drini into two branches is mentioned in the fifteenth century and the middle of the nineteenth century. As it was said, the entire Drini was artificially diverted into the Buna River. Its feeding waters consist of rain and snow and groundwater. The flow of the river is irregular. Average annual flow in the estuary is 352 m³/second, annual flow module 24.8 l/sec/km2, and the flow coefficient 0.64 (Pano 2015). The river is used to harness electricity but also for navigation, irrigation, etc.

The Buna River (44 km) is the only lowland river in the country, with a difference level of only 5 m and a decrease of 0.11‰. It originates from Shkodra Lake. The catchment area of the Drini and Lake Shkodra is a ruggedly mountainous, and it consists mainly of karstic limestone. There is heavy rainfall (over 1500–2000 mm/year) and scarce vegetation. Due to the small slope and solid discharge of the Drini, the Buna River has changed its bed several times, which has numerous meanders. The river ends with a large delta in the sea, where there are two alluvial islands: Ada (5 km²) and Franz Joseph. Its hydrological regime is determined by the regime of Lake Shkodra and the Drini river. The average flow of the Buna at the exit point in the lake is 320 m³/second, while at the estuary, it is 680 m³/second, coming third in the Northern Mediterranean, after the Rhone and the Po river, while the flow module is 61.7 l/sec/km² (Instituti Hidrometeorologjik 1987). It is navigable by small boats only at the downstream section. In the past, small tonnage ships used to come to Shkodra, but the filling of its bed by the solid discharge of the Drini river hampers navigation in the middle and upper sections of the river (Photos 7.2 and 7.3).

Besides these, there are other rivers: the Mati (144 km), the Shkumbini (181 km), the Semani (281 km) formed by the branches of the Devolli and the Osumi. The Vjosa River (272 km) is considered the only "wild" river in Europe, because no hydropower plants have been built in this river yet.

7.6 Groundwater

The country is rich in groundwater. Their distribution, reserves, and regime vary by region. There are three hydrogeological regions (Tafli 2002): (1) the Ionian area, Krasta and Kruja, with 12 aquifers consisting of quaternary alluvial deposits, with 27 m³/sec usable reserves, eight carbonate aquifers with large water reserves (Bistrica springs 18 m³/sec, Poçemi springs, etc.), and several aquifers composed of strong rocks with medium water; (2) Mirdita and Korabi areas have several large aquifers (the large artesian basin of Korça in Plio-Quaternary deposits, the basin of Bilishti, and the Black Drini with average water content in porous formations); six carbonate masses of abundant water, which are main drinking water supply of cities; and nine magmatic masses, with average water content (up to 20 l/sec) and non-uniform; (3) the Albanian Alps area with several aquifers, where large sources of carbonates emerge: the Eye of Shegani, 1000 l/sec, the Vrella of Shoshani, the Vukli Spring, the Valbona, etc. (Eftimi et al. 2019).

Photo 7.3 The Vjosa River, the only wild river in Europe (P. Qiriazi, 2020)

Depending on their genesis and the hydrogeological conditions of its collection, groundwater is divided into phreatic or shallow water and artesian or deep water (Fig. 7.3).

Karst waters are found in about 25 karst regions, where there are numerous springs with high discharge, especially in tectonic-lithological contacts between permeable karst and impermeable formations (Figs. 7.4, 7.5, and Photo 7.4) (Eftimi 2000).

Submarine springs emerge on the high coast from Vlora to Ksamili. The ones distinguished here are the springs in the bay of Borshi and Himara, etc.

Usable groundwater reserves, according to the formations where they are located, are differentiated into gravel aquifers, with reserves 1 km^3/year; carbonaceous massifs, with reserves 4.8 km^3/year; and molas and magmatic rocks, with reserves of 1.2 km^3/year. The total usable reserves reach up to 7 km^3/year. There are large reserves in soluble formations, which, explicitly expressed, reach 252 m^3/second or 68% of the total usable reserves. Only 3.3 m^3/second or 2% of these reserves are used.

Although the exploitable reserves are large, the utilization rate is still small. There are many problems related to the use of these reserves: their uneven geographical distribution, which does not match the larger concentrations of population and water demands, their variable regime, the pollution in many regions, and so on.

7.7 Mineral and Thermo-Mineral Waters

The country is rich in these waters. According to their mineral composition, they are divided into the following: (1) Sulfur waters, which have a high content of sulfur gas (5–10 mg/l), mineral salts (1000–1700 mg/l), etc. They are associated with depth carbonates that "float" on the evaporites of the Korabi and Ionian area. (2) Sulfide waters, which mainly contain sulfur dioxide (H$_2$S) and other gases as well as mineral salts. They have high temperature. (3) Iodobromide waters, with high iodine and bromine content. They are found in Ardenica, Patosi, Semani, etc. They are connected to oil storage structures. (4) Water with low mineralization (up to 1 gr/l), as the Spring of Glina, which has curative qualities and is also used as bottled water. (5) Steam explosions (near Leskoviku), containing sulfur gas and mineral salts, are related to tectonic fissures (Eftimi and Frashëri 2018b; Dakoli 2000; Frashëri and Çermak 2004).

Some of the thermo-mineral resources are used for curative purposes. Among them are distinguished: The Spas of Elbasani, Peshkopia, Bënja (in Përmet), etc. Thermo-mineral resources are the solid basis for the sustainable development of curative tourism (Figs. 7.6, 7.7 and Photo 7.5).

7.8 Lakes

7.8.1 Small Lakes

As a result of suitable climatic conditions, large tectonic subsidence, karst development, and low coast dynamics, the country has many lakes and lagoons. There are also many artificial lakes. Lakes vary in their altitude, flowing or non-flowing character, geological composition and climatic conditions of the watershed, morphometric features, water balance, hydro-chemical, and optical regime. Since the origin of the lake pit is one of the main factors which deter-

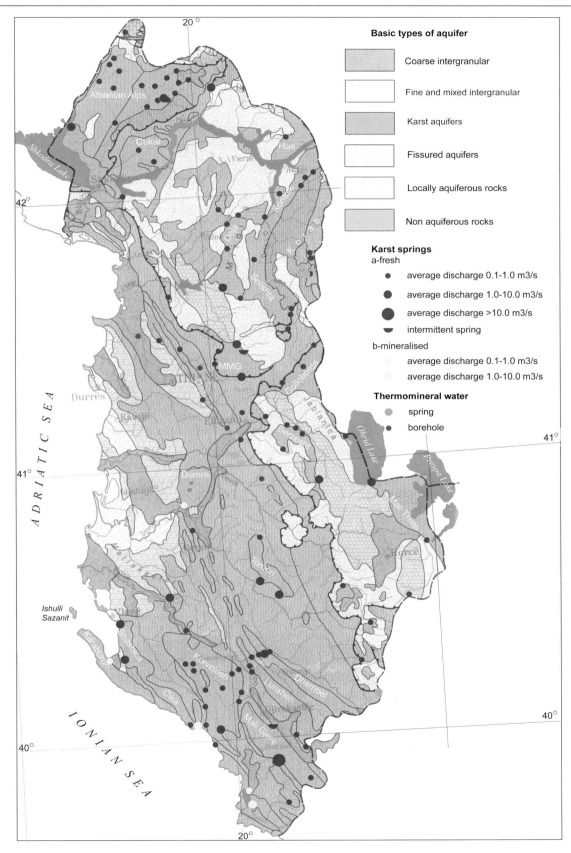

Fig. 7.3 Simplifield hydrogeological map of Albania (Eftimi R, Zojer H (2015) Human impakts on Karst aquifers of Albania. Environ Earth Sci (2015) 74–57-70. DOI https://doi.org/10.1007/s12665-015-4309-7, p. 58)

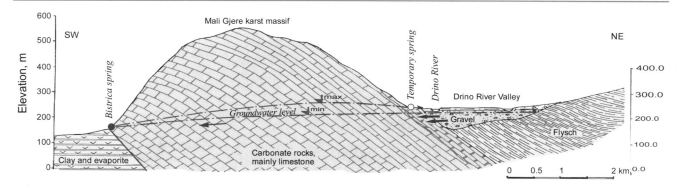

Fig. 7.4 Cross section in Mali Gjerë karst massif (Eftimi R, Amataj S, Zoto J (2007) Groundwater circulation in two transboundary carbonate aquifers of Albania; their vulnerability and protectio. In: groundwater Vulnerability Assessment and Mapping International Conference (2004: Ustron, Poland), p. 63)

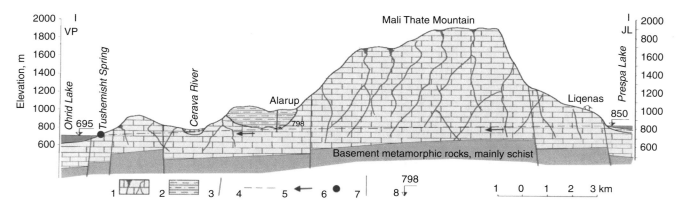

Fig. 7.5 Hydrogeological cross-section in Mali Thate Mountain. (Eftimi R, Zoto J, isotope study of the connection of Ohrid and Prespa Lakes In: International Symposium "Towards Integrated Conservation and Sustainable Development of Transboundary Macro and Micro Prespa Lake". Korça – Albania, 24–26 October 1997: pp. 32–37, page 36)

Photo 7.4 Karst spring of Blue Eye (Syri i Kaltër) (F. Plaku, 2019)

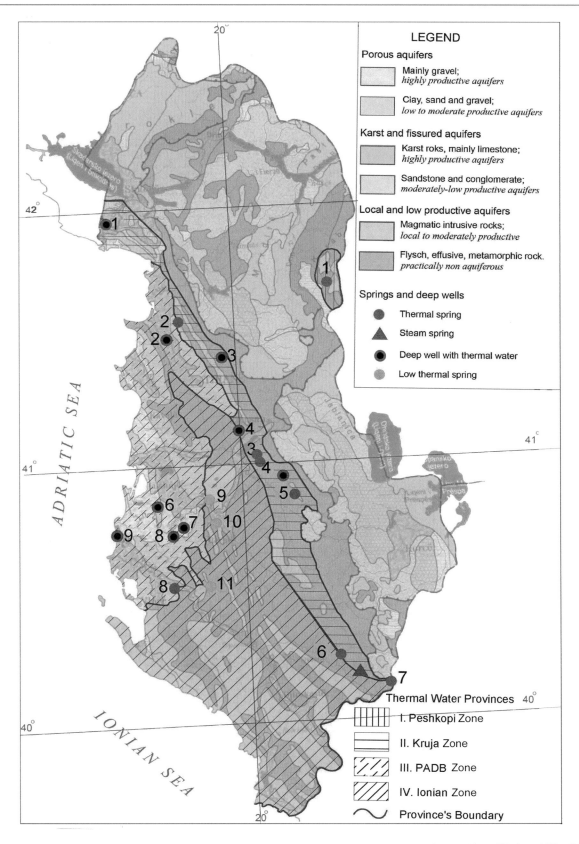

Fig. 7.6 Map of thermal water zones of Albania and main thermal water springs and deep boreholes with thermal water (Eftimi R, Frasheri A (2018a) Regional hydrogeological characteristics of thermal waters of Albania. Acta Gepgraphica Silesiana,12/2 (30) Wno, Sosnowiec, 2018, p 6 (electronic version)

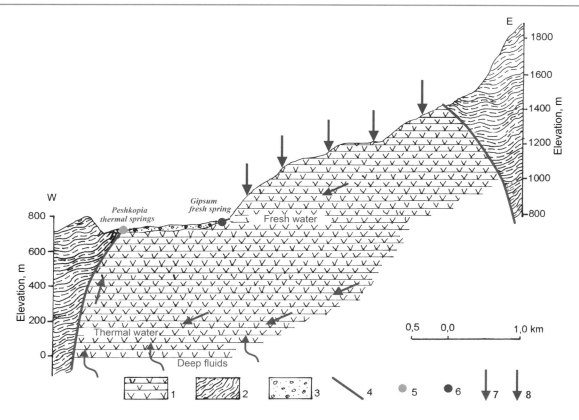

Fig. 7.7 Hydrogeological section through Peshkopia thermal spreng (Eftimi R, Frasheri A. Regional hydrogeological characteristics of thermal waters of Albania. Acta Gepgraphica Silesiana,12/2 (30) Wno, Sosnowiec, 2018c, p. 7 (electronic version)

Photo 7.5 Thermo-mineral springs of Hibrait, Elbasan (P. Qiriazi, 2016)

Photo 7.6 Lura Glacial Lakes (F. Plaku, 2018)

mines the lake's essential geographical features, lakes are grouped according to their origin.

– *Glacial lakes* are of the cirque type. They are located above 1600 m of altitude. In most cases, they are in groups and have small size (up to several hectares) and depth (up to several meters). They end in glacial valleys and troughs. A part of these lakes was dried due to river flows and the development of karst processes (Photo 7.6).

Most are lakes with no flowing leak. Their main water supply comes from the melting of snow. Therefore, they have a nival regime. In karst areas, the lake regime is nivalo-karstic. They have small-level amplitudes (flowing lakes) and large-level amplitude, the big lakes, the ones without flow. During the winter, they freeze on the surface. They are distinguished for their great purity of water, special fauna, which gives them stunning beauty and great scientific, ecological, and tourist values. Many of these lakes (Lura, Martaneshi, Shebeniku lakes) have been declared natural monuments. Among the main groups of these lakes are distinguished: the lakes of Jezerca, Lura, Balgjaj, Shebeniku, Valamara, etc. Their great values were affected using their waters for irrigation in the lower areas. Currently, their values are threatened by water pollution and landscape damage.

– *Karst lakes* are found in karst territories. There are the lakes of Dumrea, which is built of evaporites and carbonates, while its relief has the appearance of a wavy plateau with many karst forms. In the sectors where the karst has reached the maturity stage, karst lakes have been formed, which, due to the lack of connection between them, are at different hypsometric heights. There are cases when karstic underground gaps are unlocked, and wells, funnels, and suction wells are suddenly formed on the ridges or slopes of hills, even at the bottom of the lake. This is testimony to the continuation or reactivation of the karst in Dumrea (Photo 7.7).

There are about 85 lakes in Dumrea, while during the summer, there remain about 60 lakes. They are formed by (1) the collapse of the ceiling of karstic underground gaps, which is connected with the steep slope of the surrounding slopes (Lake Çestija and Mërhoja, etc.), and (2) filling with water of valleys or karstic pits, to which the oval shape is connected, from the union of karstic forms, to which the excessively elongated form is connected (Lake Dega (Branch), 1.6 km) (Sala and Qiriazi 2007) .

They are shallow, average 7 m, while maximum 61 m (Mërhoja) and without leakage. They have small values of the morphometric coefficient (3.4–6.3). Their only water source is rainfall, while water consumption is done by evaporation and infiltration and use for irrigation. Therefore, they

Photo 7.7   Dumrea karst lakes (F. Plaku, 2018)

have a Mediterranean regime of levels, with great oscillation.

The average water temperature reaches 7.5 °C in winter and 26 °C in summer. Due to the carbonate composition and lack of runoff, lake waters have high mineralization and low transparency. They have rich and diverse living worlds: hygrophilous vegetation, especially white and yellow water lilies and rich fauna (several species of fish, etc.).

Their water pollution is related to their situation without surface runoff and human activity around the lakes, with waste of agricultural and municipal chemicals. Until 1990, they were used for irrigation and fish farming, which affected their ecological and scientific values. Now the lakes are being reevaluated in ecological terms and especially as a tourist attraction.

- **Salted coastal lakes (lagoons)** are associated with the evolution of the low coast. They have a small depth (up to 2.5 m), due to their origin. The deepest is the lagoon of Butrinti (25 m), in the formation of which the tectonic sinking also acted. They have surfaces from tens of square meters (Alimura lagoon) to several square kilometers. They are connected to the sea through natural canals, in which, due to the ebb and flow, the water changes direction every 6 h. The ratio between the water entering the lagoon and the water coming out of it varies according to the seasons. During the winter, more water comes out of the lagoon. This is related to winter rainfall and the greatest evaporation in the lagoon during the summer. The level of salinity in the lagoon is also related to these changes: During the summer, the eastern part (near the ground) has maximum salinity (over 45‰, even in Karavasta 65‰ and in Narta 70‰). During the winter, the opposite happens. This part of the lagoon has minimal salinity (15–30‰), which is related to the fresh water that flows into the lagoon. In Butrinti lagoon, there are two water layers with unequal salinity, the upper layer 13–26‰ and the lower layer about 33‰. That makes this lagoon with floors.

Photo 7.8 Landscape from Karavasta lagoon (F. Plaku, 2018)

The water temperature of the lagoons varies from 5–7 °C (December–January-February) to 23–24 °C (July, August). In the deep lagoon of Butrinti, consisting of two water masses, the temperature distribution in depth changes. The lower mass has a higher density than the upper one. Therefore, the vertical water mixture goes up to 7–8 m. In the upper mass, the temperature varies from about 25 °C to 12 °C, while in the lower, it is constant, about 16 °C.

The mixing of water in the lagoons, the development of plants, and the constant exchanges between the sea and the lagoon enrich their waters with oxygen. In the Butrinti lagoon, during the summer, the amount of oxygen on the surface is lower (7.2 mg/l) than in the winter period (10.2 mg/l). Deep oxygen delivery is mainly related to the vertical partial mixing and the permanent presence of hydrogen sulfide gas (H_2S) in the bottom layer, where the oxygen content decreases, and hydrogen sulfide gas appears. At depths from 7 m to 25 m, the oxygen content varies according to the seasons from 1 mg/l to 2.5 mg/l. In the upper layer, rich in oxygen, plant and animal life thrives, especially mussels, while in the lower layer, due to the reduction of oxygen, and, in most cases, either it's the lack or the large increase in hydrogen sulfide gas, life is lacking, with the exception of aneroid bacteria (Puka and Selenica 1982).

At depths over 20 m in the Butrinti lagoon, the sulfur dioxide content often reaches 60 mg/l. This high content is related to the activity of sulfur-reducing bacteria and the powerful sources of this gas at the bottom of the lagoon. This is evidenced by its increase during and after strong earthquakes, which reactivate tectonic cracks, from which gas comes out (Photo 7.8).

The lagoons have great scientific and ecological values for the evolution and dynamics of the accumulating coast, the functioning of the wetland ecosystem, and so on. The Lagoon of Karavasta is a well-known one, part of the Divjaka-Karavasta National Park, which ranks among the lagoons of great importance as an area for migratory bird species and the habitat of over 20,000 wintering birds. The lagoon is known for the large colony of the Dalmatian pelican (*Pelecanus crispus*), which preserves the vitality of the lagoon's wildlife and enables the tourism of biological spectacles. The ban on hunting for a period has led to a significant increase in the colony of pelicans and even flamingos (Phoenicopterus roseus). The Karavasta, Butrinti, and Viliumi lagoons are listed in the IBA (Important Bird Area), or Ramsar Wetland (Qiriazi 2019).

The lagoons are rich in many types of fish: cod, sea bass, mullet, eel, etc. They are used for fishing, for cultivating mussels (in Butrinti), and for extracting salt. The drying of their special parts (Karavasta, Narta, etc.), before 1990, damaged the ecological, scientific, and tourist values. In recent years, the importance of wetlands is being assessed.

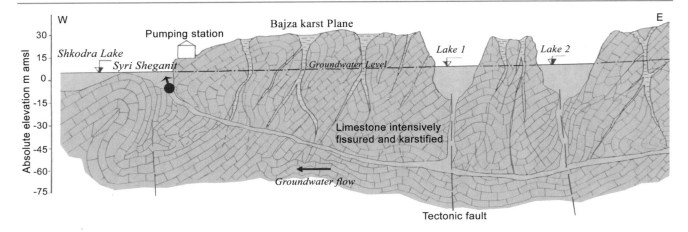

Fig. 7.8 Syri Shegani spring (Eftimi R, Andreychouk V, Szczypek T, Puchrjda W. Karst springs of Albania and their management. Acta Gepgraphica Silesiana, Wno, Sosnowiec, 2019, p. 7 (electronic version)

7.8.2 Large Lakes

7.8.2.1 Lake Shkodra

It is the largest lake in the entire Balkans (369 km²). The area within the political borders of Albania is 149 km². It is 48 km long and 26 km wide. It collects water from a catchment area with an area of about 5179 km² and an average height of 770 m. The average depth reaches 7–10 m, while the maximum is 44 m, and the volume is 2.6 km³ of water. It is a plain lake, and its end constitutes cryptodepression. The Buna River originates from this lake.

There are various, and often contradictory, opinions about the origin of this lake. The most well-argued opinion is that which describes it as tectonic-karst lake. As this argument puts it, the lake is in a large tectonic pit, which was subsequently modeled by karst processes. The tectonic sinking of the lake pit is related to the sinking of the entire Shkodra graben, in which the Shkodra Lowland was formed.

The western shores of the lake are limestone and full of rocky islands, while the eastern shores are low-lying accumulative-type plains. Temporary streams end there with deltas. Once upon a time, when the lake stretched to the foot of the Alps, these shores had the same character as today's western shores. The lake is thought to have gone through three stages: (1) In the first stage, the lake filled a closed Plio-Quaternary graben pit and reached as far as the western foot-hills of the Alps. (2) In the second stage, because of the rupture of the Buna gorge and massive accumulations, the surface of the lake gradually subsided, leaving behind first swamps and then to the dry land. Today's features of this lake were formed not so long ago. According to some authors, still late, in historical time, the size of the lake was very small. (3) In the third stage, the lake began to grow, and this growth continues even today.

The greatest growth began in the middle of the last century, when a branch of the Drini was formed and it flowed into the Buna, and especially after 1963, when artificially the entire Drini river was diverted into the Buna. The Drini river is filling the bed of Buna with its alluvium, making the latter increasingly unable to discharge water. The area increase is also associated with slow decrease of land (its shores), with the last filling of the lake from the stream alluvium flowing into the lake. Some proof of this increase is the reduction of the Vranina peninsula into an island, numerous archeological discoveries, underwater dwellings, and claims of local inhabitants.

Lake Shkodra is characterized by great-level oscillations. According to Montenegrin data, the lake surface at the maximum level reaches up to 542 km², while the water volume is 4.4 km³. The amplitude of the water levels reaches 4.25 m, while the maximum recorded between the lowest level (9.10.1928) and the highest level (27.02.1941) was 9.4 m. In January 2011, the lake deluged several neighborhoods in the city of Shkodra. These large oscillations are related to the features of the lake water balance and to the hydrographic node of the Lake-Buna-Drini (Qiriazi 2019).

As far as the lake's key water source is concerned, the main role is played by surface running water and groundwater. The Moraça River (outside Albania's border) and many mountain streams with irregular regime flow into the lake. The lake basin is mountainous, composed mainly of karstic limestone, with little vegetation and a lot of snow. The lake is the lowest level of the area. Therefore, many large springs (like the Eye of Shegani, Viri, Krevenica, etc.) appear in the lakebed or in the area nearby (Fig. 7.8).

The precipitation regime in its catchment area is very erratic (in each dry month of the year falls 1.7% of annual rainfall, while in each wet month falls about 12–18% of them), which bring large amplitudes of levels, which are also

Fig. 7.9 Elements of the regime levels of Lake Shkodra temperature regime, and precipitation regime in its watershed (N. Pano, 1984). Source: Mediaprint P. Qiriazi Gjeografia fizike e Shqipërisë, 2019, p. 196

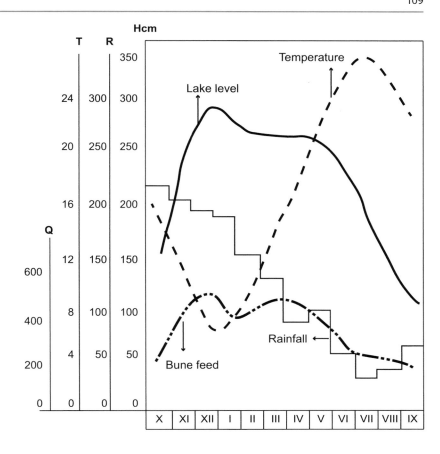

affected by the large morphometric coefficient (about 14) and the flow coefficient (up to 0.90) (Pano 2015). But the main cause of large-level amplitudes is the hydrographic node mentioned above. The strong current of the Drini at the time of its maximum discharge prevents the flow of the Buna. Moreover, the strong Shiroku wind slows down the flow of the Buna. Under these conditions, the lake's water level rises a lot. Large aquifers of hydropower plants on the Drini river, by dispersing its sediments, have somewhat influenced the regulation of the Drin flow, reducing the Drini impact on the swelling of Lake Shkodra. At the same time, for the same reasons, the solid materials that the Drini deposits into the Buna bed have been reduced. Therefore, the accumulation of this bed is made more slowly than before. The levels of the lake, to some extent, follow the levels of the Drini and are mainly conditioned by rainfall, so the lake is of the tropical type.

The waters of Lake Shkodra are warm: average temperature on February 4–6 °C, August 25.9 °C, and annual 16.3 °C. The difference between surface temperature and depth is very small (around 2 °C), due to the shallowness of the lake (Fig. 7.9).

As a result of the shallow depth, the lake waters are saturated with oxygen from the surface to the lake bottom. Due to the large amount of solid matter brought by torrential rivers and the great development of microorganisms, its water is turbid, with little transparency, in the green color or even dark blue.

Lake Shkodra is distinguished for its rich flora and fauna, one of the most interesting in Europe. It serves as a breeding ground for many species of fish. Therefore, there are influxes of their passing from the sea to the lake and vice versa. There are 20 species of fish living in its waters, some of economic importance: carp, mullet, eel, oyster, pike, kubla, etc. Its waters are used for fishing, irrigation, and tourism.

For its special scientific, ecological, and tourist values, this lake has been granted the status of a protected area: It is a national park in Montenegro and a natural park in Albania. However, there are many problems. The lake continues to be polluted by both countries, especially the Podgorica aluminum plant. The sewage of the inhabited centers around it are dumped into the lake. These are damaging values and rapidly depleting many living organisms, including endangered species. The protection of this lake requires joint strategies and programs with the Montenegrin government.

7.8.2.2 Lake Ohrid

Lake Ohrid, together with Lake Prespa and the Black Drini, forms another complicated hydrographic complex. Lake Ohrid is one of the largest lakes in the Balkans. The average area of Lake Ohrid is 362 km², of which 111.4 km² are within the borders of Albania. It is 30.4 km long and 14.5 km wide.

Photo 7.9 Ohrid Lake (A. Cmeta 2020)

The lake's watershed, together with that of Lake Prespa, amount to 2788 km², while the average altitude is 1080 m.

Lake Ohrid has a tectonic origin. But the karst processes have also acted in the modeling of its bed, especially in its eastern part. Based on archaic fauna, there are views that consider it as "the oldest lake in Europe" (formed at the end of the Miocene). But geological and geomorphological data show that the lake graben basin was formed later (Plio-Quaternary), at the same time as the other southeastern graben of the country.

Based on the lake terraces at altitudes up to 960 m, it is thought that during the Pliocene, there was a large lake, the surface of which covered the graben pits of Korça, Bilishti, Prespa, and Ohrid and the graben pits of Doborca in the north (outside Albanian border). This large graben pit was formed because of differentiating movements, from the end of the Pliocene, while during the Quaternary, the graben basin of the great lake was involved in differentiating neotectonic movements. Today's graben basin of Ohrid was involved in sinking land movements with much more amplitude larger than the graben base of other southeastern pits, inside and outside the state border. As a result, the lake conditions in the Ohrid graben pit were preserved to this day, while the lakes formed in other neighboring graben pits were

dried up. The lake was also preserved in the Prespa graben pit, but it is much shallower. In Lake Ohrid, apparently, the conditions were created for the continuation of the life of the archaic Miocene fauna, which makes it a real museum of living fossils. Due to the land movements with greater basement amplitude, the end of this graben pit is much lower than the ends of other neighboring graben pits (Qiriazi 2019).

The surface of Lake Ohrid was constantly shrinking. The proof for this is the remains of lake terraces and cavities of karst springs which are found at different hypsometric levels of Mali i Thatë (Dry Mountain) and Jablanica. These springs once flowed to the lake shore or bottom, just as current springs do. At the level of 850 m, the connection with other lakes was cut off. At 720 m, the lake still covered the Ohrid and Struga plains in the north and the Buçimas plain in the south. The further lowering of the lake's water level made the island of Lin into a peninsula, and the plains north and south of the lake came out of the water (Photo 7.9).

Even though its surface is shrinking, it should be noted that it is a lake with a very long life, which is associated with neotectonic sinking land movements with large amplitude of its graben base, of which, among other proofs, are the movements of frequent seismic. The evolution and tectonic origin are closely related to the morphological features of this lake:

Fig. 7.10 Elements of Lake Ohrid's level regime, temperature regime, and precipitation regime in its watershed (N. Pano, 1984). (Source: Mediaprint P. Qiriazi Gjeografia fizike e Shqipërisë, 2019, p. 199)

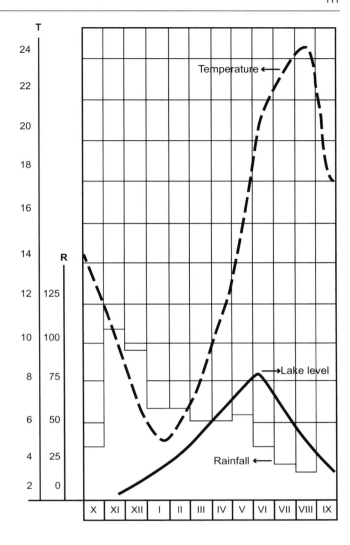

(1) It's surrounding to the east and west by mountains about 2000 m high which fall with a steep slope over the lake, while in the north and south, there are low structural thresholds that separate it from other homologous graben pits; (2) it is a very deep lake (maximum 295 m and average up to 138 m), the deepest in the entire Balkan Peninsula and one of the deepest in Europe, which conditions the large volume of water (50 km³); (iii) the shoreline is almost a straight line (meandering coefficient 1.3).

A good part of the lake shore is of the high or abrasive type. Its ridges are mostly of the tectonic-abrasive type, which are related to the new tectonic faults and their scanty modeling by the lake's ripple process. To the north, south and partly to the west of the lake, the shore of the low accumulation type is found.

The catchment area of Lake Ohrid has mountainous relief, built of karstic limestone, terrigenous and magmatic. On average, it receives 813 mm of rainfall per year and a lot of snow. Small streams and the waters of numerous karst springs flow into the lake. The Black Drini originates from this lake (Fig. 7.10).

The hydrological regime of the lake has small average annual amplitudes of water levels (30–40 cm) and perennial ones of up to 135 cm. This relates to the character with leaks, small morphometric and lake flow coefficient (2.6 and 0.53, respectively), and the large role of groundwater, expressed in the almost constant regime of springs (flow 16 m³/sec), whose waters come from Lake Prespa and the karst ridge of Mali i Thatë (the Dry Mountain) (Eftimi and Zoto 1997), with the less marked disproportion of precipitation and the significant role of snow in its water balance. The regime of its water levels is more related to the temperature: the maximum levels in spring, while the other lower maximum, in autumn, which is related to the heavy rainfall of this season. From this indicator, this lake is of the average latitudes type.

There are two main layers in the lake waters: the activity layer or epilimnion (from the surface to a depth of 130 m), which is subject to vertical mixing, and the lower layer of

Photo 7.10 Lake Prespa (A. Cmeta, 2020)

homothermic or hypolimnion (from 130 m at the bottom of the lake).

The average water temperature of Lake Ohrid fluctuates from about 2 °C (in January) to 21 °C (in August). These average temperatures are 1.4 °C higher than the air temperatures. This is proof of the positive role of this lake in the climate of the surrounding area.

The waters of Lake Ohrid have a low mineralization content, which fluctuates in very narrow limits (200–250 mg/l). Calcium and magnesium bicarbonates predominate. These waters are rich in oxygen (8–9 mg/l), especially in the epilimnion layer (Pano 2015).

The small volume of solid flow and the small development of microorganisms condition the great transparency of the waters of this lake. The "Secchi disk" can be seen about 20–25 m deep in the lake. This fact makes the lake rare in world hydrography.

The lake has diverse flora and fauna, including endangered species worldwide, endemic (about 200) and relics (over 20 species, especially mollusks.) Therefore, as it was already said, it is called "a real museum of living fossils." The endemic fish of *koran* and *belushka* (local names of fish) are famous for their rare taste. Prof. Christian Albrecht called Lake Ohrid the "the Holy Grail for biologists from all over the world." (See World Heritage from Albania.) The lake was declared a World Natural Heritage Site in 1979 (the Macedonian part) and the Albanian part was declared in 2019. After that, this lake was included in the Mix Region of Ohrid, which includes several cultural heritage sites.

Together with Lake Prespa and their surrounding regions, it was declared a Cross-Border Biosphere Reserve. The lake is used for fishing, irrigation, and tourism.

The lake is threatened by activity without anthropogenic criteria. Waste and agricultural chemicals of rural and tourist centers to the west is still dumped in the lake. Other problems are caused by the massive, spontaneous, and chaotic urbanization on its shores; overfishing, especially of the *koran*; the filling of some parts of the lake to make way for the reconstruction of the Lin-Pogradec Road that have denaturalized its shores; the project of the municipality of Ohrid for the inert and concrete filling of 75 ha of the lake; the construction of a tourist resort, etc. This activity will irreversibly destroy the important ecosystem of the lake, endangering the status of world heritage. Therefore, researchers, biologists, environmentalists, and citizens have protested in defense of the lake's values.

7.8.2.3 Lake Prespa

This lake consists of Great Prespa and Small Prespa, with an area of 285 km^2 and 44 km^2, respectively, out of which 49.5 km^2 are part of Albania, while the rest is divided between North Macedonia and Greece. The maximum length and width of Great Prespa Lake is 26.3 km and 20.6 km, while for Small Prespa, it is 10.6 km and 6.6 km, while their watershed is 1424 km^2. Great Prespa Lake has numerous inlets and two islands: Golem Gradi in North Macedonia and Maligradi in Albania (a natural monument), while there are three islands in the Greek part of Small Prespa (Photo 7.10).

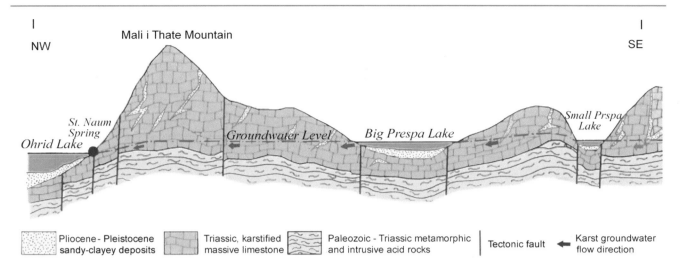

Fig. 7.11 Cross-section in Prespa-Ohrid karst massif (Eftimi R, Amataj S, Zoto J. (2007) Groundwater circulation in two transboundary carbonate aquifers of Albania; their vulnerability and protectio. In: groundwater Vulnerability Assessment and Mapping International Conference (2004: Ustron, Poland), p. 207)

Once, these two lakes were one single lake. But due to the receding water level, they were divided into two separate lakes, but they are connected through an artificial canal. This lake has the same origin and evolution as Ohrid. However, there is greater karst activity here; therefore, it is considered tectonic-karstic lake. Unlike Lake Ohrid, the Prespa graben basement sank at a smaller amplitude, while in the second half of the Quaternary, this basement underwent uplifting movements. Therefore, Prespa is a shallower lake (from 2.5 m to 54 m, with an average depth of 18 m), while the water volume is 5.2 km³. The lake is at an elevation of 853 m above sea level, 157 m higher than Lake Ohrid (Fig. 7.11).

The lake's area has been constantly shrinking. The main evidence for this shrinkage is found in the traces of the lake's earlier shore positions. In the period 1986–1995, there was a very large and rapid decrease of the lake water level, amounting up to 13 m. The most well-argued view concerning the cause of this water level decrease in the lake is related to the underground discharge of the lake waters through karstic gaps. As a result of karstic processes occurring inside the Mali i Thatë massif (Dry Mountain massif), underground gaps can be unblocked. In such cases, more water is discharged than the feeding groundwater. This causes the water level to drop and the lake's area to shrink. This is what happened in the period in question. After 1995, the lake water level drop ceased (Fig. 7.12) (Cavkalovski 1997).

Its drainage basin consists mainly of karstic and magmatic limestones. The amount of rainfall falling in this basin is about 800 mm per year. Several mountain streams flow into the lake, most of them are seasonal stream. The lake is distinguished for its great fluctuating water level amplitudes during the year (with an average of 1.54 m and a maximum of over 3.5 m). This fluctuation is related to its underground flow (underground discharge cavities allow only a certain amount of water to go through, despite the added water supply in the lake). Prespa water level regime is typical of the average latitude lake types, the same as in Ohrid Lake (Pano 2015).

The waters of Lake Prespa are colder, and the temperature drops almost to 0 °C during the winter, whereas the annual average temperature of surface water is 13.9 °C and 23.6 °C in August. As a result of the vertical mixing, the lake waters are always saturated with oxygen, while the mineralization of its waters is 250 mg/liter, and the water transparency is low (Puka 1997).

The Little Prespa Lake is located on the graben bearing the same name is much shallower. This lake is going through a swamping phase. Water reeds and other water plants grow in it. Consequently, the lake water transparency is very low. The lake's waters have been widely used for irrigation through motor pumps. For this purpose, the waters of the Devolli River have been artificially diverted into Little Prespa Lake. This has filled the lake with solid deposits from the Devolli, thus severely damaging its rich plant and animal world.

The lake has great tourist potentials and it is also used for fishing. Considering the great ecological importance, Lake Prespa and Lake Little Prespa and their surrounding territory have been declared a national park area. This national park along with its respective parts in North Macedonia and Greece now form a cross-border natural park. However, illegal hunting, mismanagement of fish resources, introduction of unsuitable fish species, and overfishing using prohibited fishing methods remain issues that are causing damage to the park (Jonovski 1997). Unlicensed construction and increased traffic within the park also pose a threat.

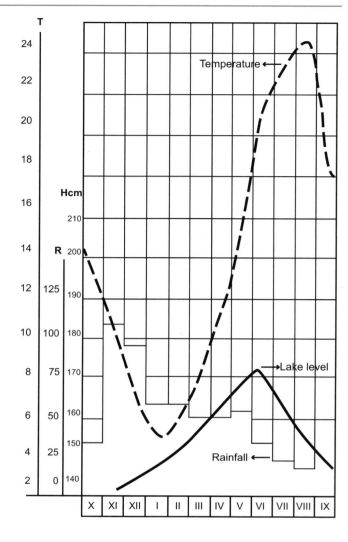

Fig. 7.12 Elements of Lake Prespa water regime levels, temperature regime and precipitation regime in its watershed (N. Pano, 1984). Source: Mediaprint P. Qiriazi Gjeografia fizike e Shqipërisë, 2019, p. 203

7.8.3 Artificial Lakes

These lakes (about 700) were created in the 1960s and 1990s to store water supplies for irrigation, electricity, and use in residential centers. The lakes of Fierza (72.6 km² and 128 m deep), Vau i Dejës, Komani, Ulza, and Banja are noteworthy for their hydropower and Lakes Thana, Kurjan, Gjanç, Doftia, etc., for irrigation. Their tourism potential has been recently reevaluated.

7.9 The Seas

The country has a long coastline. The sea has been an inseparable feature of the Albanian nature, with its great influence on the geographical landscape. A large part of the economic activity of Albanian society is related to the sea. The Albanian spiritual ties with the sea are both early and strong.

- **The Adriatic Sea**, as one bay of the Mediterranean Sea, penetrates deep into the European continent (45°9' lati-

tude in the north). The southernmost geographical boundary of the Adriatic Sea is the line drawn between Karaburuni (in Albania) and Santa Lucia de Luka (in Italy). The length from Trieste onto its southernmost border is 850 km, the width ranges from 74 km to 218 km, and the total area is 130.000 km². The Adriatic Sea lies in the intermountain tectonic pit between the Apennines and the Balkan Peninsula. The Adriatic sea's graben bed continues to decline, creating ingression shores in Dalmatia and Albania. The bottom of the sea forms a sloping trough from northwest to southeast. In most of the Adriatic Sea, the depth is less than 200 m. The northwestern and southeastern part of the Adriatic (in the Albanian coast) is well known for its shallowness. The maximum depth (1589 m) is marked near Kotorri, on the line Shengjini (Albania) to Monopol (Italy) (Pano 1973).

- **The Ionian Sea**, as one part of the Mediterranean Sea, lies in between the Balkan and Apennine Peninsulas and the islands of Crete and Sicily. To the north, it connects to the Adriatic via the Strait of Otranto, while the Strait of Messina to the west connects the Ionian to the Tyrrhenian

Sea. This sea is also formed in its graben basin and is distinguished for its great depths (3000–4000 m, whereas the maximum depth is 5121 m, the deepest in the whole Mediterranean). Isobath 100 m passes about 10 km from the Albanian coast.

7.10   The Global Water Crisis and Its Problems in Albania

Water plays an important part in human development and quality of life. As such, water is an invaluable national asset, pivotal in the development of a country. To meet the growing needs of human society, the amount of water used has steadily increased over six times in the past century. With over 7.7 billion inhabitants around the world, humanity now uses 54% of its accessible water reserves, while by 2025, it is estimated to use 70% of the water reserves. Water is constantly circulating in nature. There are still water reserves unused by man.

However, the world is facing a major global water crisis, which manifests itself in several forms: (1) Water is scarce in many countries, and large amounts of water have been polluted beyond permissible norms; (2) unequal distribution, rising water demand, and pollution are exacerbating the water crisis, making the water provision increasingly alarming (by 2030, human needs for food will increase by 55% and, consequently, the human need for colossal amounts of extra water to meet this challenge); (3) the global water crisis is also being exacerbated by global warming, and the future seems much bleaker for some regions of the world (like the Mediterranean, Africa, the Middle East, Central Asia, etc.), where water reserves are running out; (4) overexploitation and pollution of groundwater, rivers, wetlands, and the like have also taken their toll, and over 20% of their organisms have gone extinct, and ecosystems filtering the polluted water are also being destroyed. People are not aware of the causes and the risks of water crisis. Finding a solution to this crisis, which threatens our own survival, is one of the challenges of humanity at the dawn of the new millennium, but also a UN objective for sustainable development.

Humanity has the capabilities to solve the global water crisis. This is based on (1) the Earth's plentiful water resources, some not yet known or fully exploited, even in deserts (the case of the Sahara, where large aquifers have been discovered), groundwater still has unknown reserves, the major source water are polar ice caps and mountain ices, and seawater desalination; wastewater treatment and recycling, and so on; (2) the possibilities of changing the attitude and behavior of human society toward water, finding new technologies that consume less water; and (3) improving the effective management of water resources and the legal and institutional support for water use, which should be consid-

ered a "strategic element" of top importance and seen as a global problem (OKB 2015).

7.10.1 Aspects of the Water Crisis in Albania

The country's water resources, calculated per unit area and per capita, are among the highest in Europe. However, being part of the Mediterranean region (which is extremely sensitive to the effects of warming and drying climate) and being affected by the global problems and crises, the country has also been affected by the water crisis. This is especially evident in the significant shortages in the water use: Only 66% of Albanian households have access to running water; only 50% of these households have time-scheduled running water inside the house, while in rural areas only 25% of households have running water inside their houses. There are numerous cases of water pollution and misuse. The tendency is to emphasize these water problems because water has not yet been declared a "strategic priority," and there is no complete legal framework for the use of water treatment as a strategic or highly important national asset.

Solving the problems arising from the water crisis would require an accurate assessment of the country's overall water resources and a new conception and strategy for its management. The foundation of the water management strategy should be the concept of integral and sustainable management of water resources, including all types of water resources (surface, groundwater, etc.), all sectors of the economy depending on water, all the basic national objectives and constraints (social, legal, institutional, financial, environmental), the spatial distribution of water resources and the demands for this asset, etc.

Water quality and management should be considered a "strategic priority" and a major problem of XXI century (National Water Strategy of Albania 1996). New water reserves must be constantly discovered, which must be carefully collected and utilized for current needs and for future generations. Above all, water must be protected from industrial, agricultural, and urban waste. Measurements show that some of the main water bodies (the Shkumbini, Gjanica, Osumi, Fani, Kiri, etc.) and some coastal sectors have been polluted beyond the allowed limits.

The country has all the potential to solve the problems of the water crisis. This is based on the large water resources (of which only about 30% have been used), sufficient to ensure the sustainable development for the present and the future; the cooperation on global water strategies and programs, etc.

References

BCEOM and others (1996) National Water Strategy of Albania
Cavkalovski (1997) Hydrology of Prespa Lake. International Symposium, Towards Integrated Conservation and Sustainable

Development of Transboundary Macro and Micro Prespa Lakes, Albania

Cmeta A (2000–2021) Personal photo archive

Dakoli H (2000) etj. Ujërat minerare e termominerare të Shqipërisë. Kongresi i 8-të shqiptar i gjeoshkencave

Eftimi R (2000) Vështrim mbi ujërat karstke në Shqipëri, Kongresi i 8-të Shqiptar i Gjeoshkencave

Eftimi R, Frashëri A (2018a) Map of thermal water zones of Albania and main thermal water springs and deep boreholes with thermal water. Regional Hydrogeological characterisics of thermal waters of Albania. Acta Geographica Silesiana, 12/2 (30) WnoS UŚ, Sosnowiec, 2018, ISSN 1897–5100

Eftimi R, Frashëri A (2018b) Overflowing deep thermal water well Kozani-8 in Kruja Zone. Regional Hydrogeological characterisics of thermal waters of Albania. Acta Geographica Silesiana, 12/2 (30) WnoS UŚ, Sosnowiec, 2018, ISSN 1897-5100

Eftimi R, Frashëri A (2018c) Regional Hydrogeological characterisics of thermal waters of Albania. Acta Geographica Silesiana, 12/2 (30) WnoS UŚ, Sosnowiec, 2018, ISSN 1897-5100

Eftimi R, Zoto J (1997) Isotope study of the connection of Ohrid and Prespa lakes. International Symposium "Towards Integrated Conservation and Sustainable Development of Transboundary Macro and Micro Prespa Lakes, Korça – Albania, 24–26 October 1997

Eftimi R, Ahmetaj S, Zoto J (2007) Groundwater circulation in two transboundary carbonate aquifers of Albania; their vulnerability and protection. In: Witkowski AJ, Kowalczyk A, Vrba J (eds) Groundwater vulnerability assessment and mapping. Selected Papers on Hydrogeology, 11. Groundwater Vulnerability Assessment and Mapping, International Conference, Ustron, Poland, 2004. Taylor & Francis, London

Eftimi R, Andreychouk V, Szczypek T, Puchejda W (2019) Syri Shegani spring, Karst springs of Albania and their management. Acta Geographica Silesiana. WnoZ, Sosnowiec, pp 39–56. ISSN 1997–5100. A Cross section of the spring (modified after Kalaja, Rudi 1966); The connection of the spring orifice with two karst collapse lakes located about 250 m east to the spring is verified by speleodives; B. Syri Shegani Spring flows to Shkodra Lake

Eftimi R, Zojer H (2015) Simplifeld hydrogeological map of Albania, Human impakts on Karst aquifers of Albania. Environ Earth Sci 74–57-70. https://doi.org/10.1007/s12665-015-4309-7, p. 58).

Frashëri A, Çermak V (2004) etj, Atlasi Gjeotermal i Shqipërisë, UPT

Hoxha F (1999) Ndërhyrjet e njeriut në Drinin e Lezhës dhe rrjedhimet e tyre mjedisore. Gjeomorfologjia e Aplikuar, Mjedisi dhe Turizmi Bregdetar në Shqipëri

Insittuti i Hidrometeorologjik (1985) Hidrologjia e Shqipërisë

Insittuti i Hidrometeorologjik (1987) Hidrologjia e Shqipërisë

Jonovski K (1997) Ecological conditions of Prespa lakes and consequences for tourism, Internatonal Symposium, Towards Integrated Conservation ...... of Transboundary Macro and Micro Prespa Lakes, Albania

OKB (2015) Programi "Uji për jetën"

Pano N (1973) Ligjshmëria e depërtmit të ujërave të Jonit në detn Adriatk

Pano N (1984) Veçoritë e rrjedhjes ujore dhe të ngurtë. Studime Meteorologjike e Hidrologjike. nr. 10/1984

Pano N (2015) Pasuritë ujore të Shqipërisë. Botm i Akademisë së Shkencave të Shqipërisë

Plaku F (2000–2021) Personal photo archive

Plaku F (2018) Homeland Albanian's hidden gems

Puka V (1987) Varësia e mineralizimit nga prurjet e lëngëta të lumenjve, Studime Meteor. e Hidrol 12/1987

Puka V (1994) Regjimi hidrokimik i lumit Vjosa. Studime Meteor dhe Hidrol, nr. 9/1994

Puka V (1997) Water quality of Prespa lake. Internatonal Symposium, Towards Integrated Conservation and Sustainable Development of Transboundary Macro and Micro Prespa Lakes 1997

Puka V, Selenica A (1982) Regjimi hidrologjik dhe hidrokimik i liqenit të Butrinit, Studime Meteor e Hidrom, Nr. 8.1982

Qiriazi P (1980–2021) Personal photo archive

Qiriazi P (2019) Gjeografa fzike e Shqipërisë, Mediaprint

Sala S, Qiriazi P (2007) Banorët dhe mjedisi në pllajën e Dumresë

Tafli I (2002) Gjendja e ujërave nëntokësore të Shqipërisë. Strategjia e zhvillimit të qendrueshëm

Soils

8

Perikli Qiriazi

Abstract

The different types of soils are related to the diverse pedogenetic conditions and factors. The geographical distribution of soils is distinguished for its vertical stratification, mainly conditioned by the hilly-mountainous relief. Human activity also affects the soil features.

The classification of soils based on the genetic-agronomic system distinguishes zonal soils (mountain meadows soils, dark mountain forest soils, Cinnamon mountain soils, Cinnamon meadow soils, gray cinnamon soils), intrazonal soils (hydromorphic soils, saline soil), and azonal soils (alluvial soils).

Albania has limited land potential for agricultural production: Only about 25% of the land is suitable for agriculture, and part of it has rather limited fertility; about 10% of the land is in cold regions above the altitude of 1600 meters; about 55% of the lands are located on steep slopes with intense erosion activity, etc. Increasing the agricultural land area, if not impossible, is extremely limited. In some areas, soils have been severely damaged by erosion and pollution.

Keywords

Soils · Pedogenetic conditions · The genetic-agronomic system · Zonal soils · Intrazonal soils · Azonal soils

8.1 General Features

The pedogenetic conditions and factors of the country are distinguished for their diversity: The chemical and mineralogical composition of the parent formations vary; the morphological features of the relief also vary – climatic conditions and hydrographic and especially biotic factors (tall plants, microorganisms, and soil fauna). These changes are observed in the horizontal variability and particularly in the vertical variability. This brings about unequal action and features of the soil-forming factors. Consequently, different land types have been formed, whereas their geographical distribution is distinguished especially for their vertical stratification, mainly conditioned by the hilly-mountainous relief. Besides the natural soil-forming factors, the soil characteristics and their respective changes are also influenced by human activity, especially in the cultivated soils.

The genetic-agronomic system has been used for soil classification, considering soil formation conditions and morphological features of the soil profile, differentiated by pedogenesis and agronomic features that soils acquire during their agricultural use. According to this classification, soils are divided into zonal, intrazonal, and azonal soils (Gjoka and Cara 2003).

8.2 Zonal Soils

They stretch in the form of belts from sea level to the mountain heights. They are like or the same with the lands of Mediterranean mountainous countries. There are five types: mountain meadow soils, dark mountain forest soils, Cinnamon mountain soils, Cinnamon meadow soils, and gray cinnamon soils, which have respectively developed under the four main plant belts: alpine pastures, beech, conifers, oak, and the Mediterranean shrubs. There is a decrease in the height of these belts from the south to the north and from the west to the east, which is related to the differing climate conditions and vegetation in these areas (Figs. 8.1 and 8.2).

– *Mountain meadow soils* (cambisol and leptosol) extend over altitudes of 1600 m and make up about 270,000 hectares or 9.4% of the territory. The most common ones are

Fig. 8.1 Vertical zonation of soils. (Gjoka and Cara 2003, p. 26)

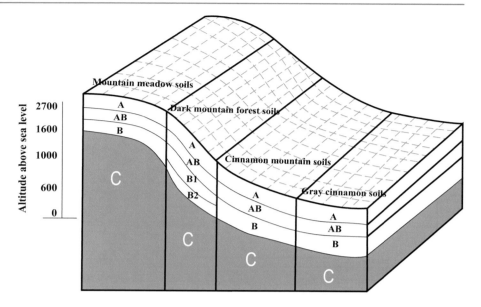

found in the north and east. These are mainly formed on limestone, sandstone, and metamorphic and magmatic parent rock, in very heterogeneous relief and with great fragmentation and slope. This type develops in areas with the harshest climatic conditions in the country, with cold winter, heavy rainfall, and little evaporation and snowfall predominance, while the summers are cool, with considerable rainfall. The herbaceous vegetation of alpine pastures, with great development during the short vegetation period, from late spring to autumn, when the temperature drops, and the snow begins, a large mass of humus is formed under these conditions (17–25% on the surface) (Gjoka and Brahushi 2007).

The humus composition in the mountain meadow soils of Southern Albania is lower (up to 17%), which is explained due to the scarcer vegetation cover and the longer and drier summers, which translates into greater mineralization. As summers grow longer and drier from the mountain heights toward the plains, the mineralization process is greater, which in turn reduces the humus. Besides the frequent reduction of decomposing organic matter, the reduction in the humus content in brown forest soils, and particularly brown and gray-brown soils, is also due to this mineralization process.

These soils have a profile about one-meter thick, with a well-defined, thick, and elongated humus horizon. Humus content decreases with depth, from about 20% to 1–1.5% on horizon B. On the surface, there is a thin layer (of 3–15 cm) permeable to the roots (barren horizon), which forms a porous mass. The humus horizon (A), up to 40 cm, dark brown to black, is brittle and airy, with well-balanced granular structure. Horizon AB reaches up to 70 cm and B up

to 100 cm. This horizon is denser, with a brownish-yellow color, without structure, and dusty.

They have medium-sized mechanical clay composition in all their profile (clay composition is 30–50% of soil) and acid reaction (pH 4–5.6), while on limestones, they are neutral or slightly acidic. This reaction is related to the high content of organic matter and the low activity of the perspiration process, conditioned by the cool climate. Depending on the conditions of their formation, and their morphological and chemical features, mountainous meadow soils are divided into four subtypes: typical, black, dry, and meadows of barren forests. The dense vegetation cover that grows on mountain meadows is used as pastures during the summer. During the communist period, some of them were used for the cultivation of potatoes, rye, and barley.

– *Dark mountain forest soils (of beech and pine forests) or cambisol, leptsol, and phaeozem* lie at altitudes between 1000 and 1600 m above the sea level. In the north and southeast of the country, they descend up to 900 m or even lower (the Albanian Alps, etc.). They occupy 14.8% of the country's area. They are more widespread in the north and east and less so in the south. They develop on terrigenous rocks, limestone, magmatic and metamorphic rocks, in rugged relief, in the humid and cold Mediterranean pre-mountainous and mountainous Mediterranean climate, under the beech and coniferous forests with sensational development, and herbaceous vegetation that requires little light (Gjoka and Brahushi 2007).

The profile of these soils, 80–140 cm thick, begins with a layer up to 5 cm thick (Ao) made of undecomposed debris and then continues with the humus horizon, up to 30 cm thick, dark brown or brown in color, medium granular struc-

Fig. 8.2 Soils map of
Albania. (Gjoka and Cara
2003, p 121)

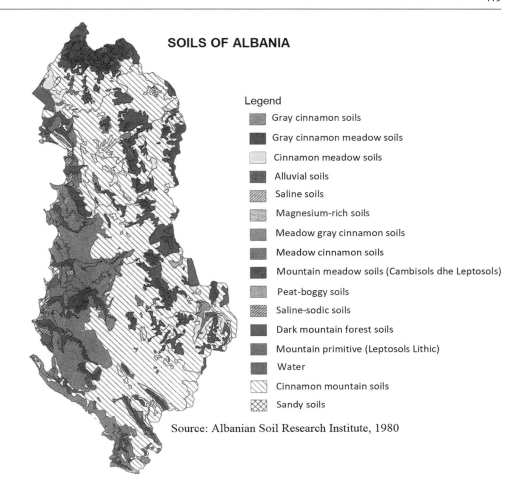

SOILS OF ALBANIA

Legend
- Gray cinnamon soils
- Gray cinnamon meadow soils
- Cinnamon meadow soils
- Alluvial soils
- Saline soils
- Magnesium-rich soils
- Meadow gray cinnamon soils
- Meadow cinnamon soils
- Mountain meadow soils (Cambisols dhe Leptosols)
- Peat-boggy soils
- Saline-sodic soils
- Dark mountain forest soils
- Mountain primitive (Leptosols Lithic)
- Water
- Cinnamon mountain soils
- Sandy soils

Source: Albanian Soil Research Institute, 1980

ture, and sub-clay composition, and then with horizon B having yellowish brown color and granular structure, up to
pleats.

Beech and conifer forests develop in the soils of this belt.
However, rye, wheat, potatoes, apples, cherries, plums,
chestnuts, etc., can also be cultivated in these lands. Many of
these soils have been abandoned, running the risk of degradation and desertification by erosion.

Depending on the conditions of their formation and their
morphological and chemical features, these soils are divided
into seven subtypes: typical forest brown soils or *humic
cambisol*, reddish forest brown soils or *humic nitosols*, black
brown forest soils, rinsed brown forest soils, carbonate-
humus soils or *rendzinc*, brown forest meadow soils or *humic
cambisol,* and shtoi soils, brought to the Shtoit plain of Upper
Shkodra (20–26 m above sea level) by waters from the slopes
of the Albanian Alps.

- *Cinnamon mountain soils and Cinnamon meadow soils*
  (cambisol, luvisol, leptosol, vertisol) have a greater extent
  (42.1% of the territory), and they are located at an altitude
  of 600–1000 m above sea level, while in the north of the
  country, they descend up to about 200 m in altitude and

formed on sedimentary and magmatic rocks, under the
rare xerophilous oak forests in relief with significant slope
with intense erosion, in climates with annual rainfall
between 600 and 1600 mm, with not too harsh winters
and generally hot summers. In winter, soil leaching
occurs, while in summer, salt accumulation occurs due to
high evaporation. Consequently, on the middle horizons
of the earth, pseudomycelia and new spots of carbonate
formations appear which condition the reaction around
the neutral value. They have a granular structure, a deep
profile up to 1.20–1.5 m, and humus surface composition
of 4–9%, while the depth is not more than 1% (Gjoka and
Brahushi 2007).

Depending on the conditions of their formation and their
morphological and chemical features, these soils are divided
into five subtypes: typical soils or *eutric cambisol*, brown
limestone or *calcium luvisol*, reddish brown or *rhodic nitosol*, black soils or *eutric vertisol*, and brown meadow soils or
luvic phaeozems.

Cereals, industrial plants, vines, fruit trees, etc., are cultivated in brown soils.

– *Gray cinnamon soils* are a characteristic of the Mediterranean shores. In Albania, they lie up to altitudes between 200 and 600 meters. They amount to 48.5% of the country's arable land. They are mainly formed on sedimentary rocks, mostly in plain relief in the typical Mediterranean climate, with summer droughts lasting up to about 3 months, accompanied by strong physical and chemical alteration, leading to the formation of secondary minerals, inhibiting the development of vegetation, consisting of Mediterranean shrubs and herbaceous vegetation, resulting in the large mineralization of organic matter and, consequently, in the small humus content (2–4% humus). They have a slight basic reaction, which is more related to the perspiration process, elongated humus horizon, mechanical sub-clay composition, and granular structure. Horizon B is saturated with bases, especially carbonates in the form of pseudomycelia and white spots. Depending on the conditions of their formation and their physical and chemical characteristics, gray cinnamon soils are divided into four subtypes: typical brown gray soils, loamy soils or *calcic luvisol*, red soils (*terra rosa*) or *rhodic nitosol,* and gray brown meadow soils or *eutric cambisol* in flat fields (Gjoka and Brahushi 2007).

Gray cinnamon soils are used for the cultivation of subtropical crops, cereals, industrial crops, vegetables, fodder, vines, fruit trees, etc. (Veshi 1986).

8.3   Intrazonal Soils

These soils are found in several vertical belts, and they are mainly related to the action of the main local factor in their formation, such as a water phreatic layer near the surface, which creates conditions for the formation of swampy soils (hydromorphic), and salty waters which form saline soils.

– *Hydromorphic soils* (*vertisol, cambisol, castanozem, gelisol*) are found in many areas of the country, but they are more common in the western plains and on the shores of lagoons, where they are often connected with saline soils. Until the 1950s, these soils took up about 38.000 hectares. The drying of swamps and the massive land reclamation in the following years turned most of these soils into fertile lands (the swamps and marshes of Maliqi, Tërbufi, Vurgu, Roskoveci, etc.), turning them into suitable land areas where they are located. Currently, they amount to 9970 hectares (Gjoka and Brahushi 2007).

The swamping process occurs when there is excess moisture in the soil, which leads to a decrease in oxygen in the soil, an increase in acidity, and a decrease in microbiological activity. Therefore, the decomposition of organic and mineral matter has to take place under the conditions of lack of oxygen, which creates little mineral bonding. As a result of incomplete decomposition of abundant organic matter resulting from rich hygrophilous vegetation, a thick layer of peat is formed in these soils. In Tërbufi, this layer reaches a thickness of 1.2 m; in Maliqi, it is 8 m, etc. They are divided into three types: swamp meadow soils, which are formed on alluvial and lake deposits in the vicinity of swamps; peatlands, which are found in former swampy territories (Maliqi, Tërbufi, Kakariqi, Roskoveci, Thumana, etc.); peatlands located mainly in Maliqi, Tërbufi, Kakariqi, Roskoveci, etc.

Due to recent preservation and drainage, these soils have been improved. They now have high yields in some agricultural crops, such as fodder, industrial plants, etc.

– *Saline soils (solonetz)* contain over 1300 mg/liter of salts. Their general area is constantly shrinking, because of artificial reclamation and desalination, after which they evolve into gray-brown soils. In 1960, there were about 35000 hectares, while in 1985, they amounted to about 16000 hectares. These soils lie in the coastal belt of the Western Lowlands and in the Vrina plain. They are formed under the conditions of high perspiration regime, because of which salts accumulate in the upper layers of the soil. During wet winters, the opposite process occurs. These soils are found in flat areas and in some cases in areas below sea level and have poor halophyte vegetation: glasswort (*Salicornia herbacea*), marina, etc (Veshi and Leka 1997).

Soil salinization is associated with the deposition of salts in geological formations which rise to the surface. It relates to the sea salt infiltration into the light soils of the coastal belt, with phreatic waters with a high degree of mineralization near the surface and with excessive irrigation, which results in the mixing of the irrigation water with the salty groundwater. It is also associated with artificially or naturally dried salty swamps or marshes. There are cases when they are related to salty springs. Most salts are made up of sodium chloride. According to their salt content, these soils are divided into poorly saline soils (0.2–0.5% salts), moderately saline soils (0.5–1% salts), and highly saline soils (more than 1% salts) (Qendra e Studimeve Gjeografke 1990; Veshi and Leka 1997).

To make these soils suitable for cultivation, they must be desalinated, their water regime adjusted, their physical and chemical qualities improved, etc.

8.4 Azonal Soils (Alluvial or Fluvisol, Cambisol, Regosol, Arenosol, Gelisol)

These are soils that form on river deposits. They can be carbonate or acid, sand, silt, or clay. They are usually formed in flat areas, and they are young. The most typical types are found in the major river valleys. They are also formed on marine, lake, swamp deposits, which are usually clay and sand, etc. Due to repeated alluvial deposits, they are composed of layers, which replace their genetic horizons, while the scarce vegetation is usually herbaceous, and in rare cases, it is forest with poplar, willow, etc.

They generally have sub-clay and clay compositions in river estuaries or in the lower and sub-sandy and sandy streams in the direction of their middle and upper course. They are unstructured soils, with basic reaction, poor in organic matter (up to 2% humus). They occupy about 160,000 hectares. About 95,000 hectares are cultivated land. According to the conditions of their formation and the degree of their development, we distinguished alluvial meadow soils, located near the bed of the lower river flows; gravel soils, which form in gravels and sands in which the pedogenetic process has just begun; undeveloped sandy soils, in beach sands; and sandy soils of old little developed dunes (Qendra e Studimeve Gjeografke 1990).

Alluvial soils are suitable for all agricultural plants, but they must be carefully treated to maintain their moisture and biological health.

8.5 Land Resources, Condition, and Damage

The total land potential for agricultural production in Albania is limited. Only about 25% of the land area is suitable for agriculture, while a part of it has limited fertility. About 10% of the land is in cold regions above 1600 m in altitude, about 55% of the lands are located on steep slopes with intense erosion, etc. Expansion of the agricultural area, if not impossible, is extremely limited. In some areas, the soils have been severely damaged by erosion and pollution.

Soil erosion is very intense due to the natural factors and exacerbated by centuries-old human activity on their vegetation cover. As a result, soils were impoverished, depleted, degraded, even desertified in many areas, and abandoned altogether.

Soils are polluted by industrial waste (mining industry of chromium, copper, iron-nickel, coal, extraction, processing of oil, etc.), urban waste, and agricultural chemicals (pesticides and chemical fertilizers exceeding standard norms, etc.).

Soil preservation and protection from degradation and pollution requires studies and effective measures (reforestation of baren areas, terraces, improvement of irrigation, tilling and cultivation technology, processing of polluting waste, and so on).

References

Gjoka F, Cara K (2003) Soils of Albania (Monograph). Migeralb, Tirane, 136 pp

Gjoka F, Brahushi F (2007) Tokat, natyra, cilësitë shpërndarja dhe përdorimet e tyre

Instituti i Studimeve të Tokave, Harta e Tokave të Shqipërisë, 1999

Qendra e Studimeve Gjeografke. Gjeografia fizike e Shqipërisë, pjesa e parë, 1990

Veshi L, etj. Pedologjia, 1986.

Veshi L, Leka I. Kripëzimi dhe dukuritë e tj në Shqipëri. Studime Gjeografke, nr. 11/1997

Veshi L, Leka I. Tokat acide në Shqipëri. Studime Gjeografke, nr. 9/1997.

Flora and Fauna

9

Perikli Qiriazi

Abstract

Albania is well known for its diversity of ecosystems, both terrestrial and aquatic habitats, which are part of the Mediterranean and Balkan chain of natural ecosystems. Many of them are preserved and intertwined in natural or almost natural state, such as marine, coastal, wetland, lake, estuary, river, plain, hilly, and mountain ecosystems; shrub ecosystems; deciduous, coniferous, and mixed-type forests; subalpine and alpine meadows and pastures; and high mountains.

These ecosystems are home to a large variety of plants (3651 species, accounting for 52% of Balkan plants and 36% of European plants) and wildlife (340 species of birds, 85 species of mammals, accounting for 42% of all mammals of Europe, etc.). But the country's biological diversity also includes the creation of several indigenous plants and livestock breeds. This is a legacy of great value to preserve but also to improve the production and quality of agricultural and livestock products. The living world of the country is of special importance even on a European and world scale: relict, endemic, and subendemic plants and many species of plants and animals globally endangered, such as Dalmatian pelican (Pelecanus crispus), pygmy cormorant (Microcarbo pygmaeus), koran (Ohrid trout, Salmo letnica), linden tree (tilia), etc.

Over the centuries, human activities have damaged the country's living world and its biodiversity.

Keywords

Diversity of ecosystems · Holarctic region · Mediterranean floristic subzone · Vegetation migrations · Endemic · Subendemic and relict species

9.1 General Features of Albanian Flora

- *Geographical position and composition.* The country's vegetation is part of the larger Holarctic region, of the Mediterranean floristic subzone of this region, and it consists of elements of the mid-European (boreal), North Balkan, Alpine-Carpathian, and Eurasian flora, which dominate the upper plant strata (floors). Some of the representatives of these subregions are beech (*Fagus sylvatica*), European ash (*Fraxinus excelsior*), maple (*Acer campestre, A. pseudoplatannus*), hornbeam (*Carpinus betulus*), (Greek) juniper (*Picea excelsa*), pine (*Pinus sylverstris*), bog pine (*Pinus mughus*) etc. Besides these, Mediterranean plants have an important place. Due to the dry and hot summers, these plants have a xerophilous character, especially plants developed in karst rocky terrains where summer drought is more intense (Qiriazi 2019).

The areas of Mediterranean vegetation and the vegetation in the north of the Mediterranean (mid-European, etc.) are divided by the western slopes of the mountain ranges that lie east of the Western Lowlands, in the northeast and east of the Middle Devolli valley, and the Miçani, Qelqëza, and Melesini Mountains from where it continues abroad. Mediterranean vegetation lies almost entirely to the west and southwest of this border, and together with the Adriatic and Greek species, it makes up 35% of all plant species in the country. This vegetation is represented by Mediterranean shrubs and trees. In the north and east of this border, starting from the oak stratum (floor), there are plants mainly of Central Europe and elements of other regions: Eurasian, North Balkan, Alpine-Carpathian, etc. All these make up 65% of the number of plants of the country's flora. This vegetation is represented by broad-leaved deciduous trees and subarctic conifers. A special place is occupied by the Illyrian-Pannonian vegetation: beech, white fir, oak (bunga), oak (*Quercus macedonica*), Turkey oak (Quercus cerris), etc (Baldaçi 1932–1937; Mitrushi 1954).

Studies on the flora have identified several layers repre-sented by old members, remnants of the tertiary tropical flora (relics), isolated in small areas, such as laurel (*Laurus nobilis*), etc., which have shown significant vitality to spread over a wider area than by members of the flora of other countries, which have the limit of their distribution in Albania (Qendra e Studimeve Biologjike 1988, 1992).

The vegetation of the country has its strongest floristic connections with the northern subregions of the large Holarctic floristic region (European, Eurasian, North Balkan, Alpine-Carpathian) (Qendra e Studimeve Biologjike 1988, 1992). More than 550 species of these regions have their southern border in Albania. These plants are found mainly in the beech belt and Alpine pasture belt. They have their largest extension in the Albanian Alps, and while gradually decreasing, they extend as south as Tomorri. In this area, species of northern origin predominate over species of other origins. Some of the northern plants are beech (*Fagus sylvatica*), black fir (*Abies nigra*), birch (*Betula alba*), etc.

The connection of our flora with that of the southern regions is the weaker, and about 150 species plants of these areas have their northern border in Albania. These plants are from the Greek flora and other elements of the Mediterranean. They are most prevalent in the maquis shrubland belt and become predominant compared to plant species of other origins.

Plant species coming from the Eastern Mediterranean (about 20) are found throughout the country, while those coming from the interior of the Balkans are found in the east and northeast of the country. There are 50 such species of plants, among them is the (Greek) juniper (*Juniperus excelsa*), etc.

The floristic connections of our vegetation with the Western Mediterranean are very weak: Only 40 species of western plants have their eastern border in Albania. They are found mainly in the belt of Mediterranean shrubs in the western part of the country and not so much in its interior parts. Some such plants are *Phillyrea angustifolia L, Plantago serraria L, Centaures sonchifolia L*, and, in the western inland part also, *Centaures nigra L*. They are mainly concentrated on the stratum (floor) of shrubs, Mediterranean forests, and oaks.

– *The evolution of flora.* The vegetation of Albania, like the whole Mediterranean vegetation, has its origins from the Paleogene (Demiri 1973). In the Oligocene deposits of Southern Europe and even the Eocene, plant remains of the same category with today's Mediterranean vegetation have been found. In the Oligocene, plants that are like today's species emerged, such as mastic tree (*Pistacia lentiscus oligocenica*), *Olea europea praxima*, etc. During this geological period and the Miocene, there are the evergreen tropical plants and deciduous plants, which represent the beginnings of medium (boreal) vegetation. Marine transgressions and regressions and the frequent

and profound climatic changes brought about great changes in the flora of the country, manifested in the reduction of the tropical elements and the expansion of the Mediterranean ones. There were many species in the Pliocene, which are still present. The Mediterranean flora, in general, took on its present features during the late Neogene and especially after the last glacial period.

Due to the extremely cold climate during glacial periods in the Quaternary, tropical and subtropical vegetation was almost destroyed or greatly changed. In the most protected places, only the species with the greatest adaptive abilities survived. They are relict plants, such as eight-petal mountain-avens (*Dryas octopetala L*), net-leaved willow (*Salix reticulata L*), and notch-leaf willow (S. retusa L.).

As a result of these profound climate changes, a new redistribution of vegetation occurred. During this time, the horizontal and vertical boundaries of the plant areas shifted. The plants that migrated to the south were replaced by the plants which came from Central Europe at the same time. Such plants were the conifers and some broad-leaved species with temporary greenery. There was also a shift in the vertical dimension. The mountain plants moved to the hills, the hilly plants moved to the plain, and the field plants disappeared and were replaced by the hilly plants. Fluctuations in the horizontal and vertical boundaries of vegetation and migrations of floristic complexes or special species also occurred in the postglacial period due to climate change (Gracianski 1971; Di Castri 1973).

During vegetation migrations, plant crossings also occurred. Thus, the crossing of the European silver fir (*Abies alba Miller*) with the Greek fir (*Abies cephalonica*) resulted in the Macedonian fir (*Abies borisii-regis Mattf*), which covers the altitudes of 1200–1800 m in the south of Albania, in Greece, and North Macedonia, replacing the beech (*Fagus sylvatica*), which, moving to the south in Albania, extends as far down as the Tomorri-Gramozi line (Qendra e Studimeve Gjeografke 1990).

The process of plant evolution continues even in historical time. The proof for this is the changes in the areas of some plants. In Albania, as in all the Mediterranean shores, man has changed the vegetation cover so much that, in its natural form, the primitive natural vegetation is preserved only in certain parts of the country (beech groves, etc.). This action was carried out in three ways: through destruction and change of natural vegetation, especially forests (the cutting of oaks give rise to the development of shrubs); through the cultivation of trees of economic importance, especially olive (one of the oldest trees of the country), citrus, and many other species; and through the creation of new types of agricultural plants and the cultivation of plants from other areas.

From the ancient to the modern times, there has been great deforestations, especially of the coastal areas. Deforestations started with the Illyrians, who were not only skilled sailors but also well-known builders of merchant ships and warships. However, it was the Roman invaders and others after them (the Byzantines, Venetians, and Ottomans) who intensified the deforestation of these areas for shipbuilding and other purposes. Forests suffered extensive damage during First World War and Second World War. Even in the second half of the last century, plenty of forests were cut down, which were cleared for arable land. After 1990, the rates of deforestation increased. Forest protection measures have recently been stepped up.

– *Floristic diversity and wealth of Albania.* Flora is submarine and terrestrial. The former consists of vegetation that grows in the seabed, near the seashores, at a depth of 1–35 m, and in clean marine waters, it is also found at a depth of 40–45 m. About 60 species of marine herbs or plants are known, of which only five species grow in the Mediterranean. Underwater vegetation consists mainly of two species of kelps (*Zostera marina L, Z nana Roth*), other herbs, and especially (Mediterranean tapeweed) Neptune grass (*Posidonia oceanica Delile*), which forms underwater meadows on the seabed. *Posidonia oceanica* is the only endemic flowering plant of this sea. Underwater meadows make a great biological wealth, with a very positive role for the environment. But they have been damaged by natural and especially human factors (like polluted water, etc.). *Posidonia oceanica* is considered an endangered species in the Mediterranean and is protected by the Barcelona Convention, where Albania also adheres to.

The terrestrial flora is distinguished for its extraordinary diversity not only in species but also in the large number of endemic, subendemic, and relict species, which rank Albania among the richest countries with plant species in the Balkans and Europe (Fig. 9.1).

Albania is characterized by a rich and interesting flora. The indicators of this plant variety are numerous. It comprises ca. 3629 vascular plant taxa, belonging to 960 genera and 175 families, thus constituting nearly 30% of the European flora (Vangjeli 2015, 2021). Out of these, 25 species and about 150 subspecies are endemic to Albania. About 160 other plant taxa share the distribution area between Albania and the neighboring countries, and ca. 450 species are Balkan endemics. The main floristic elements are *Mediterranean* (24% of the total), *Balkan* (22%), and *European* 18%, but other elements, such as *Eurasian, Euro-Siberian, and Alpine-Carpathian,* are also present (Meço and Mullaj 2015; Paparisto et al. 1988; Vangjeli 2016, 2021). Besides ordinary plants, the Albanian flora has about 300 species of medicinal and aromatic plants, about 40 species of fodder plants, about 35 species of forage or dye, about 50 species of honey-making plants, and about 70 species of plants used as food (Ministria e Mjedisit 2015).

– *The high rate of endemism* is another indicator of the diversity of the country's vegetation and a characteristic feature of the Mediterranean flora and of the southern peninsulas of Europe, related to the former land connections between Europe and Africa, which conditioned the floristic unity, while the subsequent segregation of special areas of the Mediterranean from the seas and high mountains differentiated the flora, bringing about the birth of endemic plants. Moreover, the southernmost peninsulas of Europe, being warmer during the Quaternary glacial period, served as shelter for plants coming from the colder northern regions. This created large numbers of endemic plants.

In the Balkans, Albania is distinguished for the large number of endemic plants, which make up about 1% of the Albanian flora, while subendemic plants make up about 4%. Besides the highly suitable conditions throughout the Mediterranean and the southern peninsulas of Europe, this phenomenon is related to our geographical position, being on one of the routes of floristic migration and the transition from Mediterranean to continental climate. It is also related to the extremely rugged relief which has created an array of microclimates, which, in turn, have conditioned the great diversity of ecological and edaphic conditions, which give rise to relict endemism and active new species formation. The Albanian Alps are particularly distinguished for this (Photo 9.1).

The endemic plants of the country are divided into relict (*Wulfenia baldacci Degen, Forsythia europaea Degen et bald*, etc.) and neo-endemic (most of them) formed late in their present habitat (isolated and in special ecological and edaphic conditions, especially in the Alps Albanian), where are found *Lunaria telekiana jav, Crepis bertiscea Jav, Aster albanicus subsp. Paparisto, Ligusticum albanicum Jav,* etc (Qendra e Studimeve Biologjike 1988, 1992; Vangjeli 2016, 2021).

Such great plant diversity is related to the long and complex evolution, the country's position on the European subtropical Mediterranean belt, where the influence of tropical latitudes is intertwined with that of medium latitudes, whereas the location on the shores of the two Mediterranean seas (the Adriatic and the Ionian) has conditioned the combination of maritime and continental influence, which makes the nature of the country much more diverse. This great variety is also related to the features of the relief, the predominantly hilly-mountainous character with its sharp fragmentation and contrasts. It is also related to its climatic, hydrographic, lithological variety and soil diversity as well as the country's position at the intersection of important routes of floristic and faunal migration, human activity, etc.

– *The cycles of the Albanian flora.* The growing season of plants, due to changing climatic conditions, does not start,

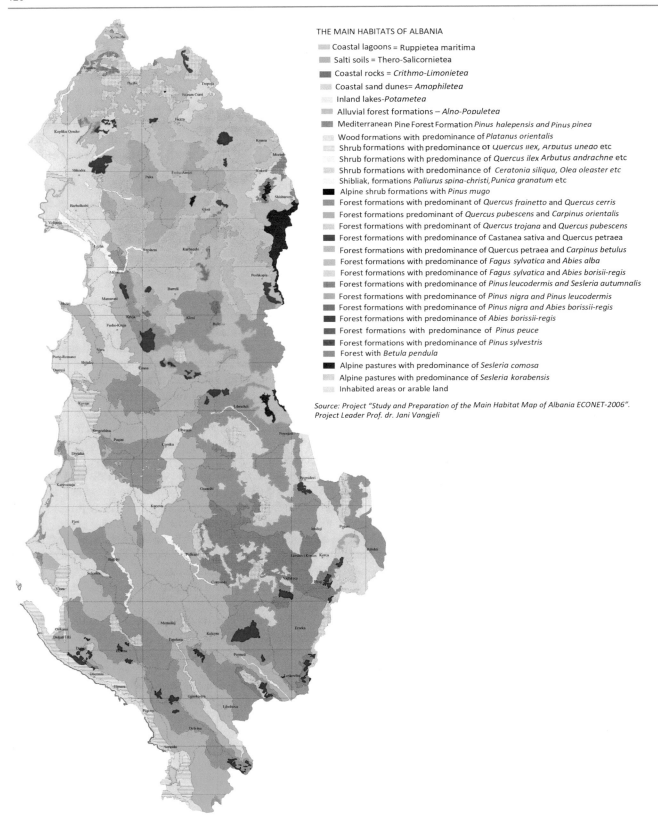

THE MAIN HABITATS OF ALBANIA

Coastal lagoons = Ruppietea maritima
Salti soils = Thero-Salicornietea
Coastal rocks = *Crithmo-Limonietea*
Coastal sand dunes= *Amophiletea*
Inland lakes-*Potametea*
Alluvial forest formations – *Alno-Populetea*
Mediterranean Pine Forest Formation *Pinus halepensis* and *Pinus pinea*
Wood formations with predominance of *Platanus orientalis*
Shrub formations with predominance оf *Quercus ilex, Arbutus uneao* etc
Shrub formations with predominance of *Quercus ilex Arbutus andrachne* etc
Shrub formations with predominance of *Ceratonia siliqua, Olea oleaster* etc
Shibliak, formations *Paliurus spina-christi, Punica granatum* etc
Alpine shrub formations with *Pinus mugo*
Forest formations with predominant of *Quercus frainetto* and *Quercus cerris*
Forest formations predominant of *Quercus pubescens* and *Carpinus orientalis*
Forest formations with predominant of *Quercus trojana* and *Quercus pubescens*
Forest formations with predominance of Castanea sativa and Quercus petraea
Forest formations with predominance of Quercus petraea and *Carpinus betulus*
Forest formations with predominance of *Fagus sylvatica* and *Abies alba*
Forest formations with predominance of *Fagus sylvatica* and *Abies borisii-regis*
Forest formations with predominance of *Pinus leucodermis* and *Sesleria autumnalis*
Forest formations with predominance of *Pinus nigra* and *Pinus leucodermis*
Forest formations with predominance of *Pinus nigra* and *Abies borissii-regis*
Forest formations with predominance of *Abies borissii-regis*
Forest formations with predominance of *Pinus peuce*
Forest formations with predominance of *Pinus sylvestris*
Forest with *Betula pendula*
Alpine pastures with predominance of *Sesleria comosa*
Alpine pastures with predominance of *Sesleria korabensis*
Inhabited areas or arable land

Source: Project "Study and Preparation of the Main Habitat Map of Albania ECONET-2006".
Project Leader Prof. dr. Jani Vangjeli

Fig. 9.1 The main habitats of Albania. Project "Study and Preparation of the Main Habitat Map of Albania ECONET-2006." Project Leader Prof. dr. Jani Vangjeli

Photo 9.1 Endemic Albanian plants (Hoda 2020). 1. Astragalus autranii Bald. 2. Gymnospermium scipetarum E. Mayer & Pulevic. 3. Forsythia europaea Degen & Bald. 4. Hypericum haplophylloides Halascy & Bald

does not end, and does not have the same duration (Demiri 1973, 1985). There are three plant cycles: (1) Continental plant cycles are found in Northern, Northeastern, and Eastern Albania. There are two vegetative periods, the first break, the result of physical drought (lack of water), in July and August, while the second, the result of low temperatures in winter. (2) Plant cycle of high places, above 1800 m, has a short vegetative period, about 5 months, and begins with the melting of snow, usually in early May, when it begins to grow rapidly, reaching optimal development at the end of June to the beginning of July, when the mountainous landscape takes on the appearance of a multicolored carpet, where the flowering of *crocus veluchensis* (*Crocus veluchensis Herbert*), snowdrop (*Galanthus nivalis L*), etc., can be seen. From the end of July to the beginning of August comes the time of summer, which does not occur every year. In September to October, the plants bear fruit and release the seeds. The first frosts and frosts of autumn mark the beginning of the long period of winter dormancy. (3) The Mediterranean cycle has a vegetation cycle longer than the first two, since early autumn, with the first September rains, the vegetation immediately begins to flourish, which, in most cases, continues into winter. In this season, the colchicum (autumn crocus) (*Colchicum autumnale L*) and sternbergia lutea (winter daffodil) (*Sternbergia lutea Ker-Gawl*) produce their flowers, forming a distinct landscape. Only in January and February, in some areas with Mediterranean plant cycles, where the temperature drops below 3–5 °C, there is a short interruption of the vegetative cycle. From the end of February to the beginning of March, the first flowers of early spring appear, such as hellebore (*Helleborus odorus, Waldst et Kit.*), primrose (*Primula vulgaris Huds.*), violet (*Viola odorata L*), etc. In April, new vibrant flowering reaches its peak in the first 10 days of May, as early as June, and especially in July, with the decrease of humidity and increase of temperature, the vegetative cycle ends, the yellow color prevails, and everything falls into calm to resume again with the first rains of autumn. In these 2 months, the xerophilous plants live.

9.2   Vertical Zones of Vegetation in Albania

In accordance with the ecological and edaphic conditions, plant communities have their own places in nature, which is expressed in their vertical stratification. Their vertical structure is mainly conditioned by temperature. As temperatures in the north are lower than in the south at the same altitude, the zones of vegetation start and end at a lower altitude than in the south. For the same reason, the extent of these zones of vegetation also decreases from west to west.

The classic of vertical zones of vegetation for Mediterranean countries can be seen on the Dajti Mountain: At the bottom comes the subzone of evergreen shrubs, and above that is the subzone of deciduous shrubs; above this stratum comes the oak zone and above that the beech zone, while in the higher mountains, above the beech zone come the pine groves and the alpine pastures. Besides this prevailing stratification, there are some changes in the placement of plant zones due to local natural and anthropogenic reasons: (1) In southern and southwestern Albania, the beech area is occupied by the Macedonian fir, while the oak layer is occupied by mixed forests. (2) In some cases (in Çarshova, on the western slope of the Melesini Mountain), the plant layering passes directly from the Mediterranean bushes to the Mediterranean fir zone. (3) In the pits of Thethi and Valbona, the oaks appear over the conifers, which is related to the deposition of cold air at their bottom. (4) There are cases of vegetation spreading from one zone to another, as a consequence of the law of compensation.

There have been several views regarding the plant stratification of the country. Margrafi distinguished four (strata) zones (Mediterranean shrubs, oaks, beeches and conifers, and alpine pastures). Later, Ilia Mitrushi distinguished five (strata) zones. To Margrafi's layers, he added pines, which form a belt on the beech zone. In the work *The Flora of Albania* (1992), four plant zones are distinguished again, although with slightly different names for them (first floor). The vegetation zones is treated here based on this seminal work (Fig. 9.2) (Margraf 1927, 1932; Mitrushi 1954, 1956, 1966; Qendra e Studimeve Biologjike 1988).

- *The zone of Mediterranean shrubs and forests* lies mainly in the western and southwestern part of the country and penetrates to the east along the transverse river valleys. In the north, this layer climbs to an altitude of 400–600 m, while in the south up to 800–900 m, rarely up to 1000 m.

This plant stratum is mainly the result of centuries-old human activity. The forests that covered the western part of the country have been cut down and burned by man. Due to the Mediterranean climate conditions, their renewal is very difficult and slow process. These shrubs mainly represent the plants which constituted the former understory (sub-forest). On the steepest and most eroded slopes, the bushes are natural species because they can develop in the poor soils of these slopes. Further degradation of the vegetation cover of the Mediterranean shrubs through cutting, burning, overgrazing has led to the formation of underbrush (coppice) which are shrubbery up to 1.5 m high composed of plant communities of sage (*Phlomis fruticosa L*), sage (*Salvia officinalis*), heather (*Erica verticillata fox*), etc. On occasions, even the underbrush is a natural community which develops in dry and poor rocky soils. In the southwestern part of the country, the underbrush consists of xerophilous plants that endure the long, dry summers. Further natural degradation, particularly artificial degradation in some sectors, has almost completely wiped out plants and, in some sectors, completely bare rock emerges.

The zone of the Mediterranean shrubs consists of two subzone: the macchia subzone, with evergreen plants, and the shibliak subzone, consisting of plants with temporary greenery.

The macchia subzone covers the lower parts of the layer in question and occupies an area three times larger than the shibliak subzone. It is the most typical formation of Mediterranean vegetation consisting of several dozen drought-resistant species of shrubs and bushes, such as strawberry tree (*Arbutus unedo L*), heather (*Erica arborea L*), mastic tree (*Pistacia lentiscus L*), myrtle (*Myrtus communis L*), laurel (*Laurus nobilis L*), kermes oak (*Quercus coccifera L*), Spanish (rush) broom (*Spartium junceum L*), etc. At higher altitudes inland, there are other evergreen shrubs that make up the pseudo-macchia, resistant to low temperatures (below −16 °C): common boxwood (*Buxus sempervirens L*), cade juniper (*Juniperus oxycedrus L*), etc. (Qendra e Studimeve Biologjike 1988).

The Shibliak subzone is found above the macchia sublayer, especially along the major river valleys. It consists of hornbeam (*Carpinus orientalis*), *Paliurus spina-christi Miller*, *Cotinus coggygria Scop*, *Forsythia europaea* (endemic plant), etc. Here is also found the wild pomegranate (*Punica granatum L*), which grows in limestone terrains, especially on the hill of Renci and Kakariqi, etc. Xerophilous herbaceous formations grow in both sub-layers forming steppe meadows.

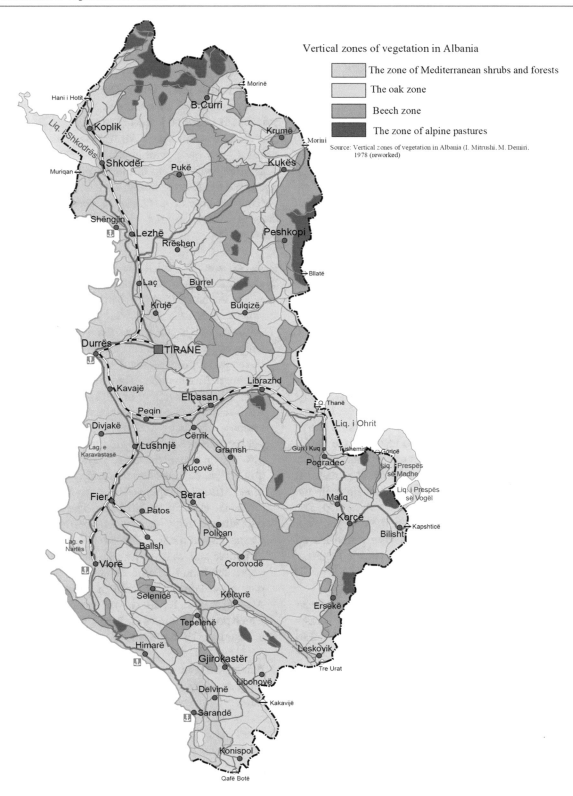

Fig. 9.2 Vertical zones of vegetation in Albania (Mitrushi and Demiri 1978 (reworked). Source: IDEART Gjeografia 11 (2019), p. 62)

Mediterranean shrubs are used for grazing, firewood, in industry (cashews, apricots, heather, strawberry tree), and as medicinal plants. In recent decades, these shrubs have been replaced by subtropical plantations and vineyards in some areas (Qiriazi 2019).

This plant zone also includes the Mediterranean forests which have a more limited extent: (1) stone (umbrella) pine forests (*Pinus pinea*) and Aleppo pine (*Pinus halepensis*) which make the famous Divjaka forest (1100 hectares) that lies in the belt coastal between the Shkumbini and the Semani estuaries. The forest stretches on sand dunes which rise between the ditches (canals, water ponds). The entirety of these microenvironments has conditioned the high biodiversity. It is a relic forest, a remnant of the former forest that probably stretched along the entire Albanian Adriatic coast. It is an ecosystem with vibrant biodiversity, special structure, and rare beauty. Its vertical structure begins with the dominant tall crowns of pines, evergreen Mediterranean oak, elm (*Ulmus minor Miller*), ash (*Fraxinus angustifolia Vahl*), poplar (*Populus alba*), black alder (*Alnus glutinosa L*), etc., and it continues with dense macchia bushes and creeping lianas (*Smilax aspera L,* etc.) and to the herbaceous vegetation, mainly hydro-hygrophilous in swamps (struga), xeromezophiles, and psammophytes. (2) Holm oak forests (*Quercus ilex*), which is the most typical Mediterranean tree. It has a smaller spread in the northeastern hills of Karaburuni and in some gorges of the Ionian coast, in the valley of the Pavlla and the Shushica, etc. (3) Macedonian oak forests (*Quercus trojana Webb*), which form simple or mixed oak groves. (4) Valonia oak forests (*Quercus macrolepis*) mainly in the coastal mountains with special spiritual values. (5) The pubescent oak forests (*Quercus pubescens Wild*) are the most widespread in the country. (6) Forests with broad-leaved essences mixed in special areas (karst, etc.) (Qendra e Studimeve Gjeografke 1990).

– *The oak zone* occupies about 36% of the country's forest area. It lies over the Mediterranean shrubs and forests stratum (or in some cases even inside them) from 400 to 1250 m above sea level. It consists of several types of oaks: bunga, bungëbuta, shparthi, qarri, and bulgari. Besides these, there are also black pine (*Pinus nigra arnold*), Hungarian oak (*Quercus frainetto Ten*), Turkey oak (*Quercus Cerris L*), the sweet chestnut (*Castanea sativa Miller*), shrubs like hornbeam (*Carpinus betulus L*), etc. Some of these plants can also descend below and climb higher than the oak stratum.

Mesophilic oaks of the Central European region and the northern Balkans predominate in the north of the country (Hungarian oak (bungë) and Turkey oak forests), and in the central part, meso-xerophilous oaks (Hungarian and Turkey oak), while in the south, xerophilous oaks predominate. They

have quite rich low vegetation. They are mostly mixed with two to three types of oaks. The most important forests of this floor stratum are (1) oak forests (*Quercus petraea, Mattuschka, Liebl*), mainly in the north and east of the country; (2) Italian oak forests (*Quercus frainetto Ten*) found almost everywhere, especially at altitudes between 500 and 900 m; (3) chestnut forests (*Castanea sativa Miller*) either cultivated or natural ones; (4) Austrian oak forests (*Quercus cerris L*) are quite widespread, and (5) hornbeam forests (*Carpinus betulus L*) at altitudes between 200 and 1550 m (Qendra e Studimeve Biologjike 1988).

The oak forests are close to densely populated areas; therefore, they are the ones most damaged by human activity.

– **Beech zone includes the beech, fir, and pine forests**. Beech forests lie at altitudes between 800 and 2000 m, where the beech (*Fagus sylvatica*) finds its optimal conditions and is associated with white fir (*Abies alba*). In Albania, the beech forms the southernmost tip in the European area, and it is even found in the form of a highly developed forests as far down as the mountains of the Southern Mountainous Province. This is related to the drier climate prevailing in the southern part of the country. The beech tree grows on fresh slopes. Significant differences in their floristic composition are observed between the beech forests growing on limestone and those growing on magmatic stone. The beech forests are generally simple but are often associated with silver fir (*Abies alba Miller*), maple (*Acer pseudoplatanus L*), ash (*Fraxinus angustifolia Vahl*), etc. The beech forests mixed with silver fir have been more numerous, but human activity has hewed more of the latter. Within the beech areas, in poorer and warmer places, black pine forests (*Pinus nigra*) are found. The beech is associated with the black pine only temporarily, until the beech has outgrown it. After that, the beech crowns create suffocating shade which no longer allows the black pines to develop (Qendra e Studimeve Biologjike 1988).

Black pine (Pinus nigra) is the most widespread species, and it forms large forests. The black pine association with the forsythia, an endemic plant, is of great ecological importance. In this plant stratum, there are also birch forests (*Betula alba L.*), although with limited extension. Recently, this plant's area has been reduced, due to global warming but also due to the great damage from hewing, especially during the communist period. It has also been used as an ornamental plant because of its special beauty (Hoda 1989).

In the lower part of the beech forests, there are Meso-European-type shrubs, made up mainly by hazelnuts (*Corylus avellana L.*). Due to the shade, the sub-layer and herbaceous vegetation is poor. They are important for timber and are home to much wildlife.

The fir forests are Mediterranean and subarctic type. In this plant layer, the former is represented by the Macedonian fir (*Abies borisii-regis Mattf*), which is found in the mountains of Southern and to some extent Southeast Albania. It grows at altitudes between 1000 and 1700 m above sea level, but it can also grow even at lower or higher areas. The xerophytic species of the Mediterranean flora predominate like oak, hornbeam, etc. It is often accompanied by black pine. The areas with Macedonian fir forests used to be larger, but indiscriminate cutting has reduced them. As far as the subarctic conifers are concerned, European spruce (*Picea excelsa*) is distinguished. In between the village Ragami and Valbona (in the Valbona valley), it is the largest spruce forest in the country. It often grows in community with beech, pine, white fir (Abies alba), aspen (*Populus tremula*), and Balkan pine (Pinus peuce). European spruce trees can reach up to 40 m in height. This kind of forest has been damaged by hewing.

Pine forests are of three main types: Bosnian pine (*Pinus leucodermis Antoine*), Balkan pine (*Pinus peuce Griseb*), and Heldreich's pine (*Pinus heldreichii Christ*). Pine forests grow higher than all other trees (1600–2300 m). There are cases where these forests grow also at 800–1000 m. They have their optimal development at altitudes between 1500 and 2000 m, where they are simple, because low temperatures, especially late frosts, hinder the growth of other trees. The sub-layer vegetation, especially the herbaceous one, is quite rich in this stratum (Qendra e Studimeve Biologjike 1988).

– *The zone of alpine pastures* has primary origin. All the other herbaceous formations are secondary, i.e., created after the extinction of the forest vegetation. They form a special stratum above the beech layer up to the greatest heights of vegetation extension. The vegetation here is of the Alpine and subalpine type, typical of the Mediterranean mountainous climatic zone. In the northern part of the country, there are pastures with Central European vegetation, which are denser and more developed, while in the south, the pastures have Mediterranean vegetation and xerophilous features (Buzo 2001).

Pastures consist of many herbaceous plant communities, mainly represented especially from grasses: some species of festuca (*Festuca sp. div.*), sesleria (*Sesleria sp. div.*), etc., and in some places shrubs like juniper (*Juniperus nana Willd*), dwarf mountain pine (*Pinus mughus*) in Northern Albania, etc. The Alpine pastures stratum is used as summer pastures for livestock.

9.3   General Features of the Fauna

The geographical position of the country at the crossroads of migration routes for bird, bats, and insects, diversity of ecosystems (marine, coastal, lagoons, swamps, marshlands, lakes, rivers, Mediterranean shrubs, and deciduous, coniferous mixed forests, but also meadows and Alpine pastures, etc.), the rugged relief full of hidden recesses, climate and water changes, etc., have created the conditions for the significant diversity of wildlife.

Although it is part of the Mediterranean region, the Albanian fauna has its special characteristics. Its faunal spectrum consists of Holarctic elements (Eurasian, Mediterranean, Balkan, etc.). There are 3250 species of mollusks, 520 species of insects, 4000 species of marine fish, 249 species of freshwater fish, 64 species of amphibians, 15 species of reptiles, 340 species of birds, and 85 species of mammals, which make up 42% of all mammals in Europe. Out of them, 79 live on land, while 6 are aquatic mammals (Dragoti 2012).

Many endemic and relict species are part of the Albanian fauna. Lake Ohrid is notable for its faunal endemism. Insects known to date offer 16 species of relicts. Further knowledge of the country's fauna will reveal other endemic species.

The large number of subendemic terrestrial species with Greece and the territories of the former Yugoslavia, as well as the Adriatic marine endemics, testifies to the importance of Albania for the preservation of these taxa in the Balkans and the Mediterranean. The coastal lagoons and the large inland lakes serve as wintering areas for migratory birds, with an average of about 70 species of waterfowl, with about 180.000 birds wintering in these wetlands. At least four of them (the Karavasta Lagoon and Lakes of Shkodra, Butrinti, and Prespa) are considered as areas of international importance for birds, known as IBA (Important Bird Area), or as Ramsar sites, with over 20.000 wintering water birds each (Bino et al. 2006).

In Albania, there are globally endangered species: 72 globally threatened species of vertebrates and 18 species of invertebrates have a part of their habitats and populations in the country's territory, species like Dalmatian pelican (*Pelecanus crispus*), pygmy cormorant (*Phalacrocorax pygmaeus*), koran (Ohrid trout) (*Salmo letnica*), sturgeon (*Acipenser sturio*), and Albanian water frog (*Pelophylax shqipericus*). Depending on its natural habitat, the animal world is divided into terrestrial and aquatic fauna.

9.3.1   Terrestrial Fauna

In terrestrial environments, the following faunal complexes are distinguished:

- *The fauna of Mediterranean shrubs and evergreen forests* consists of several species of mammals, like the jackal (*Canis aureus L*), fox (*Vulpes vulpes L*), beech marten (*Martes foina L*), European polecat (*Mustela putorius*), otter (*Lutra lutra*), weasel (*Mustela nivalis*), hare (*Lepus europaeus, Pall, L. capensis L*), wild boar (*Sus scrofa*), roe deer (*Capreolus capreolus L*), etc. Among the birds live the gray partridge (*Perdix perdix L*), quail (*Coturnix coturnix L*), pheasant (*Phasianidae*), hooded crow (*Corvus cornix*), magpie (*Pica hudsonia*), some species of sparrows, and other common birds. In this faunal complex, there are amphibians and reptiles – turtles, lizards, snakes of different species such as whip snake (*Coluber jugularis*), Aesculapian snake, and four-lined snake (*Elaphe longissima* and *quatuorlineata*) – whereas the insects world is even richer (Qendra e Studimeve Gjeografke 1990).

- *The fauna of (friganas) and rocky terrains* lives in baren areas or areas with sparse vegetation, severe drought, high summer heat, and generally mild winters. It consists of several species of mammals which are predominantly small rodents, while large ungulates like roe deer, goat-antelope (*Rupicapra rupicapra*), etc., rarely descend from the higher areas. There are also larks and starlings, quail (*Scolopacidea*), rock partridge (*Alectoris graeca Meish*), and hawk (*Accipiter*), and many reptiles like lizard (*Lacerta muralis*), Aesculapian snake (Elaphe), horned viper (*Vipera ammodytes V*), etc. As for insects, the grasshopper (*Orthoptera*) is common.

- *The fauna of mountain (oak, beech, conifers) forests* is the richest in animal species, especially mammals and birds. These animals live in conditions of often harsh and rugged relief covered in high dense forest vegetation, in conditions of harsh winters and stable snow cover, and in territories with greater extent of protected areas. The mass exodus of the population from mountainous areas after 1990 has created the conditions for the increase of wildlife in these areas. Some animals of this faunistic complex are brown bear (*Ursus arctos*), wolf (*Canis lupus L*), wild boar (*Sus scrofa*), pine marten (*Martes martes L*), beech marten (*Martes foina L*), red squirrel (*Sciurus vulgaris*), fat dormouse (*Glis glis*), and hazel dormouse (*Muscardinus avellanarius*); weasel (*Mustela nivalis*), found in cavities, etc. Less common are the lynx (*Lynix lynix*), the wild cat (*Felis silvestris*), the wild goat (*Rupicapra rupicapra*), and some rare birds, such as capercaillie (*Tetrao urogallus*), which are beautiful natural wildlife species. More common are rock pigeons (*Columba livia*), Eurasian jay (*Garrulus glandarius*), falcons, etc. (Photo 9.2).

- *The fauna of the mountain heights* is poorer. Only animals adapted to the harsh climate conditions, and the extremely rugged terrain (wild goats) lives there. Some predators like wolves and (rarely) bears in search of prey also climb at these heights. As for the birds are distinguished: the flocks of Alpine chough (*Pyrrhocorax graculus*), the bearded vultures (*Gypaetus barbatus*), the black vulture (*Gypus fulvu*), and the mountain eagle (*Aquila chrysaetos*), which build their nest on the steep rocks; Egyptian vulture (*Neophron percnopterus*); and raven (*Corvus corax*).

- *The fauna of caves and underground cavities*, due to the isolated and extreme living conditions in the eternal darkness of the underground world, it is distinguished for its special features and endemism. There are 14 species of bats, some of which are endemic. They belong to the families of *Rhinolophidae, Vespertilionidae, Molossidae*, etc. There are also only Mediterranean species, such as Khul's bat (*Pipistrellus kuhli kuhli*). Several caves with large colonies of bats have been discovered like the caves of Baruti, Ketri (in Gramshi), the cave of Mark Shytani (in

Photo 9.2 Lynx lynx balcanicus (IDEART, Gjeografia 11, 2019, p. 63)

Photo 9.3 Pelecanus crispus
(Cmeta 2019)

Photo 9.4 Flamingos (Phoenicopterus roseus) in Karavasta Lagoon (IDEART, Gjeografia 11, 2019, p. 63)

Upper Curraj), etc. Bio-speleological studies will identify many other cave creatures (Hanak and Lamani 1961; Lamani 1972; Bino et al. 2006).

– ***The fauna of lagoons, river deltas, and coastal swamps*** is the most special and the richest complex of the country and the whole Adriatic coast. Although many wetlands were drained during the second half of the last century, some not-so-small sectors have remained in their natural state to this day, especially in the river deltas. The combination of aquatic and terrestrial environments is of great importance, especially the combination of salty and fresh

waters, of rich hygrophilous vegetation with herbaceous, shrubby, and forest vegetation. The low and ebb tides and the maximum and minimum inflows of rivers also have an impact, during which entire surface areas are periodically released or covered by water. These conditions provide suitable feeding environment for wildlife but also shelters for nesting (Photos 9.2, 9.3, 9.4 and 9.5).

This faunal complex is particularly distinguished for its variety and wealth of waterfowl and partly of its terrestrial birds. The large flocks of the Dalmatian pelican (*Pelecanus*

Photo 9.5 Cygnus cygnus
(Cmeta 2020)

crispus) in the Karavasta Lagoon and the rosy pelican (P. onocrotalus) in Lake Prespa are well known. About 6% of the world population of Dalmatian pelicans live in Albania. The pelican population is declining worldwide. But only two species, among them the Dalmatian pelican, are considered endangered. The bird world consists of common species, such as shoveler and native wild ducks, pheasants, herons, winter, and summer migratory geese. The lagoons and river deltas are home to rare birds, such as whooper swans (*Cygnus cygnus*), goldeneye (*Clangula clangula*), red-breasted goose (*Branta ruficollis*), common eider (*Somateria mollissima*), white stork (*Ciconia ciconia*), once a widespread species, but the drying up of these wetlands reduced their numbers. In the coastal wetland complexes and the river deltas covered with bushes and reeds, there are several species of mammals like otters (*Lutra lutra*), foxes, jackals, rabbits, etc. (Zeko 1963; Bino et al. 2006).

9.3.2 Aquatic Fauna

A wide variety of wildlife lives in different aquatic environments (marine, lagoon and coastal wetlands, rivers, lakes and freshwater wetlands, etc.). There are several fauna complexes.

- *The fauna of the coastal waters* forms a special complex represented by several species of seagulls, little terns (*Sterna albifrons*), small shearwaters (*Puffinus*), Mediterranean cormorant (*Phalacrocorax aristotelis desmarestii*), etc.
- *The freshwater fauna in the country's interior* is very rich with species of fish, mollusks, gastropods, bivalves, arthropods, aquatic insects, amphibians, and salamanders and, in wet places, birds, some mammals (otters), etc. There are 52 species of fish, of which 35 species are

autochthonous, 10 migratory, and 7 brought from other countries. Lake Shkodra, together with Buna, are well known, and they constitute a "fish mine." In Lake Ohrid, 18 species of fish grow, half of which over are endemic: koran (Ohrid trout *Salmo letnica*) and belushka (*Salmothymus ohridanus* Stend). Glacial lakes have poor fauna, which is related to very low temperatures of water and their freezing during winter. The fauna of rivers is poorer than fauna of lakes. It is mainly found in the lower river sections, especially in their estuaries, while the water complex (Shkodra Lake, the Buna River, the sea) is an "aquarium" of aquatic fauna, particularly of ichthyofauna. Trout (*Salmo trutta*) grows in rivers and mountain streams with clean fresh water. In fresh water, there are also reptiles (turtles, two species of snakes), amphibians, shellfish, crustaceans, etc. (Raka N. *Peshqit e Shqipërisë* 2000).

- *The fauna of the Adriatic and Ionian seas* consists of about 110 fish families, which is associated with the river flows that bring food. The pollution of the rivers has reduced marine life. Marine fish are sedentary (autochthonous), such as sea bass, mullet, common dentex, sea bream, red porgy, but also migratory, such as tuna, marlin, etc. Sardines (*Sardina pilchardus sardinia*) are distinguished for their industrial importance, gray mullet (*Mugil cephalus*), sea bass (*Disentranchud labrax L*), sea bream (*Sparus aurata*), common dentex (*Dentex dentex*), common smooth hound (*Mustelus mustelus*), which is a species of Mediterranean shark, common torpedo (*Torpedo torpedo*), etc. As far as depth predator fish are concerned, there are Mediterranean moray (*Muraena helena*), stingray (*Trygon pastinaca*), etc. On hot days, dolphins (*Delphinus delphis*) also appear, and rarely, the cachalot (Physeter catodon), a whale from the Atlantic, also appears; beaked whales (*Ziphius cavirostris*), Mediterranean monk seal (*Monachus monachus*), especially on the rocky shores of the Karaburuni peninsula and the Sazani island, rarely on the Adriatic coast, etc. (Raka N. *Peshqit e Shqipërisë* 2000).

As far as the invertebrates are concerned, there are mussels (*Mytilus galloprovincialis*), which are cultivated on the shores of Saranda and Butrint; flat oyster (*Ostrea edulis*); pinna (*Pinna*); cephalopods, such as cuttlefish (*Sepia*); sponges; coelenterata; and thorny skin (*echinoderm*), such as urchins, starfish, etc. From gastropods, there are common limpet (*Patella*), murex (*Murex*), etc. Among the crustaceans, there are the shrimp (*Penaeus*), spiny lobster (*Palinurus vulgaris*), etc.; from reptiles: water snakes, leatherback sea turtles (*Dermochelys coriacea L*), and rare and very large (up to 600 kg) and loggerhead sea turtles (*Caretta caretta*).

In regions less affected by human activity, the terrestrial and aquatic fauna is much richer and far more diverse, especially in the protected areas.

9.4 Damage to Vegetation and Wildlife and the Protection of Biodiversity

The Albanian environment is renowned for its high diversity of ecosystems and habitats. Forests occupy about 33% and pastures about 15% of the country's territory. However, vital productive human activities have damaged the living world and biodiversity (Strategjia e plani i veprimit për biodiversitetin 1997–1998; Vangjeli 1999). Among these human activities are extensive and intensive development of agriculture (mass drying and reclamation of swamps, deforestation of large areas clearing them for arable land, the use of pesticides, particularly the phosphorus-organic types); faulty tillage and flood irrigation technologies; overuse of natural lake and river water for irrigation and electricity generation, going beyond the subsistence level of their living organisms; the use of old industrial technologies which emit gases into the environment and life-threatening waste; and overexploitation of forests beyond the regenerative capacity during the communist period, even to their complete disappearance in some areas. After 1990, intensive fishing and hunting; expansion of the urban centers in recent years following the movement of rural population toward cities or the plains and coastal areas. Consequently, forests, shrubs, and pastures were reduced and degraded, which in turn also affected the faunal communities.

Even in recent decades, biodiversity is being severely damaged. The existence of 122 vertebrate species is endangered, out of which 27 mammals (the Mediterranean seal, lynx, wild cat, etc.), 95 species of birds (Dalmatian pelican, vultures, birds of prey group, etc.), while 17 species of birds no longer nest in the territory of the country. There are also 6 species of fish, 4 species of amphibians (Epirus frog, etc.), and 12 species of reptiles that are endangered. Since the end of XIX century and the beginning of XX century, two species of plants and four species of mammals have disappeared, such as the red deer and fallow deer (*Cervus elaphus* and *C. dama*). Also, four plant species are endangered as they have lost over 50% of their population. The habitats of river ecosystems are being endangered, particularly where hydropower plants have been built or are being built (Qiriazi 2019).

Restricting people's movement and their socioeconomic activity, especially transport, during the COVID-19 pandemic in 2020, has significantly reduced pollution and damage to the environment, and this has created the conditions for its regeneration and for an increase in the living world.

As a country aiming to join the EU, Albania has the responsibility to carry out concrete actions in the protection of biodiversity.

All this natural diversity has created numerous special values of our natural heritage, which should be evidenced and studied to distinguish the categories of the protected areas. This is the best way to preserve and protect biodiversity.

References

Baldaçi A (1932–1937) Studime të Posaçme Shqiptare
Bino T, Zoto H, Bego F (2006) Shpendët dhe Gjitarët e Shqipërisë. Shtëpia Botuese "Dajt 2000"
Buzo K Pasuritë florike të Shqipërisë, shoqërimet bimore të rrezikuara. Studime Gjeografike, nr. 13/2001
Cmeta A (2000–2021) Personal photo archive
Demiri M (1973) Gjeografa e bimëve
Demiri M (1985) Flora Ekskursioniste e Shqipërisë
Di Castri F (1973) Mediterranean type ecosystems (origin and struktur), New York
Dragoti N (2012) Raport për biodiversitetn
Gracianski N (1971) Priroda Sredizemnomorja, Moskva
Hanak V, Lamani F etc. Përhapja e lakuriqëve të natës në Shqipëri. Bul UshT Seria Shk Nat, nr. 3/1961
Hoda P (1989) Studim gjeobotanik i formacionit të pishës së zezë (Pinus nigra) në vendin tonë
Hoda P (1990–2021) Personal photo archive
Lamani F (1972) Lakuriqtë e natës në vendin tonë. Buletni i USHT, Ser. Shkenc. të Nat
Margraf F (1927) Pflanzengeographie von Mittelalbanien
Margraf F (1932) Pflanzengeographie von Albanien
Meço M, Mullaj A (2015) Phenological aspects of Albanian flora. In: Proceedings of international conference on soil 04–06 May 2015, Agricultural University of Tirana, p 164
Ministria e Mjedisit (2015) Dokument i Politikave Strategjike për Mbrojtjen e Biodiversitett
Mitrush I (1954) Drurët dhe shkurret e Shqipërisë
Mitrushi I (1956) Dendroflora e Shqipërisë
Mitrushi I Konsiderata mbi vegjetacionin lidhur me klimën. Bul Për Shk Nat nr, 4/1966
Mitrushi I, Demiri M (1978) Harta e bimësisë të R P S të Shqipërisë. Shtëpia Botuese e Librit Shkollor
Paparisto K, Demiri M, Mitrushi I, Qosja Xh (1988) Flora e Shqiperise. vol 1, Tiranë, p 457
Qendra e Studimeve Biologjike, Flora e Shqipërisë, pjesa e parë dhe e dytë, 1988, 1992
Qendra e Studimeve Gjeografke (1990) Gjeografa fzike e Shqipërisë, Pjesa e parë

Qiriazi P (2019) Gjeografia fizike e Shqipërisë, Mediaprint

Raka N (2000) Peshqit e Shqipërisë

Strategjia e plani i veprimit për biodiversitetin, Raport kombëtar për biodiversitetn (1997–1998)

The main Habitats of Albania. Project "Study and Preparation of the Main Habitat Map of Albania ECONET-2006". Project Leader Prof. dr. Jani Vangjeli

Vangjeli J etc, Libri i Kuq – Bimë të rrezikuara, 1999

Vangjeli J (2015) Excursinflora of Albania, Koenigstein, Germany

Vangjeli J (2016) Atlasi i florës së Shqipërisë, vil I. Akademia e Shkencave e Shqipërisë

Vangjeli J (2021) Flora eskursioniste e Shqipërisë

Zeko I Invetarizimi i shpendëve të Shqipërisë. Buletni i USHT Ser Shkenc të Nat, Nr. 1/1963

Albanian Natural and Cultural Heritage

10

Perikli Qiriazi

Abstract

The highly diverse nature of the country, often described as "a great natural museum," has conditioned many natural phenomena and objects with extraordinarily unique values on a global and national scale, also considered natural heritage sites.

Two natural sites in Albania have been registered in the World Heritage List: Lake Ohrid (part of the mixed site of Lake Ohrid Region) and the ancient and primeval beech forests of Rrajca and Gashi. There are four Albanian wetlands in the Ramsar List. The Prespa-Ohrid Transboundary Biosphere Reserve is part of the World Biosphere Reserve Network. Some protected areas are part of the European Green Belt, the Emerald Network, the Parks Dinarides Network, and the Mediterranean Marine Protected Areas Network.

The national natural heritage network consists of sites of various categories of protected areas, municipal sites, and the Aos-Vjosa eco-museum.

The rich and diverse Albanian cultural heritage is related to the ancient population and the social, economic, and cultural development throughout the centuries. Cultural sites from Albania are also registered in the World Cultural Heritage List: the ancient city of Butrinti, the city of Gjirokastra and Berati, as well as partly the mixed site of the Lake Ohrid Region. The two codices of Berati have been registered as monuments of world culture in the UNESCO World Memory register, while the Albanian iso-polyphony was proclaimed "Masterpiece of the Oral Heritage of Humanity." The list of national cultural heritage also includes archaeological, historical, architectural, artistic, ethnographic, religious, and environmental monuments (like traditional bazaars and museum cities). The material and spiritual heritage in unity with the geographical environment manifests the Albanian culture and identity.

Keywords

World Heritage List · Albania Sites in the UNESCO World Heritage List · Transboundary Biosphere Reserve · Ramsar Convention · Great natural museum · *The national natural and cultural heritage* · The municipal natural parks · *Cross-border eco-museum*

10.1 World and National Natural and Cultural Heritage

World heritage sites are protected by the International Heritage Convention as property of the respective country and of all humanity. However, they are also protected by national laws. Albania is among the 195 member countries which are part of this convention. The study, assessment, protection, and management of natural and cultural heritage has emerged as an urgent necessity to the major challenges and detrimental consequences of recent developments around the world, such as globalization, rapid population growth, wide-spreading urbanization (in most cases chaotic), environmental concerns, and rapid development of tourism. The studies of the UNESCO-University Heritage Forum (FUUT) have not only helped to clarify and expand the concept of natural heritage, to include, in addition to biodiversity and other rare natural and cultural values, but have also helped in providing the legal and institutional support to heritage and in determining the criteria for the selection of the sites. On this basis, due to their rare values, many sites were identified and proposed for the world and national heritage status. New concepts and models of their management were defined, strategies for increasing the protected area were prepared, and management plans of heritage sites were turned into a source of scientific information, didactic laboratories of ecological and patriotic education, and models of sustainable development.

10.2   Albanian Sites Declared World Natural Heritage

Albanian natural and cultural sites are among the sites registered in the World Heritage List (1154 sites, July 2021).

- *The primeval (virgin) and ancient beech forests of Rrajca* (2200 hectares) are in the Shebeniku-Jabllanica National Park along the border with North Macedonia, in the Balkan part of the European Green Belt. The territory of this site consists of magmatic and limestone. They lie in a rugged relief, over 1000 m altitude, dominated by harsh winters and cool summers. The territory has clear streams and there are glacial lakes. This is a cross-border area with great biodiversity values. These primeval beech forests (*Fagus sylvatica*) with giant trees, sometimes over 300 years old, are simple or accompanied by rare species (*Pinus sylvestris, Pinus peuce, Pinus leucodermis*, white fir, ash, birch (*Betula alba*)), wetland habitats and special vegetation of karst landscapes (*Juncus, Carex*, etc.), but also subalpine and alpine meadows. There are rare and endangered plants, such as the Dukagjini violet (*Viola ducadjinika*), and endemic plants, such as Albanian rudithi (*Brachypodium albanicum*), genista (*Genista hassertiana*), and subendemic plants like *Dianthus jablanicensis*, the Albanian lily (*Lily albanica*), etc. These forests constitute the nucleus and bio-corridor for large, rare mammals like bear, lynx, chamois and wild boar, roe deer, etc., which give these forests the Balkan value. There is also the eagle, hawk, and capercaillie, and in cold running waters, there is the trout. The site also has several cultural heritage monuments like the Rrajca castle, Skanderbeg (Pass) Stairs, etc. The popular memory is also rich in symbols, legends, folklore, and toponyms from the period of Skanderbeg. The rich folklore is well known for the famous Rrajca dance (Photo 10.1) (Qiriazi 2020).

- *The primeval and old beech forests of Gashi* lie in the catchment area of the Gashi River between the altitudes of 430–2312 meters, along the border with Montenegro, in the Balkan part of the European Green Belt. They have an Alpine nature and are mainly consist of magmatic and limestone. The relief is Alpine, broken, and fragmented, with glacial forms. The climate is also Alpine, with short and cool summers and harsh winters when snow reaches over 2 meters in thickness. They have rich alpine hydrography, glacial lakes, rivers with nival regime, rapid flows, and clean, cold waters. This is a cross-border site, with rich habitat diversity, virgin forests, mainly with beech trees of over 150 years old, and pine, spruce, fir forests, but also with alpine meadows; endemic and subendemic plants; rich animal world: large mammals (brown bear, wolf, lynx, chamois and wild boar, roe deer, otters, globally threatened species); rare birds (capercaillie, eagle); and fish, like trout. The popular memory is rich with symbols, legends, folklore, and toponyms, from the Albanian titans, fairies, and patriots, like Mic Sokoli, Bajram Curri, etc. The folklore is also rich, with the famous Tropoja Dance. Women's and men's clothes are also of rare beauty (Photo 10.2) (Qiriazi 2020).

Photo 10.1   The primeval and ancient beech forests of Rrajca (F. Plaku)

Photo 10.2 The primeval and old beech forests of Gashi (F. Plaku, 2019)

– **Lake Ohrid** is part of the mixed site of "Lake Ohrid Region." This site has a very rich biodiversity, which promoted it to a world natural heritage site, also called the "Galapagos of Europe." There is a wide variety of living species, including globally endangered ones, and about 200 endemic species and over 20 relict species, especially mollusks. This rich world of living species has also given this lake name "the real museum of living fossils." There are 17 rare species of fish, belonging to the families *Salmonidae, Cyprinidae, Cobitidae, and Anguillidae* (Ohrid trout, locally known as korani and belushka, are endemic, famous for their taste), but also squids, crabs, Ohrid endemic sponge, etc., and some endemic species of plankton, etc. On the lake's shores, there are various species of birds: the Dalmatian pelican, the ferruginous ducks (*Aythya nyroca*), spotted eagle (*Aquila clanga*) and imperial eagle (*Aquila heliaca*), swans, etc (UNESCO 2019). Based on all these elements, Professor Christian Albrecht describes this lake as the "Holy Grail Cup for biologists from all over the world" (Christian 2008).

The shores of this lake have been populated since the ancient times. Therefore, there are numerous cultural heritage values on its shores, like the city of Ohrid (world cultural heritage), the early settlements of the palafitte type (in Lin), cult and artworks, castles, etc.

The lake is endangered by human activity, especially by two projects, in Albania and Northern Macedonia (see section on hydrography) (Naumoski et al. 2007).

10.3   Albanian Sites in Other World Categories of Protection of Natural Values

– **Ohrid-Prespa Transboundary Biosphere Reserve** joined the World Biosphere Reserve Network in June 2014. It includes Lake Ohrid, the two lakes of Prespa, and the territories around them in Northern Macedonia and Albania. This site is distinguished for its rich biodiversity with rare and globally endangered species. Due to the genesis and the evolution, the endemic species are numerous here. It is administered by the Ohrid Basin Management Committee (OWMC), with representatives from all the countries.

– **Sites from the list of wetlands of the Ramsar Convention.** This list also includes four Albanian wetlands, which take up 98.180 hectares or 3.4% of the country's area: (1) *the Karavasta Lagoon-Divjaka Pine* (20,000 hectares) which is the most important area for waterfowl nesting, with the Dalmatian pelican (*Pelicanus crispus*) being the most important, and about 6% of its world population is found here, flamingos (*Phoenicopterus roseus*), rare migratory birds, like *Cignus cignus, Clangula clangula*, etc.; (2) *the Çuka Canal-Butrinti-Cape Stillo* (13.500 hectares) which is distinguished for its great diversity of ecosystems and wintering birds; (3) *Shkodra Lake-the Buna River* (49,562 hectares) with great variety and wealth of migratory and sedentary birds; and (4) *Prespa Lakes* (15.118 hectares), also known as the European and Global Biodiversity Hotspot (Qiriazi 2020).

10.4 European Efforts on Natural Heritage and the Cooperation of Albania

These efforts have brought about concrete results materialized in special laws, adopted by the European Parliament and in the activity of state and civil society. Some noteworthy efforts are as follows: (1) "Nature 2000" Project is the ecological network of protected sites in the European Union. As an aspiring country to join the EU, Albania has started working for this network. (2) The European Green Belt replaced the Iron Curtain separating the democratic Western and communist Eastern Europe. This belt also extends along the Albanian land border, where there are several categories of protected areas. (3) The Emerald Network, besides species protection, the network also aims to protect particularly endangered natural habitats and migratory species. As a party to this network, Albania has declared its own Emerald 25 sites. (4) The Dinaric Arc Parks Network includes regional and national parks. The network aims at exchanging information and expertise for their protection and sustainable regional development. (5) Mediterranean Marine Protected Areas Network (MedPAN) ensures cooperation between marine-protected areas in the Mediterranean basin (Qiriazi 2019).

10.5 Albanian Cultural Sites in the World Heritage List

The 869 world cultural heritage sites (as of 2019) include several cultural sites from Albania: Butrinti, Gjirokastra, Berati, and partly the mixed site of Lake Ohrid Region (World Heritage List, July 2021).

– **The ancient city of Butrinti** is in the southeastern corner of the Ksamili peninsula, on the shores of the lake bearing the same name. It has a naturally protected position as it is surrounded on two sides by Lake Butrinti and the Vivari Channel, and it is connected to the Ksamili peninsula by a narrow strip of land. Located on the eastern shore of the Corfu Channel, its position is of a great strategic importance (Photo 10.3).

In its motivation as a "World Cultural Heritage" site (1992), Butrinti was described as "the microcosm of Mediterranean history." The city appears in the Mediterranean history from the First Peloponnesian War (fifth century BC) to Napoleon Bonaparte and Ali Pasha Tepelena (nineteenth century). The city's Illyrian, Greek, Roman, Byzantine, Anjouan, Venetian, and Ottoman monuments reflect this long and complex history. There is evidence about the city being populated since the 12 century BC. The city was part of the Illyrian tribe of the Kaons and the state of Epirus. There are myths about its establishment as a Greek colony by the displaced Trojans. Around the IV century BC, the settlement was fortified and turned into a cult center dedicated to Asclepius. In the year 167 BC, it was conquered by the Roman Empire and in the Middle Ages by the Byzantine and Ottoman Empires. The ruins of numerous monuments testify to the flourishing of the city in antiquity and in the Middle Ages: the city's agora, the theater with 1500 seats, the 3-kilometer-long aqueduct dating back to first century BC, gymnasium, nympheus, public baths, baptistery (fifth century); the Paleo-Christian basilica of the Acropolis (end of the fourth century), paved with mosaics and the Great Basilica (sixth century), as the most magnificent Paleo-Christian building, etc (Ceka 2002, 2005; Hodges and Logue 2007; Hodges et al. 2011).

– **The city of Gjirokastra** is in the south of Albania. There are human traces of the Early Iron Age. The city was declared a museum city in 1963, and in 2005, it was declared a UNESCO World Heritage Site. In 2008, the World Heritage Committee described Gjirokastra and Berati as "... carved historic centers of the rare surviving examples in the Balkans of typical Ottoman-style architectural trading towns, continuously inhabited from ancient times to the present days, and evidence of the richness and diversity of the urban and architectural heritage of this region." (Photo 10.4) (Boardman 1982; Komiteti i Trashëgimisë Botërore 2005, GCDO History part, History of Gjirokaster 2010)

Gjirokastra, also dubbed the "Stone City," today, the city is the heart of cultural heritage tourism.

– **The city of Berati** is one of the oldest cities in Albania (dating back from 2600 to 1800 BC) and one of the few cities where life began in ancient times and continues uninterrupted to this day. As one of the earliest centers of Christianity, the city is distinguished for the coexistence of different religious and cultural communities over the centuries. Life in this city began before the Bronze Age. It is considered as an early Illyrian settlement, c. sixth to fifth century BC, and in third century BC, it turned into a fortress called Antipatrea. Later, the city had different names.

Photo 10.3   The ancient city of Butrinti. (Photo P. Qiriazi, 2010)

"Berati" is mentioned for the first time in 1018. Its development declined at the beginning of XV century when the city was conquered by the Ottomans. However, thanks to its position as crossroads center with intensive road communication, the city developed and expanded beyond the castle into three characteristic neighborhoods: Kala, Mangalem, and Gorica. The great Albanian painter Onufri and his son worked here. Berati took on its present appearance in seventeenth and nineteenth centuries, after the earthquake of 1851. In the late 1800s, it was the most prosperous economic, commercial, and cultural center in the south of the country, where 23 different types of crafts were practiced (Photo 10.5) (Dvornik 1958; Karaiskaj 1981).

Berati has numerous Illyrian, Roman, Byzantine, and Ottoman monuments of precious value, which are evidence not only of the rich cultural, historical, and artistic heritage but also of life and works, masterfully done generation after generation by its inhabitants. Among them, we can mention (1) *the castle*, with its Illyrian foundations and rebuilt several times, is one of the few still inhabited castles; (2) *the Gorica Bridge* (eighteenth century) one of the main symbols of the city; and (3) *the cult buildings* (Paleo-Christian churches, etc., the Mosque of Lead and Singles, Helvetian

Tekke, etc.) and the civic buildings (complexes of Mangalem, Kala e Gorica) Neighborhoods, pearls of medieval architecture, where houses are built together in the form of terrace belts. Berati is also called "the city of one-over-one windows."

In 1961, Berati was declared a museum city and a world heritage site on July 8, 2008. ICOMOS considers Berati "a special example of the architectural and urban values of the Balkan region, in the context of a magnificent historical continuity; rich with old churches dating back to XIII century with magnificent murals, icons and wood carvings; well known for Muslim architectural works."

– *Lake Ohrid Region.* The shores around this lake have been populated since ancient times. Therefore, the cultural heritage sites of great value are numerous around the lake's shores. In the Albanian part are distinguished (1) the palafitte settlement of Lin of the early Neolithic age; (2) the prehistoric settlement of Zagradian in Lin, where there is an Illyrian settlement (fifth century BC); (3) traces of the early Illyrian settlement from the Iron Age to the early Middle Ages; and (4) the ruins of the Paleo-Christian triconch on the hill of Lin (fifth to sixth centuries) (Ministria e Kulturës 2005).

Photo 10.4 The city of Gjirokastra. (Photo A. Cmeta, 2018)

Photo 10.5 The city of Berati. (Photo A. Cmeta, 2018)

10.6 Albania Sites in the UNESCO World Heritage List

The two codices of Berati were registered as monuments of world culture in 2003, and the Albanian iso-polyphony was declared "Masterpiece of the Oral Heritage of Humanity" in 2005.

- *The Purple Codex* (also called "Beratinus 1") of sixth century is written on purple parchment (tanned and dyed goat skin) with letters cast in silver. It contains two gospels: Matthew's and Mark's. The other parts of it are cast in gold. The cover of the manuscript, which is later than the work, is metallic and decorated with biblical scenes. It is one of the four oldest codices in the world, at the same time as the famous manuscripts "Petropolitaus," "Vindeoboneusis," and "Sinopencis." It is at the foundation of the ecclesiastical literature of the Eastern church and ranks among the fundamental works of evangelical literature of special importance to the culture of Christianity (Sinani 2000).
- *The Golden Codex of Anthimos* (also known as "Beratinus 2") is written on parchment in gold letters. It contains four gospels: John's, Luke's, Mark's, and Matthew's. Stylistically, it is compared to the codex preserved in the St. Petersburg Library. The German scholar Kurt Witzman determined that it is a manuscript of ninth century (Sinani 2000).

The codices of Albania are monuments of Christian culture and civilization, and they are testimony of the biblical-ecumenical space where Albanians and their ancestors have lived in. They are both a valuable source of information for bibliophiles, scripture scholars, and the church and object of study for ethnopsychology, language and history, writing technique and calligraphy, applied figurative arts, and iconography. They are encyclopedias of Christian thought. They are valued as monuments of the world heritage of knowledge, as well as scientific objects of paleography, bibliology, linguistics, and history of religious beliefs (Sinani 2000).

- *Albanian Folk Iso-Polyphony* was declared "Masterpiece of the oral and intangible heritage of humanity" protected by UNESCO on November 25, 2005. That became a historic date for the Albanian cultural heritage. As an extraordinary form of human conception and communication, Albanian folk iso-polyphony took its rightful place in the list of humanity's spiritual masterpieces, which make the foundation of world culture. This is proof of the great values found in the Albanian spiritual heritage, which is becoming the main factor of European integration and the positive image of Albania. The Albanian state has drafted and ratified new laws and is a member of world conventions in this field (Tole 2018).

Proposals for the status of "Memory of the World" of two other great Albanian values are being prepared: the Epic of the Valiant Heroes (Kreshnikeve) and the famous Tropoja Dance.

10.7 National Natural Heritage

The natural and cultural heritage is regarded and treated as sacred in every cultured country, because this heritage is associated with the national identity, balance of ecosystems, sustainable development, and their posterity.

In Albania, the identification and proclamation of protected areas began in 1940, when the coastal territory of Kune-Vain-Tale was declared a "State Hunting Reserve." Until 1990, the protected area occupied 1.8% of the country, while in 2019, it amounts to about 17.6%. Protected areas will soon occupy over 20% of the country. After 1990, based on new concepts of UNESCO programs and the World Heritage Convention, etc., groups of researchers in various fields helped to provide legal and institutional support to natural heritage, in defining the criteria for declaration, protection, and management of protected sites, and then they made proposals for many protected areas (Dragoti et al. 2007; Ministria e Mjedisit dhe e Turizmit 2019; Qiriazi 2020).

The national network of protected areas consists of national sites of all categories and municipal sites. The study and proposal of the Aos-Vjosa as eco-museum is of special value. The protected areas include almost the entire natural diversity of the country (Ministria e Mjedisit dhe e Turizmit 2019).

In the diverse nature of the country, which has rightly been described as "Great Natural Museum," there are still many other unknown objects and values, which deserve the status of a protected area. The expansion of the protected areas would help the country meet the EU's criteria in this field and advance the country's accession aspiration. The numerous concerns in the protected areas are related to both natural factors and human activity. Considering new insights and developments, the studies on assessment, protection, and management of natural heritage sites are being deepened.

The protected areas in Albania are classified into six categories (Fig. 10.1).

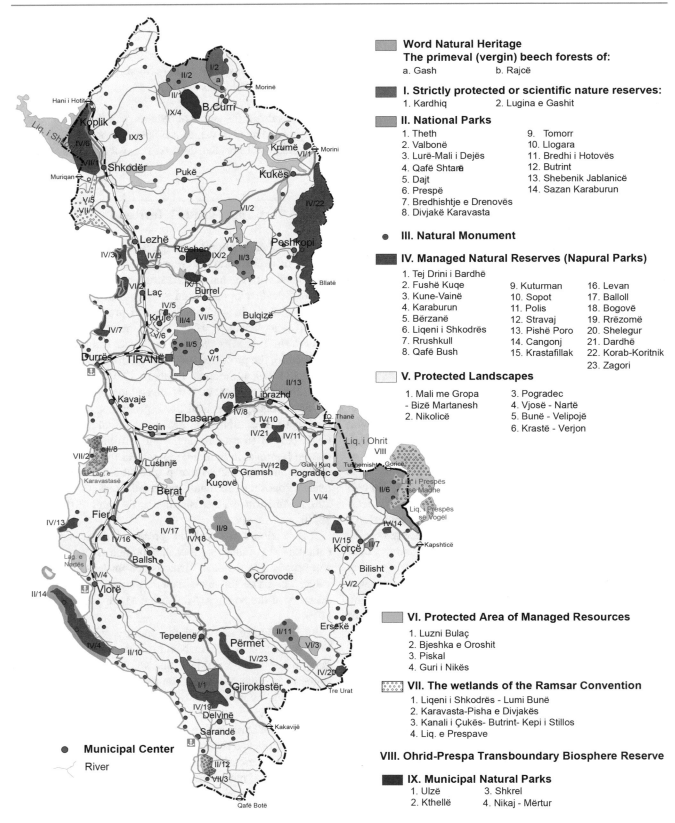

Word Natural Heritage
The primeval (vergin) beech forests of:
a. Gash   b. Rajcë

I. Strictly protected or scientific nature reserves:
1. Kardhiq   2. Lugina e Gashit

II. National Parks
1. Theth   9. Tomorr
2. Valbonë   10. Llogara
3. Lurë-Mali i Dejës   11. Bredhi i Hotovës
4. Qafë Shtarë   12. Butrint
5. Dajt   13. Shebenik Jablanicë
6. Prespë   14. Sazan Karaburun
7. Bredhishtje e Drenovës
8. Divjakë Karavasta

● **III. Natural Monument**

IV. Managed Natural Reserves (Napural Parks)
1. Tej Drini i Bardhë
2. Fushë Kuqe
3. Kune-Vainë   9. Kuturman   16. Levan
4. Karaburun   10. Sopot   17. Balloll
5. Bërzanë   11. Polis   18. Bogovë
6. Liqeni i Shkodrës   12. Stravaj   19. Rrëzomë
7. Rrushkull   13. Pishë Poro   20. Shelegur
8. Qafë Bush   14. Cangonj   21. Dardhë
                               15. Krastafillak 22. Korab-Koritnik
                                                      23. Zagori

V. Protected Landscapes
1. Mali me Gropa   3. Pogradec
 - Bizë Martanesh   4. Vjosë - Nartë
2. Nikolicë   5. Bunë - Velipojë
                                   6. Krastë - Verjon

VI. Protected Area of Managed Resources
1. Luzni Bulaç
2. Bjeshka e Oroshit
3. Piskal
4. Guri i Nikës

VII. The wetlands of the Ramsar Convention
1. Liqeni i Shkodrës - Lumi Bunë
2. Karavasta-Pisha e Divjakës
3. Kanali i Çukës- Butrint- Kepi i Stillos
4. Liq. e Prespave

VIII. Ohrid-Prespa Transboundary Biosphere Reserve

IX. Municipal Natural Parks
1. Ulzë   3. Shkrel
2. Kthellë   4. Nikaj - Mërtur

● **Municipal Center**
⌇ River

Fig. 10.1 Map of protected areas of Albania (P. Qiriazi, 2019). (Source: IDEART Gjeografia 11 (2019), p. 65)

10.7.1 Category I: Strictly Protected or Scientific Nature Reserves

These are small areas with rare values and strict protection of natural and cultural values for scientific and ecological purposes. Two areas enjoy this status:

- *Kardhiqi oak forest* (1800 ha) is in the south of the country, lying in between altitudes from 600 to 1600 meters. Its terrain consists of flysch, limestone, and gypsum, and it has rugged relief with karst forms, Mediterranean climate, and dense vegetation, and it is crossed by a network of streams. It combines a variety of rock landscapes and river ecosystems, with shrubs, primeval virgin forest, and pastures. It is a bird migration route. There is a high biodiversity of habitats and species. Oak forests predominate, accompanied by ash (*Fraxinus ornus*), maple (*Acer campestre*), and rare hardwoods: large-leaved linden (*Tilia platyphyllos*) and silver linden (*Tilia tomentosa*). There is a forest formation of Macedonian fir (*Abies borisii-regis*) and maple (*Acer pseudoplatanus*). There are endangered plants like horse chestnut (*Aesculus hippocastanum*), yew (*Taxus baccata*), yarrow (*Achillea grandifolia*), etc. There are poultry and mammal communities: chamois and wild boar, roe deer, etc. The territory around the reserve has been inhabited since the ancient times. The main proof for this is the historical traces and objects preserved to this day. The values of this reserve are well preserved (Photo 10.6) (Dragoti et al. 2007; Qiriazi 2020).

- *The Gashi River* (3.000 ha) (see section on world natural heritage sites).

10.7.2 Category II: National Parks

National parks are large areas (over 1000 hectares) with unique scientific, ecological, and educational values , and they excluded from intensive human activity. These areas are preserved and managed to protect the ecological integrity of ecosystems, as well as for scientific studies and tourism. All parks in the country are part of the "Emerald" Network and the Dinaric Arch. It is currently being worked to declare another cross-border park, which will include the territories from Albania, Montenegro, and Kosovo. So far, 14 national parks have been declared (Dragoti et al. 2007; Qiriazi 2020).

- *Dajti National Park* (29.347 ha) is the first one to enjoy this status (1960). It lies in between the altitudes of 300 m and 1827 m. Its territory consists of limestone, magmatic, and flysch. It has rugged mountainous relief, Mediterranean pre-mountainous climate, and abundant water resources. It has Mediterranean shrub vegetation; oaks and beeches mixed with conifers and alpine pastures, but also rare endemic plants and animals (Qiriazi 2019). The park serves as a source of information and genetic material for the regeneration of the degraded areas. There are several natural monuments like the Black Cave, the Vali Cave, the Gorges of Skorana and Murdhari, etc. Human traces of the late Paleolithic have been found at the western foot of the Dajti mountain. There are several cultural monuments in the area like the castles of Dajti and Tujani, the Brari Bridge, etc (Photo 10.7).

- *Divjaka-Karavasta National Park* (22.230 hectares) was declared a national park in 1966. It lies on the coastal strip between the Semani and the Shkumbini estuaries. There are several rural centers and the town of Divjaka in this area. It is the most important protected area along the Albanian coastal space and one of the most important areas on the shores of the Mediterranean. The natural values of this park are especially related to its geographical position on the low accumulation coast, in the interaction contact between the sea and the land, and in its coastal wetland system, where there are lagoons, active and dead riverbeds, swamps, marshes, saltwater and sweet-water ponds, sand dunes, and the estuaries of the Semani and the Shkumbini Rivers. This natural park is also distinguished for its significant diversity of habitats and rich biodiversity: xerophilous, halophyte, hygrophilous vegetation etc. There are Mediterranean coniferous forests, dominated by pines, macchia bushes, and lianas. The park has a rich and diverse animal world, especially birds (about 230 different species). As it was mentioned, the Karavasta Lagoon of this national park is ranked in the Ramsar wetlands. There are several natural monuments (the pelican island, oasis dunes, the Old Pine) and cultural monuments (the ancient town of Babunja, the church of St. Thanas of eighteenth century, etc.) (Photo 10.8).

- Theth National Park (2630 ha) and Valbona Valley National Park (8000 ha) are located in the Albanian Alps. Their territory lies mainly on old limestone and terrigenous and have steep Alpine relief and a variety of river and glacial relief forms, climate, and alpine hydrography. Distinguished for plant diversity are beech, coniferous and oak forests, and alpine pastures. It has many relict, endemic, medicinal, and essential oil plants. They have diverse animal world (20 species of mammals, with some rare animals among them like lynx, wildcat, bear, chamois, roe deer, etc., and 50 species of nesting birds like capercaillie, eagle, etc.). They have some several natural monuments and cultural heritage, such as villages with special architecture and structure for their dwellings, which are tower-shaped and become. The villages have an interesting organization in tribal neighborhoods, with

Photo 10.6 Kardhiqi oak forest. (P. Qiriazi, 2018)

Photo 10.7 Dajti National Park (P. Qiriazi, 2017)

Photo 10.8 Divjaka-Karavasta National Park (F. Plaku, 2018)

unique clothing, distinctive legends, diverse folklore, etc (Photos 10.9, 10.10, and 10.11).

– **Tomorri Mountain National Park** (26.106 hectares) lies on the homonymous mountain and consists mainly of limestone. It seems like up to 2415 meters high, on its surrounding hilly and plain territories. There are karst forms, with caves, deep wells (Kakruka Well is about 280 m deep), and glacial cirques. It has mountain climate, large surrounding karst springs, and significant biodiversity. The vegetation is dominated by Mediterranean shrubs, oaks, beeches, and conifers and alpine pastures. There are endemic and subendemic plants, including rare and endangered types, but also rare mammals and birds. There are several natural monuments (the Sotira Waterfall, etc.) (Photo 10.12).

Since the pagan times, Mount Tomorri has attracted the people's attention and has become a symbol and a part of their spiritual world. It has been described as a mountain "with its head into the heavens, the King of Heavens, the Throne of God." Pagan rites were performed near the mountain. In the Kulmak Gorge, there is the Tekke of Kulmak, a shrine of the Bektashi believers.

– **Prespa Cross-Border Park** (27.750 hectares) is a natural extension of the Galicica National Park in North Macedonia and Prespa National Park in Greece. It includes the Great and Small Prespa Lakes and their surrounding territories. About 15.120 hectares of this park have the status of "Ramsar Wetland." The park is a Transboundary Biosphere Reserve. The territory of the park consists of limestone and terrigenous composition, and it has diverse and rugged mountain relief, with karst forms, Mediterranean pre-mountain climate, poor surface hydrography, a variety of landscapes, and rich biodiversity (about 60 plant species of the park are rare or endangered global species on European and/or national level). There are several natural monuments in the park. Its territory has been inhabited since the Neolithic age (the Neolithic settlement in Tuminec/Kallamas and the Cave of Tren). There are several cultural monuments, like the Castle of Trojani, 11 Byzantine churches, three of which were hermit churches, carved in stone and with early paintings, etc (Photos 10.13 and 10.14).

– **Butrinti National Park** includes the ancient city of Butrinti (a world cultural heritage site), while the Butrinti Wetland Complex has been declared a Ramsar Zone. It stretches from the shores of the Ionian Sea of the Delvina

Photo 10.9 Thethi National Park. (Photo A. Cmeta, 2018)

Photo 10.10 Valbona Valley National Park (A. Cmeta, 2018)

Photo 10.11 Lura-Deja Mountain National Park (F. Plaku, 2019)

plain up to 825 m (the peak of Milesa). The park includes wetlands, shrub and herbaceous vegetation, rocks, islands, residential areas, and agricultural land. The great variety of natural habitats (including marine and coastal, wetland, hilly habitats) is the result of over 3000 years of interaction between man and nature. Their combination creates landscape, quite unique in the Mediterranean. The ecosystems of the park have a variety of habitats and rich biodiversity: meadows with Posidonia (*Posidonion oceanicae*) on the sea floor up to 30–40 m deep, about 247 species of birds, mostly aquatic species, 9 species of amphibians, and 25 species of reptiles, representing about 60–75% of the native herpetofauna and about 39 species of mammals, or about 60% of the mammal fauna of the country. The littoral waters of Butrinti are frequently visited by sea turtles (*Caretta caretta*), dolphins (*Tursiops truncatus*), and less frequently by the Mediterranean monk seal (*Monachus monachus*). Freshwater and brackish waters are populated by about 35 globally endangered species. There are several natural monuments in the park,

like the Ksamili islands, the Konispoli Cave, the Channel of Vivari, the subtropical jungle, the Butrinti forest, etc.

– ***Sazani-Karaburuni National Marine Park*** (12.427 hectares) is the first marine park in Albania. It lies in the sea belt with a width of one nautical mile around the island of Sazani and the Karaburuni Peninsula. There is a wide variety of landscapes on the shores of the marine park, like secluded beaches, capes, caves, and steep slopes, diving into the blue waters of the sea. Underwater landscapes also abound, for instance, precipices, canyons, and caves, but also steep slopes, flat surfaces, and diverse habitats, which form substrates for underwater life, like algae, sponges, mollusks, crustaceans, urchins, red coral, reptiles, mammals, etc. About 36 species of marine flora and fauna of national, regional, and global importance live in this marine park. The underwater meadows of Posidonia (*Posidonia oceanica*), which is considered the "deteriorating species" of the Mediterranean, are of special value. This area is inhabited or visited by globally endangered species of shark, sea turtle, dolphin, and Mediterranean

Photo 10.12 Tomorri Mountain National Park (P. Qiriazi, 2019)

Photo 10.13 Hotova-
Danglli Fir National Park
(A. Cmeta, 2019)

Photo 10.14 Shebenik-Jablanica National Park (F. Plaku, 2019)

monk seal (*Monachus monachus*). There are several natural monuments (the Haxhi Alia Cave, the delta of Palasa, Sazan's Cave, the Grama Cave and Bay, etc.) and underwater cultural monuments (sunken ships, from antiquity to modern times) and on the coast (ancient inscriptions in the Bay of Grama), etc.

In addition to these national parks, the country also has these other national parks: Lura-Deja Mountain (20.242 hectares), Qafë Shtama (2500 hectares), Drenova Fir (1380 hectares), Llogara (1010 ha), Hotova-Danglli Fir (34.361 ha), and Shebeniku-Jablanica (33.927 hectares).

10.7.3 Category III: Natural Monument

As of 2019, 721 natural monuments have been declared. Among these monuments are distinguished the Black Cave and the Puci Cave, the Pelican Island, the Blue Eye, the Sotira Waterfall, the Osumi and Shoshani Canyons, the Qarishta Birch Grove, the *Platanus* (Plane) of Libohova, etc. Natural monuments have scientific, ecological, cultural values. Their spiritual and religious values are being studied (Qiriazi 2020). Here are some of them:

– The plane trees in the center of the town or village, or the City Stone in Përmeti, etc., have become a symbol for their inhabitants. It is because they have served as the starting focal point for the extension of the inhabited center that many meetings, talks, and assemblies of importance to the village or province have taken place under their shade. Their values have been passed down through the generations, becoming part of the spiritual world of the inhabitants and their strong spiritual ties with the settlement. Around these central plane trees, there are places where one can relax and get informed. When the plane tree of the village of Zhulati (in Gjirokastra) fell due to its age, out of its trunk, the inhabitants artistically carved the monument of Papa Sulli (a well-known personality of their province), which now stands in the place of the centuries-old plane tree (Photos 10.15, 10.16, 10.17, and 10.18).

– *Some natural monuments* were furnished with supernatural powers due to their strange shapes and consequently were turned into objects for religious rites, such as the Cave of Sarisalltëkut (in Kruja) and the Trail of Abaz Ali (in Skrapari) for the Bektashi believers; the Cave of Llanga (in Mokra) for the Orthodox believers; the Cave of St. Andouts (in Laçi) and the Holy Stone (in Puka) for Catholic believers, etc. There is no lack of pagan rites in this respect either, like near the salt springs of Golemasi (in Kavaja), young girls and brides predict the realization of their wishes, according to the force of the explosion of "papas," as these springs are known locally.

– *The Blue Eye Spring* (6 m³/sec), located at the western foot of the Mali i Gjerë (the Broad Mountain), is a major karst spring. The powerful gurgle of its water from below has been likened to a boiling cauldron. At the bottom of the vertical well of this spring, about 60 m deep, begins the gallery through which water flows from this karst massif. In the center of the "eye," several meters in

Photo 10.15   The natural
monument of the Island of
Pelican (A. Cmeta, 2019)

Photo 10.16   The natural monument of City Stone in Përmeti (F. Plaku, 2019)

Photo 10.17   The natural
monument of Grunasi
Waterfall in Thethi National
Park (F. Plaku, 2019)

Photo 10.18   The natural monument of Platanus (Plane) of Leskovik (P. Qiriazi, 2020)

diameter, there is the dark blue "pupil." In ancient times, the Blue Eye has been the center of pagan rituals to honor the gods by making animal sacrifices and throwing wine, tools, and ornaments into the spring waters. Still people toss coins for good luck into the spring waters.

10.7.4 Category IV: Managed Natural Reserves (Natural Parks)

The area and number of sites in this category has been constantly increased. Their list counts 23 sites with a total area of 147.564 hectares. Scattered from the coast to the heights of the mountains, they include terrestrial and aquatic ecosystems, forests, meadows, pastures, and the like. Their size varies from about 30 hectares to 55.550 hectares (Korabi-Koritniku Natural Park). There are inhabited centers inside or near these areas. Noteworthy among them are the natural parks of Kulari, Rëzoma, Sheleguri, the Lake of Shkodra, etc. There are residential centers inside or near them (Qiriazi 2020).

10.7.5 Category V: Protected Landscapes

These are land and/or coastal/maritime territories over 1000 hectares in which the interaction of man and nature has created an area with special phenomena and visible aesthetic, cultural, and/or ecological values. They also include residential centers. Their management aims at the preservation and normal evolution of landscapes, by ensuring the harmony of traditional human interaction with biological diversity, the variety of terrestrial and/or marine ecosystems, cultural monuments, etc. Six protected landscapes with a total area of about 97.500 hectares have been declared (Qiriazi 2020).

10.7.6 Category VI: Protected Area of Managed Resources/Resource Reserve

They constitute protected areas managed for sustainable use of natural resources. They have a wide range and are relatively isolated. They have unmodified dominant natural systems and are either uninhabited or sparsely populated, but they may be under the pressure of greater population and more intensive use. Their management aims at protecting

and preserving biological diversity and other values, promoting sustainable production by protecting basic natural resources, and aiding regional and national development. Four reserves of managed resources have been announced with a total area of 18.245 hectares (Qiriazi 2020).

10.7.7 Municipal Natural Parks

The municipal natural parks are proclaimed by the municipal councils for the needs of the community. They are natural green areas (like forests, pastures, wetlands, etc.) or cultivated areas with special values, which fulfill the needs of the community with traditional activities. They have an area of over 50 hectares. These parks lie within the administrative territory of the municipality and they are not included in the national network of protected areas. This is a new approach that brings protected areas closer to the local community. Their management aims at protecting the environmental values and local biodiversity, as well as promoting economic development and educational activities. One municipal park (in Ulza) has already been declared and proposals for three other municipal natural parks have been made.

10.7.8 Cross-Border Eco-Museum of the Aos- Vjosa River the Municipal Natural Parks

The eco-museum is a form of protection and scientific management of the natural and cultural identity of specific areas. It is an important means by which communities protect and preserve their natural and cultural heritage and a new approach to the identification of local features and the effective management of this heritage by contributing to the revitalization of the abandoned rural areas and the forgotten traditional professions and crafts to both improve the living standards of residents and to alleviate their poverty (Maggi 2002).

This eco-museum includes 3540 km^2 of the Vjosa River, its upstream catchment area in Albania and Greece. These two border countries created the interactive eco-museum for the recognition and protection of natural and cultural heritage in situ on both sides of the border, as well as the protection of productive, environmental, and sociocultural identity.

The Vjosa River is considered one of the "wild" rivers in Europe, "a blue traveler" winding through mountains, gorges, canyons, and valleys. On the banks of this river and in the picturesque villages alongside it live thrives and the vegetation is profuse. Colors, sounds, aromas, and tastes form a diverse environment (Qiriazi 2020).

Over the centuries, along this important waterway, the Vjosa River has shaped the life and traditions of the population, and it has united civilizations by facilitating communication and being the lifeline to so many living creatures. In its watershed, there are two of the largest national parks, like the North Pindus (in Greece) and Hotova Fir Forest (in Albania) and several natural monuments, like the Zagoria natural park, etc.

10.8 National Cultural Heritage

The national cultural heritage is material and spiritual. The former consists of palpable objects, covering the various early materials created by Albanians. The latter is related to the use of the Albanian national language in literary works, to the oral folklore preserved in the memory of the people and to choreographic, instrumental, and oral folklore, to the

traditions and customs, to the beliefs and superstitions, to the traditional crafts, and the like.

The rich and diverse Albanian national cultural heritage is related to the ancient population and its social, economic, and cultural development over the centuries. Some noteworthy among these are the archaeological and historical monuments of Apollonia, Butrinti, Antigone, Lezha, Durresi, Shkodra, Berati, Zgërdheshi, Bylisi, Amantia, etc. There are architectural monuments (temples, theaters, amphitheaters, stadiums, fountains, monumental tombs, bazaars, towers, bridges, baths, etc.), artistic monuments (mosaics, sculptures, portraits, murals), ethnographic monuments (handicrafts and art products, traditional folk costumes), religious monuments (ancient churches, mosques, and tekkes), and environmental ones: museum cities (Berati, Kruja, Gjirokastra) and bazaars in Kruja, Korça, Shkodra, etc. The material and spiritual heritage and the geographical environment express the Albanian culture and identity (Photos 10.19, 10.20, 10.21, 10.22, and 10.23) (Ministria e Kulturës 2005).

Photo 10.19 Apollonia Archaeological Park (A. Cmeta, 2019)

Photo 10.20 Medieval castle of Gjirokastra (F. Plaku, 2019)

Photo 10.21 Hermit Church in Prespa (F. Plaku, 2019)

Photo 10.22 Labova Kryqi Church (thirteenth century) (F. Plaku, 2018)

Photo 10.23 The cultural monument of the Mosaic of Lin, fourth to fifth century (F. Plaku, 2019)

Like all civilized countries, Albania regards and treats its cultural heritage as sacred. Therefore, it has provided legal and institutional support for its preservation, scientific study, and management. The results of these studies are put in the service of culture, school, as well as ecological and patriotic education. The rich and interesting natural and cultural heritage of Albania is an added value for the sustainable development of the regional economy. To this end, strategies have been prepared for the protection of this heritage through legal and institutional support, through restorations, etc.

References

Boardman J (1982) The prehistory of the Balkans and the Middle East and the Aegean World, tenth to eighth centuries B.C. Cambridge University Press, Cambridge

Ceka N (2002) Buthrotum: its history & monuments. Tirana, Cetis Tirana

Ceka N (2005) Butrint: a guide to the city and its monuments (Migjeni Books)

Christian A, Thomas W (2008) Ancient Lake Ohrid: biodiversity and evolution. Hydrobiologia

Cmeta A(2000–2021) Personal photo archive

Cmeta A (2012) Tirana through lens. Albas Publishing House

Dragoti N, Dedej Z, Abeshi P (2007) Zonat e Mbrojtura të Shqipërisë

Dvornik F (1958) The idea of apostolicity in Byzantium and the legend of the Apostle Andrew. Harvard University Press, Cambridge, p 219. OCLC1196640

GCDO History part, History of Gjirokaster (in Albanian). Retrieved 1 Sept 2010

Hodges R, Logue M (2007) The mid-byzantine re-birth of Butrint, Minerva 18

Hodges R, Bowden W, Lako K etc. Byzantine Butrint: excavations and surveys 1994–1999. Retrieved 8 Jan 2011

Karaiskaj GJ (1981) 5000 vjet fortifkime në Shqipëri. Tiranë

Komiteti i Trashëgimisë Botërore (2005) Raport *Gjirokastra sit i trashëgimisë botërore*

Maggi M (2002) Ecomusei: guida europea. Allemandi editore, Torino

Ministria e Kulturës (2005) Lista e Monumenteve të Kulturës

Ministria e Mjedisit dhe e Turizmit (2017) Lista Kombëtare e ASCI për Shqipërinë

Ministria e Mjedisit dhe e Turizmit (2019) Rrjeti i zonave të mbrojtura në Shqipëri

Naumoski TB, Jordanoski M, Veljanoska-Sarafiloska E (2007) Physical and chemical characteristics of Lake Ohrid. In: 1st international symposium for protection of the natural lakes in Macedonia

Plaku F (2000–2021) Personal photo archive

Plaku F (2018) Homeland Albanian's hidden gems

Qiriazi P (1980–2021) Personal photo archive

Qiriazi P (2019) Gjeografia fizike e Shqipërisë. Botmi Mediaprint

Qiriazi P (2020) Trashëgimia natyrore e Shqipërisë. Botmi i Akademisë së Shkencave të Shqipërisë

Sinani SH (2000/6) Kodikët e Shqipërisë dhe 2000-vjetori i krishtërimit, në "Media"

Tole T (2018) Isopolifonia shqiptare

UNESCO. Report about the Lake Ohrid watershed region, 2019

Population Development

11

Dhimitër Doka

Abstract

Although Albanians, as descendants of the Illyrians, are among the most ancient peoples in the region and in Europe, the population to date has greatly changed because of various historical, political, and social factors. The establishment of present-day political borders in 1913 left outside these borders almost half of the Albanian population, which today lives in the neighboring countries. Thus, according to the first census conducted in 1923 within the current political borders of the country, about 815,000 inhabitants lived in Albania and the population density did not exceed 29 inhabitants/km² (Doka 2015).

The largest population growth occurred during the communist period (1945–1990), and it reached a maximum of almost 3.2 million inhabitants in 1989, with an annual growth of 2–3% (INSTAT 1989). Meanwhile, from the 1990s until today, there has been a significant decrease in the number of population due to two main factors: the mass emigration of Albanians abroad and the decrease in the birth rate. Consequently, 2,831,000 inhabitants live in Albania, based on the data from the last census in 2011, and the population density is 99 inhabitants/km², about 13% less than it was in 1989 (INSTAT 2011). The same trend has continued in the following years; therefore, even in the upcoming census (due in 2022), an even greater decrease of the population of Albania is expected.

Almost two-thirds of the population is concentrated in the central and western part of the country, (Tirana-Durrësi area) especially due to the large influx of population settling in this area from the other regions of the country after 1990.

In the last 30 years, great changes have also been observed in the qualitative structures of the population, such as age, social structure, residence, etc.

Keywords

Illyrians-Arbëreshët-Shqipëtarët · Population growth · Geographical distribution of the population · Migration · Structural of the population

11.1 Ethnogenesis of Albanians

Albanians are regarded as descendants of the ancient Illyrian population; therefore, the history of the population of present-day Albanian territories is very ancient. On this basis, present-day Albanian territories are considered among the oldest inhabited territories in the Balkans and Europe. There is numerous archaeological evidence that confirms the most ancient traces of the population. These numerous archeological objects found in different areas, such as in Xarra in Saranda, near Shkodra, in the Dajti mountain and other sites, are a testimony to this (Misja et al. 1987).

From the study of these archeological objects, it has been proved that these areas have been inhabited since the Middle Paleolithic period (100,000–30,000 years BC). The Neolithic period (7000 to end of the fourth millennium BC) testifies to an even denser population of these areas (Doka 2015). The archeological excavations carried out in Korça, where 12 settlements belonging to this period have been discovered, is another evidence (see Photos 11.1 and 11.2). Similar settlements have also been discovered in Cakran, Kolonjë, in the Drini i Zi (the Black Drini) valley, in Mati, Përmeti, and other places. Judging by the vast geographical extent of these settlements, we can conclude that these areas have been densely populated during that period. Among these settlements, it is worth mentioning the area of Korça, where there were also settlements built on water, as is the case Maliqi settlement (Misja et al. 1987).

The Bronze Age testifies to a further increase of the number of population and an even wider geographical distribution of this population.

Photos 11.1 and 11.2 Cave of Tren and palaeophilic settlements of Maliqi. (Source: https://www.zgjohushqiptar.info/2021/01/historia-e-lashte-e-shpelles-se-trenit-apo-grykes-se-ujkut.html; Source: https://link.springer.com/chapter/10.1007%2F0-387-36214-2_3)

After the creation of the Illyrian ethnic population, both the living space and the number of settlements of this population expanded considerably. This area already occupied almost all the western parts of the Balkan Peninsula. The Illyrian population at the time was organized in several tribes, such as the Encheleis, Taulantis, Dalmataes, Dardanis, etc. (Misja et al. 1987).

The collected evidence of the period of antiquity is proof of the same Illyrian ethnocultural continuity, as well as of the existence of a considerable population. One characteristic for this period was the continuous increase in the number of cities, where there were about 20 of them. Some cities among them were particularly important (see Photos 11.3 and 11.4), such as Scodra, Lisi, Bylis, Dyrrahu, Apolonia, Butrint, etc. (Photos 11.3 and 11.4).

The medieval period, when the country was called *Arbëria* and its inhabitants *Arbëri*, marked a further important development of the population, both in number and its geographical distribution. The population is estimated to have reached about 1.5 million inhabitants during this period (Misja et al. 1987).

Although statistical data are missing for all these periods until the Ottoman occupation, the archaeological evidence proves two very important facts: the ancient origin but also the ethnic and cultural continuity and development of the same population in all Albanian territories today.

11.2 Population Growth

The first statistical data that testify to the progress of the Albanian population belong to the period of Ottoman occupation. Thus, according to the Sanjak of Albania Land Register of 1431–1432, the population of the Albanian territories was about 800,000 inhabitants (Doka 2015).

After the Ottoman occupation and the declaration of independence, several censuses were conducted, respectively, in 1921, 1923, 1927, 1938, and 1945 (Berxholi et al. 2003).

Although very important, all these censuses until 1945 failed to rise above the level of genuine population counting. However, since 1923, the general censuses conducted would include the whole country and its population. Consequently, there are added data resulting from these censuses, which indicate not only the population number but also its geographical distribution, qualitative structures, etc.

Thus, based on the first general census of Albania in 1923, there were 814,380 inhabitants and the population density did not exceed 29 inhabitants/km^2. At the same time, 84% of the population lived in the country and only 16% lived in the city (Doka 2015).

Nine general censuses have been conducted since the end of Second World War. The first four were performed every five years, three every ten years, the penultimate after 12 years, and the last one again after 10 years.

The data from these censuses specifically reveal (see Table 11.1) an increase in the total population from 1945 to 1990, and this has happened entirely because of the natural population growth within the territory of Albania. Thus, in the general census in 1989, the population of Albania was 3,182,417 inhabitants, which means the tripling of the population within a period of 45 years.

However, the comprehensive changes that swept Albania after 1990 included not only the political and economic sphere but also the entire Albanian society. Unlike the previous period, when the demographic policy was dominated by such factors as the largest absolute possible increase in the population number, the prohibition of free movement of people, the even population distribution in the entire area of the country, etc., the demographic policy changed after 1990. Thus, family planning was also introduced in the natural population growth, free movement of people domestically and abroad was allowed, and overpopulation or depopulation of certain areas in the country became apparent.

All these new developments and their impact were immediately reflected not only on the significant decline of the total population number but also on many other indicators and parameters of demographic development of the country, such as natural population growth, population density per unit area, age and gender structure, population structure by residence, etc.

The large decline in population is evident from the data of the general census conducted in Albania in 2001. According to that census, the total population of Albania was 3,069,275 inhabitants, a figure about 130,000 smaller, or almost 4% lower than in 1989. As the data from the last census conducted in Albania (in October 2011) reveal, the population decline has continued even further, reducing the total population to only 2,831,741 inhabitants (INSTAT 2011). This declining tendency has also been dominant in the following years; therefore, an even greater decline in Albanian population is expected in the next census that is due in 2022.

Although such a decrease in the total population is not specific to Albania, as this phenomenon has been observed after 1990 in almost all former communist countries, Albania results particularly affected by this phenomenon in relative values. The main reasons for this large decrease in the total population number are related, on the one hand, to the extremely massive external emigration that involved the population of the country since 1990, and on the other hand, the decrease in the birth rate has also played its part.

Considering these two factors, it can be said that in the case of Albania, emigration abroad has played a crucial role in the population decline. This does not mean that the birth rate has not been on the decrease, but this decrease has not been negative, as has been the case in many other countries in the region.

Based on this analysis of the total population growth of Albania at different time periods, the question rightly arises:

Photos 11.3 and 11.4 Illyrian and ancient cities of Bylis and Butrint. (Photos by Dhimiter Doka, June 2017)

Table 11.1 Population of Albania 1923–2011

Years	Population Total	Male	Females	City in %	Country in %	Density inhab./km²
1923	814.380	421.618	392.726	15,9	84,1	29.0
1930	833.618	428.959	404.659	18,2	81,8	30.3
1945	1.122.044	570.361	551.683	21,3	78,7	39.0
1950	1.218.945	625.935	593.008	20,5	79,5	42.0
1955	1.391.499	713.316	678.184	27,5	72,5	48.0
1960	1.626.315	831.294	795.021	30,9	69,1	56.6
1969	2.068.155	1.062.931	1.005.224	31,5	68,5	71.9
1979	2.590.600	1.337.400	1.253.200	33,5	66,5	90.1
1989	3.182.417	1.938.074	1.544.343	35,5	64,5	110.7
2001	3.069.275	1.530.443	1.538.832	42,1	57,9	107.4
2011	2.831.741	1.421.810	1.409.931	53.7	46.3	98.0

Source: 1. Statistical Yearbook of Albania 1991, p. 361
2. INSTAT: Census Results, 2001, p. 19

11.2.1 How Many Albanians Live in the World Today?

It is difficult to give a correct answer to this question. On the one hand, this has been particularly difficult because of the many and frequent changes the country has historically undergone under different governments which have made it difficult to have accurate statistics, and on the other hand, the difficulties of life and numerous wars have historically forced many Albanians to move from their territories and scatter around the world (Misja et al. 1987).

Another reason that has made this problem very difficult has been the frequent border changes and fragmentation of the Albanian territories. Albanians are among the few peoples in the world whose territories are politically divided into six countries, namely, Albania, Kosovo, North Macedonia, Serbia, Greece, and Montenegro. This has made the statistical estimation of their total number different, according to the countries where they reside.

However, based on various data and estimates, it can be said that the total number of Albanians in the world today fluctuates between 6 and 9 million (Doka 2015). The first figure relates mainly to Albanians currently residing in Albania, Kosovo, North Macedonia, Greece, Serbia, and Montenegro, while the second figure includes the Albanian population has already immigrated to different countries of the world, both in the context of early emigration (Albanian diaspora), but also in the context of the recent emigration after 1990.

11.3 Geographical Distribution of the Population

The geographical distribution of the population is conditioned by various natural, economic, and social factors. Albania is distinguished for its diverse relief, from its coastal lowlands to the hilly terrain and the high mountains. It is self-evident that in a little economically developed country, natural factors have had a huge impact on the geographical distribution of the population. On the other hand, the development of the country has had its characteristics and priorities concomitant with the specific stages which the country has gone through. Likewise, the development of the population itself, depending on the performance of various demographic indicators (natural population growth, migration, etc.) as well as the policies pursued at certain stages, has significantly affected the distribution characteristics of the population.

The role and influence of these groups of factors in the geographical distribution of the population in Albania has been different depending on the periods in which the demographic development of the country has passed.

Until 1945, being enveloped in significant backwardness with little socioeconomic development, the natural factors played the main role in the geographical distribution of the population in Albania. Therefore, the movement and concentration of the population was directed toward those regions and areas of the country which met the best natural conditions in the terms of agricultural lands, warm climate, the availability of water, and other natural resources, whereas the influence of other (demographic and socioeconomic) factors was much less than natural ones for the period until 1945. On the one hand, the reason for this was the low natural population growth due to the high level of mortality as well as the considerable population departure from different areas of the country through external emigration. On the other hand, the economic situation was very difficult in almost all areas of the country, which meant that it could not have played a significant role in the distribution of the population.

Only a few specific areas became attractive to a certain segment of the population in search of better jobs, and they were mainly those areas where foreign capital was concentrated in investments such as in mineral extraction or road construction. Therefore, almost 40% of the population of that time was concentrated in four prefectures of the country: Berati, Durrësi, Tirana, and Vlora. Although they did not occupy more than 25% of the country, these prefectures had the best natural conditions to meet the people's needs, whereas most of the country's area had a very small population and the average population density did not exceed 40 inhabitants/km^2 (INSTAT 1989).

This situation would change during the period 1945–1990, when the increase of the total population of the country also increased the population density per unit area, from 39 inhabitants/km^2 in 1945 to 110.7 inhabitants/km^2 in 1989 (see Table 3.2). The two main factors for this increase were the high natural population growth in the whole territory of the country coupled with the legal prohibition of emigration which forced the growing population to stay within the country (Table 11.2).

Besides the increase in population density throughout the country from 1945 to 1990, differences in areas depending on the priority of their economic development were also witnessed. Therefore, in the 1960s and 1970s, a greater concentration of the population was observed in the areas and centers where numerous industrial facilities were built or where new agricultural lands were cleared and acquired. As a result of the need for labor force in these areas, there were numerous influxes of population toward them, a process that was planned and organized by the state. The Western Lowlands and its most important urban centers (Tirana, Durrësi, Elbasani, Vlora, Fieri, Shkodra) would see their population increase by fresh arrivals, which would also significantly increase the population density per unit area, reaching over 400 inhabitants/km^2 (Doka 2015). The incoming population were mainly from the areas and districts of the south, southeast, but also north of the country.

Besides the Western Lowlands, noted for its significant population attraction, almost all the country's areas and districts had a significant increase in population density brought about by the increase in their population. Even in the high mountainous areas of the north and northeast of the country,

there was a significant increase and concentration of population because of job creation (mainly in the mineral industry) and banning the free movement of people.

In the late 1970s and during the 1980s, the further increase in the population, with its dominance of active working-age population, the fulfillment of the previous labor needs for many areas and industrial centers, all these factors coupled with the ever-deepening economic crisis within the country lead the state of that time to increase the bureaucratic measures hindering the free movement of the population even within the country. Under these conditions, the population would remain "immobile" in their areas of birth and growth, until the regime finally collapsed.

Meanwhile, one of the results of the new demographic developments in Albania after 1990 onward has to do with the deepening level changes in the population between different regions and areas of the country. Unlike the demographic policy prior to 1990, which aimed at a possibly uniform population throughout the country and at hindering the free movement of people between different areas, the free movement of people was allowed in Albania after 1990.

This gave many people the opportunity to go abroad or leave their previous place of residence in search of better working and living conditions, often taking their families with them too.

The decline in the total population of Albania after 1990 has been accompanied by a decrease in population density per unit area. Thus, the indicator of the average population density per unit area in Albania in 1989 was 110,7 inhabitants/km^2; according to the 2001 census, the same indicator resulted in a decrease to 107.4 inhabitants/km^2, and it fell even further to under 100 inhabitants/km^2 in the census data in 2011 (INSTAT 1989 and 2011).

The general tendency of the population has been the departure from the high mountainous and peripheral regions of the country and the settlement in the central part and in the main cities (see Fig. 11.1).

The central part of Albania, the Tirana-Durrësi area, has seen the largest population arrivals, and this is the main factor for the significant population increase in the concentration per unit area. The population settling to this region has mostly departed from the northern and northeastern areas, but there are also residents from almost all other regions of the country.

Because the territory of the country is dominated by mountainous terrain with harsh climate, the highest population density is observed in the country's plains with their good soil structure and warm climate, which also provide better working and living conditions.

Table 11.3 and the population density map above reveals that there is a large concentration of population in the central and western part of the country where the population density exceeds 200 inhabitants/km^2, while in the region of Durrësi,

Table 11.2 Trend in population density in Albania

Years of censuses	Density (inhabitants/km^2)
1923	29.0
1938	36.0
1945	39.0
1969	56.6
1979	71.9
1989	110.7
2001	107.4
2011	99.0

Source: 1. Statistical Yearbook of Albania 1991, p. 361
2. INSTAT: Census Results, 2001 and 2011

Fig. 11.1 Population density
by grid 2011

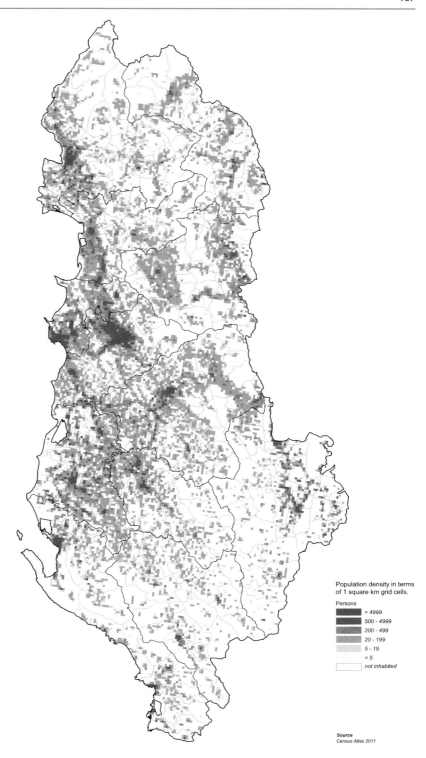

Population density in terms
of 1 square km grid cells.

Persons

> 4999
500 - 4999
200 - 499
20 - 199
5 - 19
< 5
not inhabited

Source
Census Atlas 2011

it reaches almost 400 inhabitants/km², while in Tirana, it goes over 500 inhabitants/km². The districts with the lowest population density are Dibra, Kukësi, and Gjirokastra, with 46, 32, and 22 inhabitants/km², respectively, as given in the table below.

These changes in population density have significantly deepened regional differentiation between the areas and regions of the country (see Fig. 11.2). The large flow of domestic migration has played the main part in this differentiation.

Table 11.3 Population indicators by district 2019

Districts	Population	Density (inhabitants/km²)
Berat	126.295	70
Dibër	119.963	46
Durrës	289.877	378
Elbasan	276.765	87
Fier	296.446	157
Gjirokastër	62.188	22
Korçë	209.034	56
Kukës	76.994	32
Lezhë	125.998	78
Shkodër	203.945	57
Tiranë	889.578	538
Vlorë	189.297	70
Shqipëria	2.866.376	100

Source: INSTAT – Demographic Indicators 2019

These large differences in geographical distribution have deepened even further in recent years, leading to the total depopulation of certain areas.

11.4 Natural Population Movement

The population of Albania has been traditionally distinguished for its high birth rate. However, its impact on natural population growth has varied depending on its performance and the mortality rate. In this regard, the general state of socioeconomic development and demographic policies applied at different times have had their own influence and impact.

The data on the progress of the natural movement of the population in Albania for the period up to 1945 show low rates of natural population growth. However, this was not because of the low birth rate but due to the high mortality rate, especially the infant mortality rate. Despite the high number of births, difficult living conditions, numerous diseases, and lack of adequate health care accounted for the high mortality rate. The closed agrarian economy, which viewed children as future farm hands, widespread poverty, and low cultural and educational level, stimulated the birth of as many children as possible. However, the high infant mortality rate (0–1 year of age) directly affected population growth. Also, the average life expectancy for the entire population was low (38 years old), even lower for women, which meant that a good part of them did not live their entire fertility period (Misja et al. 1987).

This situation would change significantly **during the period from 1945 to 1990**. The new demographic policy of the communist state would particularly stimulate the rapid and highest possible increase in the country's population. This new policy was accompanied by the improvement of living conditions and the establishment of a new health-care service, with special care for mothers and children. On the one hand, these combined measures fostered and sustained high fertility rates, gradually restraining and reducing the mortality rate on the other hand. Therefore, very high rates of natural population growth ensued. The data in the table below confirm this fact (Table 11.4).

The data in the table clearly show the high rates of natural population growth for the entire period from 1945 until 1990. The natural increase has been particularly high in the 10-year period between 1950 and 1960 (Misja et al. 1987). This happened due to the combined effect of the very high birth rate and the continuous decrease in the mortality rate. During the 1970s and 1980s, the tendency of gradual decline in natural population growth was observed. This fact is best illustrated by the average annual rate of population growth, as reflected in the table below (Table 11.5).

Such a thing was not the result of an increase in the mortality rate, which continued to decrease even further at the time, but because of the decrease in the birth rate. Some the most important factors that influenced this decrease, which would grow deeper in the following years, were the following:

- An increase in the age of girls entering marriage, for reasons of education and career
- A change in the function of the family, the move from a small agrarian economy that stimulated having children as labor force, toward a state-run economy, that guaranteed the provision of all the needs of the family
- A decrease in infant mortality, which also boosted parental confidence in the survival of their children
- Several psychological, social, emancipator factors, personal comfort, etc.
- The decreasing tendency in the fertility rate would deepen more and more, especially from the 1990s until today.

The shift from the previous demographic policy of stimulating the natural increase of the population to the new policy of family planning, the official allowing of abortion, and mass departure of the mainly young population through external emigration were some of the main factors, which led to this large decrease in the fertility rate after 1990. These new conditions have caused the gross coefficient of natural population growth to fall below 2% in recent years (Fig. 11.3) (INSTAT 2019).

Significant differences in the indicators of natural population growth in terms of space between different administrative units can be observed. The greatest contribution in absolute value comes from the largest districts and cities, primarily Tirana, but also Durrësi, Elbasani, Fieri, etc.

This accounts for the great concentration of the population in the central and western parts of the country, while the peripheral and mountainous areas have made insignificant contribution in this regard due to their high number of the population departures, especially young people. However,

Fig. 11.2 Population density, 2019

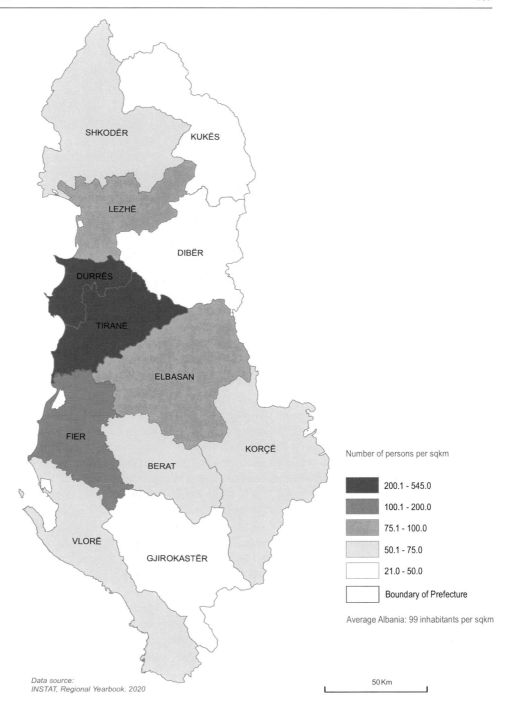

Data source:
INSTAT, Regional Yearbook, 2020

Table 11.4 Natural population growth in Albania during the period 1945–1990 (per thousand inhabitants)

Period	Fertility	Mortality	Natural growth
1945–1950	35.5	15.4	20.1
1951–1955	40.1	14.5	25.6
1956–1960	41.7	10.6	31.1
1961–1970	38.5	8.4	30.1
1971–1980	31.2	6.6	24.6
1981–1990	26.1	6.0	20.1

Source: Dh. Doka, Geografia e Shqiperise (Popullsia dhe Ekonomia). Tirana 2015

although the decline in natural population growth is quite large compared to the period up to 1990, the population of Albania compared to many other European countries (see Fig. 11.4) is nevertheless distinguished for positive natural population growth (INSTAT 2019).

Being a specific case, Albania has belatedly adopted and moved slowly toward the consumer society, which, among other things, makes the upbringing of children very competitive. With an average monthly income among the lowest in Europe, most of the Albanian society is far from a real con-

Table 11.5 Average annual population growth rate 1945–1990 (in %)

1945–1950	1950–1955	1955–1960	1960–1965	1965–1970	1970–1975	1975–1980	1980–1985	1985–1990
1,7	2,7	3,2	3,0	2,7	2,4	2,2	2,1	2,0

Source: Klaus-Detlev Grothusen (Hrg.): Albanian. München 1993. p. 466

sumer society. Considering the consumption rate level, the meeting the demands for housing, better food, and clothing is still far from the desirable, thus eroding the strong competitive advantage for the upbringing of children.

Moreover, Albanian society continues to maintain its tradition as a family-oriented rather than an individual-oriented society. This makes the upbringing of children to be considered a task not only of the parents but of the larger extended family. For a significant part of Albanian families, especially in the deep mountainous regions, having children in the difficult conditions of today's market economy is also considered as helping families to have jobs.

However, things are changing fast. On the one hand, the ever-increasing decline in the birth rate and the continuation of massive external emigration on the other, particularly affecting young people, will further reduce the impact of natural population growth.

However, the year 2020 would make an exception and worsen the situation. This is because the effects of the COVID-19 pandemic were added to the previous factors, which increased the mortality rate by 25.8% or about 7000 more people than in 2019. This made for the first time the number of births during 2020 to be lower than that of deaths and the total population on January 1, 2021, to be 0.6% lower than a year earlier (INSTAT 2021). All these together led to the reduction of the entire population of Albania, especially of the separate regions as in the following scheme (Fig. 11.5).

Thus, except for the regions of Tirana and Durres, which have had a small increase as a result of internal migration, all other regions have had a decrease in population, where the most extreme cases with more than −3% are the prefecture of Gjirokastra and more than −2% the prefecture of Dibër and Berat (see Fig. 11.5).

In these conditions, it is only the improvement of the general socioeconomic situation and Albania's fast integration in the region and Europe that can save the country from further abandonment but also stop the decline in population growth and in the total population number.

11.5 Migrations

The Albanian population is considered a migrant population. Various factors, ranging from the numerous wars to the prolonged occupation of the country by foreigner powers, the lack of political stability, and especially the difficult eco-

nomic conditions, have often forced Albanians to become part of migratory movements, both within the country and abroad. Under these conditions, both external and internal migration have played decisive roles in all the other demographic and socioeconomic developments of the country (Göler and Doka 2015).

11.5.1 External Immigration

Emigration abroad has been an Albanian historical phenomenon. The dimensions and directions of this emigration have had a great impact on the distribution and development of the Albanian population everywhere in the world. This has made the number of Albanians living outside the borders of Albania today considerable, which has had a significant impact on various socioeconomic developments. Although this is a centuries-old phenomenon, complete and detailed data on the extent of Albanian emigration are still missing. However, several cases of mass emigration of Albanians are well known in history (King et al. 2010).

One massive wave of emigration of Albanians occurred after the death of Skanderbeg in January 17, 1468, when about 200,000 Albanians were forced to leave their ancestral lands and escape to the neighboring countries (mainly Greece and Italy). Most of them ended up settling in Southern Italy. Today, there are about 80 settlements inhabited by Albanians who left then. They are known as the Arbëresh of Southern Italy (see Fig. 11.6).

The prolonged Ottoman occupation, an almost five-century-long period, forced many Albanians to flee their country. In its centuries-long natural development, this population is estimated to have nowadays reached over 4 million people of Albanian origin in Turkey alone (Doka 2015).

Similarly, huge departures of Albanians settled to Greece, where they are still known as Arvanitas. Due to their assimilation, it is difficult to give an estimate about the number of the Arvanitas population in Greece. However, it is known that at certain times, this population amounted up to 25–30% of the total population of Greece. Other significant emigration influxes of Albanians took place in Bulgaria, Romania, Croatia, and other countries (Misja et al. 1987).

The process of external emigration of Albanians continued even during King Zog's rule and the Second World War. It was during this period that Albanian emigration spread

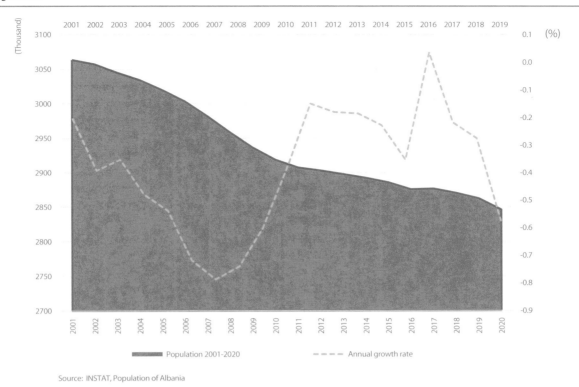

Fig. 11.3 Population of Albania and natural population growth rate

beyond the borders of Europe, heading toward the USA, Canada, Australia, etc.

All these emigration waves together created the Albanian diaspora, which has played an important and decisive role in the events in Albania, as was the case of the preparation of the country for its declaration of independence in 1912.

11.5.1.1 Prohibition of Emigration During the Communist Period (1945–1990)

During the communist rule (1945–1990) in Albania, because of the country's increasing isolation, external emigration of Albanians ceased to exist. For this entire period, external emigration only existed in the form of sporadic escapes of the political opponents of the system, who, in many cases, paid for this endeavor with their own lives (Gëdeshi and King 2018). However, it must be underscored here that it was precisely the emigration that dealt the first blows to the communist system in Albania, when in July 1990, about 5000 Albanians who sought to leave the country broke into some of the foreign embassies in Tirana. Most of them emigrated to Germany, France, and Italy (Göler 2017). This was followed in November 1990 by the departure to France of the most famous Albanian writer, Ismail Kadare, followed by other departures as the December protests swept the country the same year bringing about the collapse of the communist system in early 1991.

11.5.1.2 Mass Emigration After 1990

In the wake of the democratic changes that took place in Albania after 1990, a new phase of migration of Albanians began, especially emigration abroad: from 1991 onward, the transition of the Albanian economy from central planning to the market economy, a series of political and socioeconomic problems set in such as the immediate rise of unemployment, political insecurity, crime, and other problems. Under these conditions, emigration abroad was regarded by many Albanians as the best way for their survival or overcoming their difficult situation. In addition to the individual or small-group emigration, a massive wave of emigration broke out in March 1991, which engulfed over 100,000 people within that month alone. Most of them, about 76% went to Greece, while the rest, about 25,000 people or 24%, went to Italy (Heller et al. 2004). To this day, all Albanians and many others still remember the ships overcrowded with Albanians heading to the Italian coast in early March 1991 (Photo 11.5).

Other mass emigration flows resulted in the rapid increase of the number of Albanian emigrants abroad. During 1991–1993 alone, the number of Albanians emigrating abroad was estimated at around 300,000, with an average of 100,000 per year. This huge influx of emigration since 1993 has made Albania a country with the highest intensity of external emigration compared to many other countries of the former communist bloc (Doka 2015).

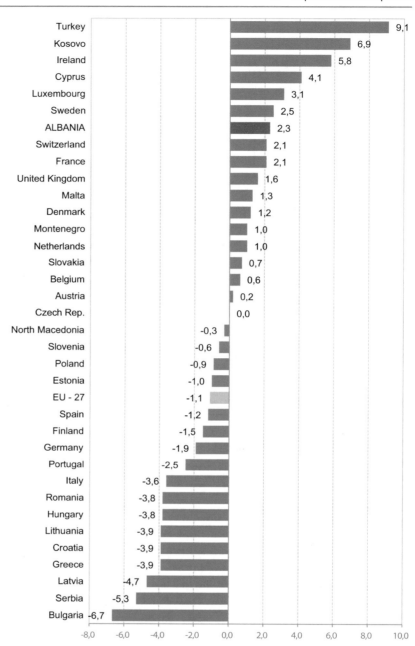

Fig. 11.4 Crude rate of natural change of population by countries in Europe (2018)

Source: INSTAT; Eurostat - https://appsso.eurostat.ec.europa.eu/nui/show.do?dataset=demo_gind&lang=en

Despite its improvements and deteriorations in the following years, the general situation in Albania remained difficult. This made external emigration of Albanians continue unabated, involving both genders and all ages.

Thus, it is estimated that ca. 1.6 million Albanians have emigrated abroad in all these years from 1990 until today. Most of them, over 600,000 are in Greece and over 500,000 in Italy, while the rest have emigrated to other countries of Western Europe, USA, Canada, etc. (see Table 11.6).

This mass emigration of Albanians after 1990 has been overwhelmingly an illegal phenomenon. Therefore, the official data do not match this massive influx. However, based on data and assessments from different directions, the progress of external Albanian emigration can be summarized in the table and map below (Fig. 11.7) (Heller et al. 2004).

In terms of the territory of the different areas of Albania, it is observed that the central and southern parts of the country have suffered the greatest loss of population, because the population of these areas has had the opportunity and tendency to emigrate to Italy and Greece, whereas for the northern areas of the country, in the first years, their population loss is primarily associated with the internal migration

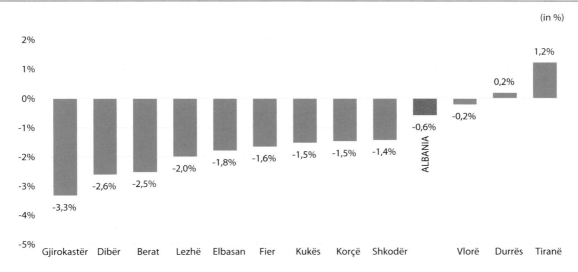

Fig. 11.5 Total change of population by prefecture, 1-st January 2021 vs 1-st 2020

through their departure mainly to the central and western parts of Albania. However, over time, this population would also be strongly involved in external emigration (Fig. 11.8) (Berxholi et al. 2003).

The motives of this mass emigration of Albanians abroad have been various. Initially regarded as protest against the communist regime and the country's long-standing total isolation from the rest of the world, the economic motives were at the core of Albanian external emigration. However, the educational motives and the hope for a better life or for political stability are not insignificant urges for Albanian emigration.

Based on various surveys conducted with Albanian emigrants in Greece and Italy, the following motives primarily stand out for the external emigration of Albanians (see Table 11.7).

The progress of comprehensive reforms and their concomitant fluctuations typical of this difficult process during the last three decades in the country have also affected the dimensions, directions, and motives of the external emigration of Albanians. A considerable portion of this emigration has taken a seasonal character, especially for emigrants heading to Greece. The structure of this emigration has also been expanded to include females to a considerable extent but also the agrarian and mountainous areas of the north and the northeast of the country (Gëdeshi and King 2018).

This type of emigration has been overwhelmingly illegal. However, in recent years, legal emigration and family reunification with relatives who had already emigrated in the first years of transition have also become important.

The shocks that the reform process has suffered in Albania have had a direct impact on the progress of external emigration of Albanians. Thus, it is worth underlining the

events of 1997, when the country was engulfed by a vortex of chaos and uncertainty because of the collapse of the Ponzi fraudulent schemes. The upshot of this difficult situation was not just the financial ruin suffered by most Albanian families, the political insecurity, and the bad image for the country, but the events became the cause for a massive new wave of external emigration. It is estimated that about 50,000 Albanians left the country by whatever means they could find during the difficult months of 1997 and 1998, further increasing the number of Albanian emigrants abroad (Doka 2015).

This situation continued to have an impact on the external emigration of Albanians in the following years, increasing the number of emigrants even more, many of whom would now aim to cross the borders of Europe toward an irreversible emigration. After 1998, two new features of the external emigration of Albanians can be distinguished:

1. The first one is related to the events of 1999, when, for the first time in its history, Albania became a host country of a massive influx of emigrants from Kosovo fleeing from the War of Kosovo. Thus, about 500,000 Kosovars would find refuge in Albania in the spring and summer of 1999 (INSTAT 2001).

2. The second one concerns the importance of legal immigration to the USA and Canada, especially after 2000. Unlike the previous type of Albanian emigration, this new kind of emigration mostly includes the qualified portion of society, and it is mainly family emigration, which means there is little hope of returning to Albania. This has exacerbated the problem of the most qualified portion Albanian population leaving the country, also known as "brain drain."

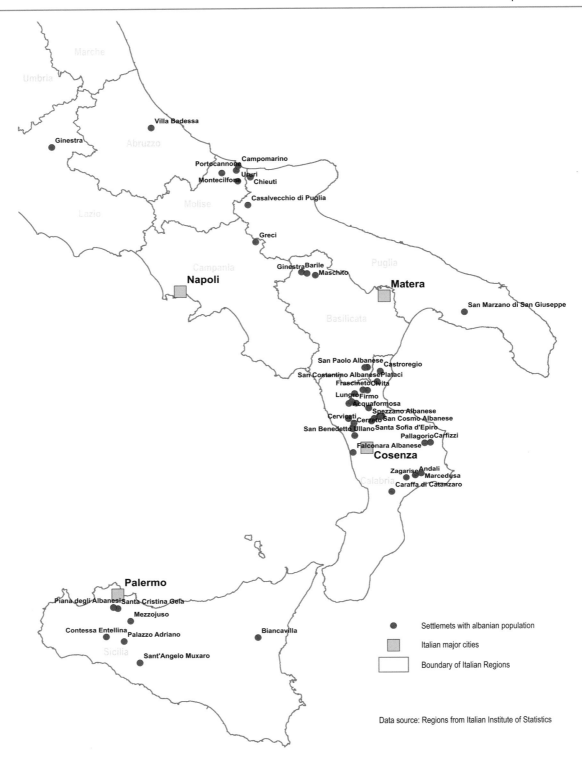

Fig. 11.6 Settlements with Arbëresh in Italy

11.5.1.3 "Brain Drain" Emigration of Albanians

According to data from scientific institutes and universities in the country, it turned out that about 40% of intellectuals from these institutions during the 1990s were already involved in the massive influx of external emigration. Compared to other former communist countries of Eastern Europe, the percentage of Albanian intellectuals who emigrated was higher (Table 11.8) (Gëdeshi and King 2018).

Photo 11.5 Albanians emigrants to the Italian in early March 1991. (Source: https://i.redd.it/h7x6wunmbdn01.jpg)

Table 11.6 The main countries of Albanian emigration after 1990

Country of emigration	Number of emigrants
Greece	650,000
Italy	550,000
USA	85,000
Canada	55,000
England	35,000
Germany	25,000
France	3000
Belgium	2000
Turkey	2000
In other countries	about 170,000
Total	**1,577,000**

Source: International Organization for Migration, UN, 2018

This process has continued in the following years, mostly to Canada and the USA, further increasing the emigration number of intellectuals. Every year, over 100,000 applications for immigration to the USA and over 20,000 to Canada are registered in Albania (Doka 2015).

The structural analysis of these emigrants, according to their academic qualification, reveals that most of them are either recent graduates or with a doctorate degree, but many are students. Therefore, they are mainly young people, who after a prior qualification in Albania and the connections they establish with the outside world, but also thanks to their academic position, they have a greater inclination to leave the country.

Most of the emigrated Albanian intellectuals are from Tirana, where most of the scientific institutions of the country are located. The rest are from the most important cities, which have universities and scientific research institutions, such as Shkodra, Gjirokastra, Elbasan, Korça, etc (INSTAT 2011).

The directions of this emigration are different. Originally aiming to reach the neighboring countries and Western Europe, it has increasingly taken on the new character of the distant emigration, mainly to the USA and Canada.

11.5.1.4 Forms and Structure of Emigration

In general, the external emigration of Albanians after 1990 has appeared in two forms:

1. Long term (individual or family emigration)
2. Temporary (or seasonal emigration)

The first form has had a wider geographical reach, and it did not just include the neighboring and European countries but also some distant countries like USA, Canada, Australia, etc.

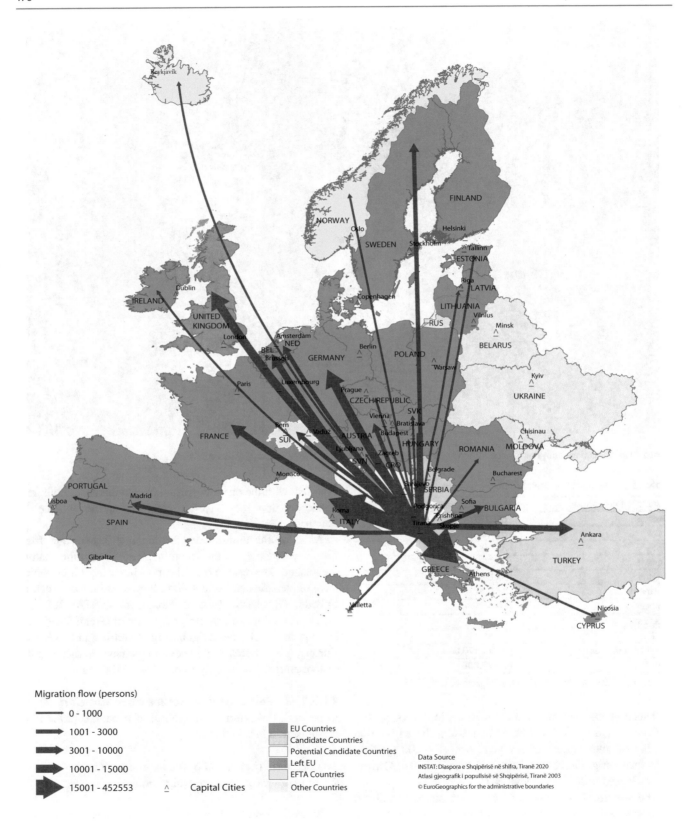

Fig. 11.7 Destination of external migration, 1990–2020

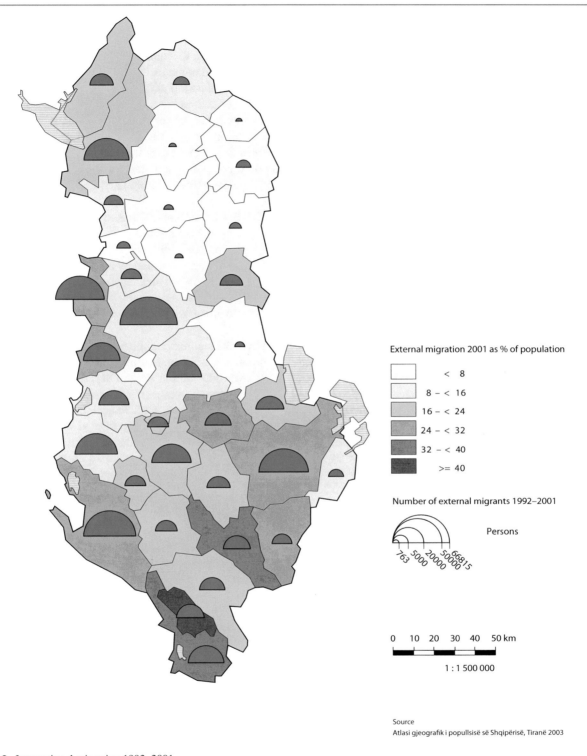

External migration 2001 as % of population

	< 8
	8 – < 16
	16 – < 24
	24 – < 32
	32 – < 40
	>= 40

Number of external migrants 1992–2001

Persons

763 5000 20000 50000 66815

0 10 20 30 40 50 km

1 : 1 500 000

Source
Atlasi gjeografik i popullsisë së Shqipërisë, Tiranë 2003

Fig. 11.8 International migration 1992–2001

The second form is mainly directed toward the neighboring countries, especially Greece and Italy. These emigrants to the neighboring countries chose to go there only for a few months, when the needs for labor increase, mainly in agriculture (Heller 1997).

As far as the treatment of external emigration is concerned, an important element is the analysis of its structure by age, gender, education, place of residence, and the like, as this is directly related to a series of socioeconomic problems that the country is going through.

Considering their age, most of the emigrants from Albania were young and of working age. On the one hand, emigration at this age has alleviated the difficult situation created by unemployment in the country, because many people who have left were ready to enter the labor market. On the other hand, a large portion of the most vital population has left, and this factor can bring about long-term socioeconomic problems, both in the country's future demographic developments and its economic reconstruction (Heller et al. 2004).

Moreover, the young age groups of the population involved in external emigration represent not only the most "adventurous" age groups, aiming to reach at all costs the beautiful Western Europe portrayed as a dreamland for them in the Italian TV channels, but also the age groups that, after getting a certain education level, enter the difficult labor market in Albania, which offers very few opportunities for them.

As far as gender is concerned, most Albanian emigrants are men. However, there has been a shift in this structure in recent years as females have become part of emigration through the process of family reunification.

Considering their education, about 2/3 of Albanian emigrants have high school certificates or higher education degrees (Gëdeshi and King 2018). Moreover, most of them already speak foreign languages which are usually an advantage for young immigrants seeking to integrate into the economic and social life of the host country. However, Albanian emigrants being overwhelmingly illegal, the linguistic advantage has been to little avail. In most cases, they must perform the most difficult and unskilled work with the lowest pay. Even many intellectuals who have opted to leave Albania have had to do similar jobs.

As far as their residence is concerned, most Albanian emigrants come from the urban areas. This has mainly happened because, on the one hand, the urban population has been better informed and well-oriented to participate in external emigration and, on the other hand, the closure of industrial activity created immediate unemployment in the urban population who, being left with no alternative employment opportunities, turned mostly to emigration. In contrast, land privatization for the rural population made it more closely linked to agrarian activities. However, there have been changes in this trend over the years, and the participation of the rural population in external emigration has been on the rise, especially toward Greece and Italy (Heller et al. 2004).

11.5.1.5 Effects and Consequences of Emigration

This massive emigration of Albanians (Doka 2015) after 1990 has been accompanied by a series of effects and consequences, both positive and negative ones.

Some of *the most positive effects* of the external emigration of Albanians are the following:

1. Financial and material income entering Albania from the remittances that emigrants send home. Thus, an average of over 800 million dollars have entered Albania every year through emigrant remittances (INSTAT 2001). Considering the difficult financial and economic conditions of the country, this foreign currency income is indeed a great help for Albania and Albanians. In certain years, this income has amounted up to 20% of the total financial income of the country. Besides the financial income, emigrants have also contributed to other material forms (like cars, equipment, furniture, and the like) for their families and relatives, thus significantly helping to improve the housing and living quality in Albania.
2. Another important positive effect lies in the impact of external emigration on the labor market in Albania. The emigration of many people of working age has significantly alleviated the problem of unemployment. If these people had stayed in Albania, the unemployment rate in the country would have been at least double the current figure.
3. Besides their financial and material gains, Albanian emigrants have also enhanced their professional capacities or have gained new experience and technical know-how. They have learned many new professions abroad, and many have applied these new skill sets in various eco-

Table 11.7 Motives for emigration of Albanians in foreign countries

Motives	Emigrants in %
1. Low wages in Albania	27.4
2. Assistance for families in Albania	29.2
3. Better working conditions in the emigrated country	15.6
4. Better opportunities for leisure activities	17.2
5. Better education opportunities	7.0
6. Political insecurity	3.4

Source: Misja et al. (1987)

Table 11.8 Emigration of intellectuals from some countries of the former communist bloc (in %)

Countries	Emigration of intellectuals	Countries	Emigration of intellectuals
Albania	38.5	Slovenia	1.7
Bulgaria	11.5	Estonia	13.8
Czech Rep.	4.0	Poland	15.0
Slovakia	11.3	Romania	3.0

Source: Brain Drain from Central and Eastern Europe, April 2001

nomic sectors of Albania, especially in the services, trade, construction, etc.

4. Emigration provided many Albanians not only with the opportunity to work but also to study and educate their children in the host country. This has further increased their educational level and cultural background, thus investing in a more professional and qualified future generation, which will ultimately be of benefit to them and to Albania.

However, besides its positive effects, the external emigration of Albanians since 1990 has had a series of negative effects, which stem primarily from the fact that this emigration has almost completely been illegal. The illegal form of their emigration has deprived Albanian migrants from enjoying their emigrant status in the host countries, or it has taken them a long time to gain that status.

Under these conditions, they have faced all kinds of discrimination and have found it difficult to integrate into the socioeconomic life of the host country. Only the portion of the emigrants who have settled in other countries together with their families can be considered the most stable part of Albanian emigrants abroad. Nevertheless, a considerable part of the external Albanian migrants remains illegal.

This illegal emigration has been very difficult and full of hardships for many Albanians, sometimes claiming their lives while crossing the land border with Greece or crossing the sea to Italy. Official data alone show that over 2000 people lost their lives during their emigration to Greece and Italy (Heller et al. 2004). However, considering the illegal nature of emigration, this figure could probably be much higher. Being mainly illegal immigration, it has often been associated with crime and prostitution, thus negatively affecting both Albanian migrants' integration in the host countries and Albania's image in the outside world.

The mass exodus of Albanians, involving over 1.5 million people since 1990, most of them at a young age, is by far one of the biggest losses of the country's human capital (Doka 2015).

Albania already feels the void left behind by the departure of this enormous human capital that is vitally important for the country's future development.

However, since 2010, besides the ongoing departure of Albanians, there has been a reverse tendency, and a significant portion of them have returned to Albania, because of the general financial crisis and other concomitant problems, especially in Greece.

11.5.2 The Return of Albanian Emigrants

This trend was observed during 2010–2013 because of the financial crisis that affected various countries of the world,

especially the neighboring Albania (Greece and Italy). It was precisely the global financial crisis in general and the Greek crisis, one of the main destination countries for Albanian migrants, which has reduced the opportunities for migrants to work there after 2010 (Göler 2014).

The crisis aggravated the financial situation of many Albanians migrants, who were left with the alternative of their return to Albania. This forced return, created by the new circumstances starting from 2010, would become part of the history of Albanians migration. According to the 2011 Census, 4.9 percent of the Albanian population, amounting to 139,827 citizens who had resided abroad, had returned home after 2010. A study conducted by INSTAT and IOM in 2014 also shows that returnees are relatively young and at a working age (Göler and Doka 2015).

Although the phenomenon started after 2009, this trend has been on the increase, reaching its peak in 2012 and 2013 with 53.4%. As far as the gender of the returnees is concerned, there is a significant dominance of males with 73.7% of them, compared to 26.3% of females. As far as origin is concerned, this is dominated by "voluntary" returns, amounting up to 94%, the vast majority is from Greece with 70.8%, followed by Italy with 23.7% and other countries, such as United Kingdom, Germany, etc. (INSTAT: 2014, p. 13.)

As it is clear from the table data, the largest percentage of returnees comes from Albanian emigrants in Greece, followed by those in Italy, which explains once again the difficult situation created in these two countries due to the financial crisis. The data (see Table 11.9) also show that the Albanians' stay in these host countries has been relatively long (an average 7 years for those in Greece and over 4 years for those in Italy). This fact shows that despite their long stay in these countries, they did not manage to put aside enough savings or reserves, which they could use in the difficult times.

Most returnees came back home with their families, 74% and 79.1% of them aimed at a long-term stay in Albania. The main reasons for their return include unemployment in the host country (88%), followed by lack of documentation and the desire to invest in Albania (INSTAT 2014).

Table 11.9 Returnees by last (emigration) host country (in %)

Countries	Percent	Average years of residence
Greece	70,8	7,0
Italy	23,7	4,3
England	1,6	4,1
Germany	1,1	6,0
US of America	0,6	4,2
Others	2,2	4,1
Total	100,0	6,2

Source: Survey form INSTAT and IOM in Albania, 2013

As far as gender is concerned, the dominant reason for men is unemployment, while for women, family reasons predominate.

The data also show that most emigrants (60.3%) have returned to the place of residence they had before leaving for emigration (Göler and Doka 2015). However, it turns out that resettlement, for a good portion of them, also depends on the progress and directions the internal migration of the population in Albania has had since their time of departure. Under these conditions, their resettlement in Albania is oriented in accordance to the current location of their family.

In many cases, there is a tendency for these returnees to resettle in those areas and administrative units with a generally improved socioeconomic development, with the hope that they can better fulfill their life plans in these areas. Thus, Tirana is the favorite region, followed by Vlora, Elbasani, Fieri, and Korça (for the returnees from Greece) and Shkodra, Lezha, Durrësi, and Dibra (for the returnees from Italy).

Considering the reasons and progress of the return of Albanian emigrants, it turns out that a good portion of them have returned to Albania with the aim of not leave again, while a significant percentage of them (32.6%) intend to emigrate again due to the difficulties in providing sufficient income for their families (Göler and Doka 2015).

Their reason for seeking another possible emigration is explained by the hardships and difficulties to reintegrate in Albania. Most of them want to emigrate again to a host country they know best.

Another important result is that the motivational factors to re-emigrate (such as unemployment, uncertain future in their country) are stronger than motivations to stay (such as new employment opportunities). The data confirm once again that reintegration opportunities for this category, along with the improved situation in the country, can greatly influence the returnees' decision to re-emigrate.

Under these conditions, the possibilities for returnees' reintegration in Albania remains an important challenge. First, for many returnees, this reintegration means finding employment an opportunity in Albania. However, their employment depends on both the opportunities offered by the labor market in Albania and their professional skills matching the few alternatives that Albania offers in terms of employment.

The data show that most of the returnees have had work experience in such sectors as construction, agriculture, services, domestic work, tourism, mechanical and electrical services, and the like. This makes them seek to find work in the same sectors in Albania, because they feel they have the appropriate knowledge and sufficient skills in these fields. Only about 10% of returnees have returned with investment plans in Albania, with some 5.4% of them already having set up a business before returning to manage it (Göler and Doka 2015).

Considering the total number of returned Albanian emigrants, the data show there is no great investment potential in Albania. This proves that most returnees are Albanians, whose income in the host country depended on their daily work and their low savings.

Under these conditions, the investments made in Albania by emigration returnees are small- and medium-sized types, supported, in over 90% of them, by their own savings in emigration, while a small portion have supported these investments in bank loans or loans taken from parents or relatives (Göler 2014).

Their reintegration process includes not only the provision of employment and investment opportunities but also many social, educational, and cultural aspects. Thus, the social reintegration of many emigrants who left Albania years ago is not easy process, given the very different living conditions between the countries of the European Union, where they have lived as emigrants, and Albania (Heller 1979).

Similarly, one of the most difficult issues of this process is related to the reintegration of the children of these Albanian returnees. Most of them were born in the emigration country and find it difficult to reintegrate into Albania life and continue their education in these new conditions. All these factors combined necessitate adopting strategies and undertaking concrete measures for the reintegration of emigrants returned to Albania in recent years (Doka 2015).

The current situation is only a momentary reflection, while the tendency is toward a continuous migration movement between Albania and other countries where Albanian emigrants reside and work.

In the current conditions, it becomes necessary to further examine and study the process of the return of Albanian emigrants in the coming years, to analyze the trends of return migration and the potential impact on returnees, their families, and the community where they settle, as well as the impact of the return migration on institutional capacities for reintegration.

Since most returnees belong to the most active labor force, it is important to further explore the impact of their return on the labor market in Albania, as well as the impact of the employment opportunities on their decision to stay in Albania or to emigrate again (Göler and Doka 2015).

A good portion of returnees are eager to invest their knowledge, skills, and financial capital gained abroad into income-generating activities. This is an indication for reintegration service providers to improve services related to starting new businesses.

All these elements combined underline the importance to address this phenomenon in the future, as many processes in the socioeconomic and regional development of the country will depend on this progress.

11.5.3 Internal Migration of Albanians. Its Performance and Features

Besides external emigration, the internal migration of Albanians has always been an integral and important part of all demographic developments in Albania.

Such factors like difficult living conditions, especially in high mountain areas, lack of infrastructure, few employment opportunities, etc., have historically forced Albanians toward internal migration, in search of better living conditions.

Until 1945, the migratory movement of Albanians appeared in two main forms: external emigration and internal migration. Although external emigration played a major role, domestic migration has also had significant impacts on the demographic and socioeconomic development of the country (Doka 2015).

In Albania, being a traditionally economically underdeveloped country, the internal migratory movement of the population has been particularly influenced by geographical-natural factors, such as climate, relief, water resources, etc. Therefore, the moving population had the tendency to settle in the lowlands, hills, valley, or plains which were distinguished for better soils, warm climate, and availability of natural water resources, etc.

Similarly, the numerous employment opportunities offered by some areas in Albania, where domestic and foreign capital has concentrated their investments in the search for minerals, road construction, etc., became compelling motives for the population to move toward these areas.

Under the conditions of that time, an important role has also played the short-term migratory movements for trade, education, health, etc. primarily toward the main cities of the country, which, in many cases, became incentives for permanent settlement in these centers, either individually or with their families.

Due to the lack of statistical data, it is difficult to estimate the extent of internal migration of Albanians in exact figures until 1945. However, it is well known that its role and importance have been significant in the socioeconomic developments of the country.

The period 1945–1990 has its own specific features in terms of internal migration of the population in Albania.

On the one hand, the internal migration was the dominant and sole form, as the external emigration of Albanians was prohibited by law at that time. On the other hand, even the internal migration was controlled, planned, and organized only by the state.

Under these conditions, the opportunity to be involved in the process of internal migration was reserved only to those individuals and families whom the state deemed that the country needed in different areas (Fig. 11.9).

It was indeed the new economic developments during the period from 1945 to 1990 that have played a crucial role in the internal migratory movement of the population, thus overshadowing the influence of geographical-natural factors. Different branches of the economy, especially industry, would also determine the volume and directions of the internal migratory movement of the population. Mostly, this population moved to the new cities and industrial areas where there was high demand for labor force (Misja et al. 1987).

However, because this migratory movement was planned, run, and controlled by the state, which then did not allow the free movement of people, its dimensions and directions for the entire period 1945–1990 did not have a major impact on the demographic developments of country, resulting in the majority of the population (about 65%) still residing in the village (Doka 2015).

The process of urbanization during that period did not aim at allowing the free movement of the population toward the cities; rather, the policy focused on "urbanization of the villages," with the aim of keeping the population in their birthplaces.

Therefore, for about 30 years (1960–1990), the urban population in Albania increased by only 6%. Such a phenomenon was unique only for Albania, as in all other countries, the urban population grew at a much faster pace (see Table 11.10).

The data in the table above clearly show the low population movement from village to town, keeping the population tied to its place of birth forcefully.

However, man, as a social being, needs to be looking for the best place to live, and when he chooses the place of residence, he not only calculates the availability of the necessary resources in terms of livelihood but also the possibility for some other important needs, such as health-care services, education, electricity, transportation, regular supply of food and water, efficient working tools, entertainment, recreation, etc.

Unavailability of all these services, to meet a satisfactory level, in many mountainous areas of Albania due to physical isolation, difficulties in using natural resources or other natural hazards, brought about stagnation in the socioeconomic development in these areas, hence the need to move away.

The main economic activity of the inhabitants in the mountainous areas is livestock, followed by agriculture, forestry, extraction of minerals and collection of medicinal plants, beekeeping, etc. If the income accumulated from all these activities together is not enough to ensure an acceptable standard of living, then the population will be looking for better working and living conditions.

Prior to 1990/1991, the demographic policy of the communist government encouraged the increase of the population as much as possible even under the conditions of the absurd restriction or prohibition of the free movement of people or under the throes of economic underdevelopment. Such a policy led to overcrowding of the country's moun-

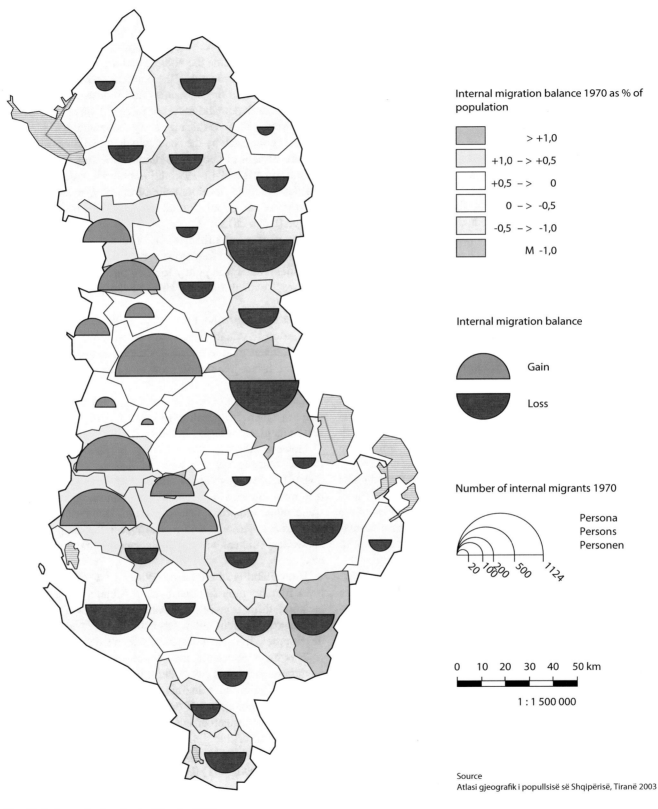

Fig. 11.9 Internal migration in Albania, 1970

Internal migration balance 1970 as % of population

> +1,0

+1,0 – > +0,5

+0,5 – >   0

0 – > -0,5

-0,5 – > -1,0

M -1,0

Internal migration balance

Gain

Loss

Number of internal migrants 1970

Persona
Persons
Personen

20 100 200 500 1124

0   10   20   30   40   50 km

1 : 1 500 000

Source
Atlasi gjeografik i popullsisë së Shqipërisë, Tiranë 2003

Table 11.10 The urban population progress in Albania, 1960–1990

Years	Urban population
1960	30,90
1969	31,50
1979	33,50
1989	35,50

Source: Statistical Yearbook 1991, p. 91

tainous areas, often beyond the acceptable population density targets per unit of arable land.

Thus, over 60% of the rural population lived in the hills and mountains, greatly increasing the population pressure on the environment, especially through the opening of arable lands on slopes. Likewise, forests were exploited at faster rates than their natural growth rates (ratio 3/1), the food base was reduced, movement of livestock was abandoned, and the result was overgrazing (Becker et al. 2005).

This hard situation, especially in the remote rugged areas with limited supply of living resources (their isolated location and total dependence on the natural resources of the area), would become the cause of a massive population movement after 1990 onward, which led in many cases to the total abandonment of many mountainous areas in the country.

After 1990, following the change of the political system and the allowing the free movement of people, internal migration, as well as the external emigration, of Albanians took on massive proportions.

Factors like the lifting of the ban on free movement of people existing until 1990, difficult living conditions especially in mountainous areas, high unemployment due to the closure of numerous industrial activities, lack of road infrastructure, etc., would be the key motivational factors that produced the huge influx of internal migration after 1990 and onward.

This migration will no longer be considered simply as a population movement from settlement A to settlement B, but as a multi-consequential factor, particularly in terms of Albania after 1990, when this migration took place on a rapid, massive, and chaotic scale (Göler 2017).

On the one hand, it can be argued as the individual need of people for a better life. However, on the other hand and in a broader sense, it was also a need to fill the structural drawbacks present in different areas of the country.

These drawbacks were related to the limited number of available jobs, low income, lack of infrastructure, poor cultural life, recreational activities, etc., and these become important motives everywhere in the world for population movement toward more suitable areas. On top of these issues, the problem of land allocation under law 7501 adopted in 1991, which, especially in mountainous areas, was not implemented and over 80% of families received their land according to the old land boundaries, leaving a considerable portion of the new families either with very little or no land at all, thus creating one more reason to move toward

settlements with better working and living opportunities (Doka 2015).

Under these conditions, in the wake of the political changes in 1990–1991, a huge influx of internal migration began, happening within a very short time and in a country unprepared for such a massive migration.

It is well understood that such a phenomenon is associated with several effects, both in remote mountainous areas and in new residential ones. The effects of this process are social, economic, and environmental. All these effects combined are reflected in the great regional or local differences or inequalities observed today, but also in the socioeconomic developments between mountainous and other areas within the country.

Under these conditions, internal migration in Albania appears with its large spatial dimensions, its multifaceted effects, and consequences in almost all administrative units (regions, districts, municipalities) of the country.

The data on migration movements clearly reveal the fact that internal migration in Albania has appeared in all its possible directions, with the movement from peripheral areas toward the center or the west of the country being the dominant one, especially toward the Tirana-Durrësi area. This migration appears to have been quite diverse, both in the form of departures and final settlements of the population, as well as in the form of seasonal or daily movements (Becker et al. 2005).

The main result of this migration movement in Albania is, on the one hand, the reduction of the resident population in the mountainous areas, especially those at high elevation, as well as the increase of the population of the plain area, on the other hand. As a result of this massive migration process, almost all mountainous areas in Albania have experienced huge population loss.

It has been observed that in the regional comparison, except for the capital and the western region, all the other regions of the country have been characterized by numerous population displacements. This abandonment has led to ruin in the mountainous areas, rural settlements, and small urban centers and overpopulation in the region of the capital, especially Tirana.

This process has been accompanied by significant changes in population between areas and districts within the country (see Figs. 11.10 and 11.11). This is reflected in the increasingly positive values of the migration balance in the districts of the central and western part of the country, such as Tirana, Durresi, Fieri, Elbasani, Shkodra, and Vlora, as well as in the continuous decline of the population in the districts of mountainous and peripheral areas, such as Tropoja, Kukësi, Dibra, Korça, Gjirokastra, etc.

This migration trend would expand further over the years, including many displaced populations from different areas of the country, such as those in the south and southeast, settling

Internal migration by prefecture of destination

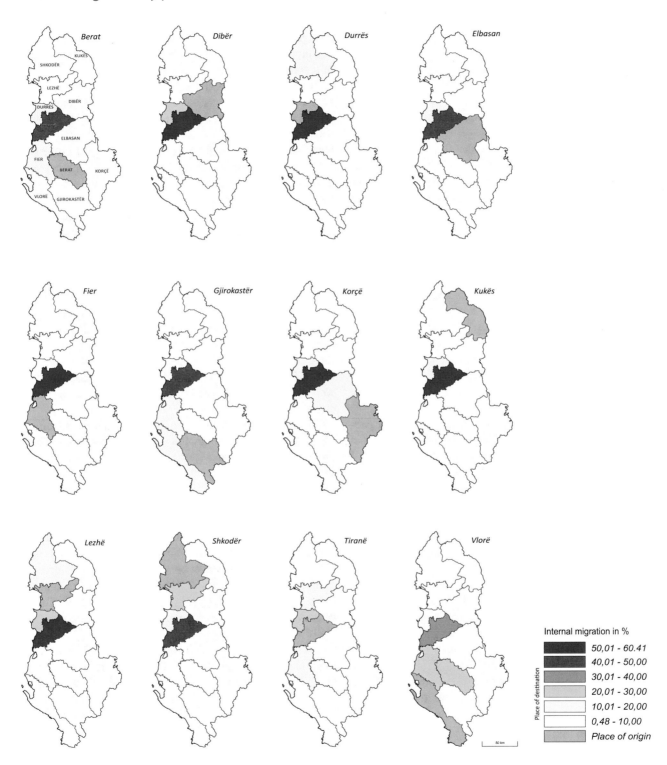

Source: INSTAT. Migration in Albani, Tiranë 2014

Fig. 11.10 Internal migration by prefecture of destination

Internal migration by prefecture of origin

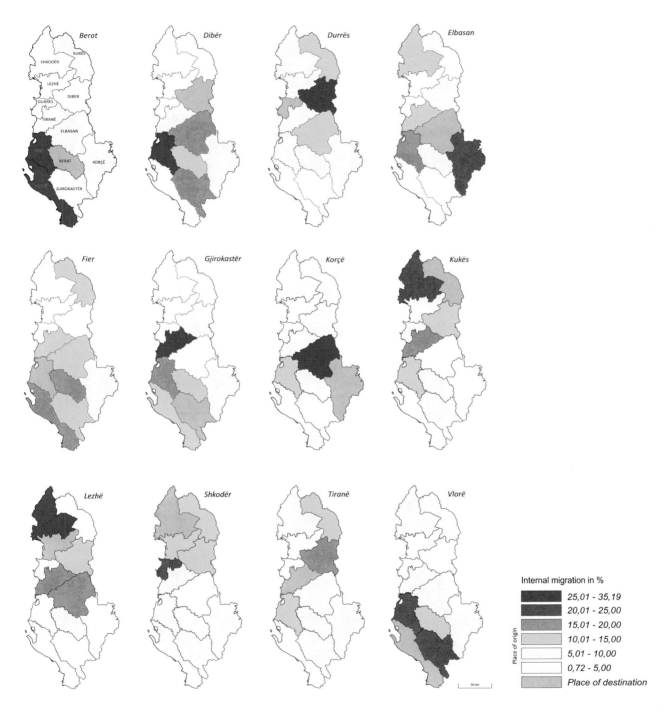

Source: INSTAT. Migration in Albania. Tiranë 2014

Fig. 11.11 Internal migration by prefecture of origin

in many cities and plain areas within their own region, besides settling in the main cities of the central and western Albania.

This has led to a significant population influx within the same district or region, mainly from rural to urban areas, thus compensating to a considerable extent the population moving away from small and peripheral cities of the country.

The effects and consequences of this massive migration process that took place in Albania starting from 1990 onward have been multifaceted. First, this process has affected the large differences in the number of population and its distribution between different areas and regions of the country. Thus, by comparing the data of the last two censuses (in 2001 and 2011) with those of the last census of the communist period (in 1989), there is not only a general decrease in the number of the population as a whole, including most of its districts, but it has also been noted that the difference in population change figures among regions has been deepening (Fig. 11.12) (Doka 2015).

This has been due to the ever-increasing role of internal migration which has made some districts of the country, such as Tirana, Kruja, Durrësi, Elbasani, etc., see a significant increase in their population, mainly by population arrivals from other districts of the country. Meanwhile, for many other districts, especially in the south and north of the country, there has been a further population decrease, sometimes at a very fast pace, and in some cases almost half of their population (as in Delvina, Saranda).

The data in the Table 11.11 clearly reveal the continuing declining trend in population for most districts of the country, not only during the period 1989–2001 but also during the decade 2001–2011. There are even districts where this decline in population has been even greater than during 1989–2001. Thus, for instance, in the districts of Mirdita, Përmeti, Gramshi, and Kukësi, the decline has exceeded the limits of − 40%, and the most extreme cases of population decline are Tropoja, with − 64%; Skrapari, with − 60%; and Tepelena, with − 52% of the population compared to 1989.

In some of these districts, the process of the further decline of the population, immediately following the political changes of 1090–1991, has continued, while in some

Population distribution in Albania by Census results

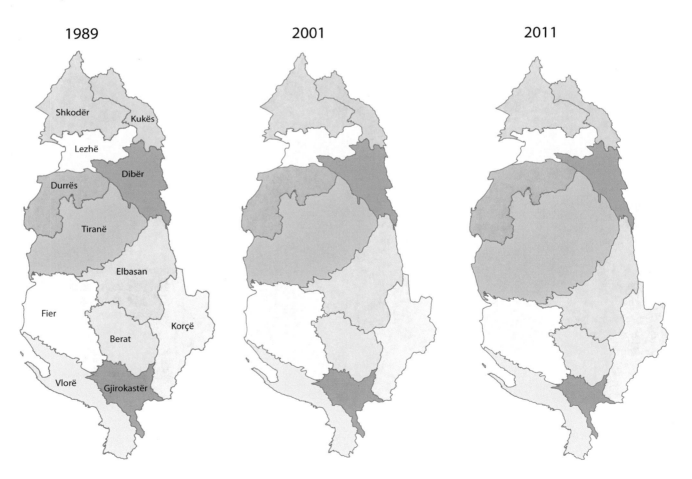

Data source: INSTAT, Population and Housing Census, 1989, 2001, 2011

Fig. 11.12 Population distribution in Albania by census results

Table 11.11 Situation in population change figures by district in 2001 and 2011, as compared to 1989 (in %)

Rrethet	Ndryshimi 2001–1989	Ndryshimi 2011–1989	Rrethet	Ndryshimi 2001–1989	Ndryshimi 2011–1989
1. Berat	−6.32	−14.22	19. Lezhë	+9.25	+24.48
2. Bulqizë	−14.55	−43.58	20. Librazhd	+0.56	−12.22
3. Delvinë	−54.74	−49.70	21. Lushnjë	+7.19	+6.69
4. Devoll	−9.06	−11.40	22. M. Madhe	−16.20	−17.58
5. Dibër	−13.76	−32.75	23.Mallakastër	−3.41	−23.39
6. Durrës	+10.44	+47.61	24. Mat	−20.20	−36.36
7. Elbasan	+4.57	+6.01	25. Mirditë	−26.54	−47.14
8. Fier	−2.48	−2.30	26. Peqin	+9.87	+3.33
9. Gramsh	−17.94	−44.39	27. Përmet	−35.19	−44.62
10.Gjirokastër	−17.67	−14.55	28. Pogradec	−1.36	−1.22
11. Has	−10.15	−20.40	29. Pukë	−29.78	−50.33
12. Kavajë	−3.50	+2.65	30. Sarandë	−45.16	−24.24
13. Kolonjë	−30.75	−42.23	31. Skrapar	−35.82	−59.42
14. Korçë	−19.32	−21.59	32. Shkodër	−3.69	−3.57
15. Krujë	+17.52	+25.25	33. Tepelenë	−35.00	−52.26
16. Kuçovë	−11.52	−12.60	34. Tiranë	+41.15	+94.83
17. Kukës	−19.69	−42.56	35. Tropojë	−37.59	−63.69
18. Kurbin	+3.00	+4.11	36. Vlorë	−16.78	−14.41
SHQIPËRIA	−3.56	−11.02			

Source: INSTAT: Results of the 1989, 2001 and 2011 Censuses

other districts, such as Gramshi, Përmeti, and Skrapari, although engulfed later in the mass migration influx, there has been very high migration rates, therefore resulting in such a huge decrease in the population number.

While the increase in population has continued for some of the districts of the country, in four of them (Tirana, Durres, Kruja, and Lezha), this increase has amounted to over 25%. Among them, Tirana stands out, with an almost 95% population increase compared to 1989, and Durrësi, with about 50%. There are also five other districts (Lushnja, Elbasani, Kurbini, and Peqini) that had a population increase in 2011, as compared to 1989. However, this increase fluctuates in the low values of 3–7% (Fig. 11.13) (INSTAT 2001 and 2011).

From the data on internal migration by region, in the regions of Dibra and Kukësi, there is a significant population loss, which amounts up to about 50% of the population residing in these regions in 1989.

The main reason for the numerous population departures from these districts is related to their particularly difficult economic situation. The evidence for this is the fact that about 40% of families in Kukësi and about 34% in Dibra districts live on social assistance (INSTAT 2001 and 2011).

Another important factor that exacerbates the current difficult situation of these districts is related to the stoppage of their industrial activity leading to the dismissal of many workers employed in this sector until 1990. Under these conditions, departure was considered the best opportunity to survive for many individuals and their families. Similar problems, although on a smaller scale, have been observed in many other regions of the country.

The largest concentration of the displaced population took place in the district of Tirana, where 60% of the entire internal migration movement has settled. The Durrësi comes second with about 20%. Thus, about 80% of the entire migration movement is concentrated in only these two regions of the country.

By the same token, it is estimated that about 30% of the population of Tirana and 20% of Durrësi region are recently settled population, mainly after 1991. This clearly proves the large concentration of the displaced population in the Tirana-Durrësi area (Doka 2015).

All these changes in the decline or increase of population in different administrative units around the country are primarily the result of the migration process, which has been very massive both in its proportions and space.

The large spatial dimension of this process can also be seen in the comparison among the main geographical regions of the country. Thus, the central and western region, which include the regions of Tirana, Durrës, Lezha, Fier, and Elbasan, has borne the main brunt in the population settlements from other regions. The data show that over 91% of the persons involved in the internal migration movement of the population have settled in the central and western regions. While in the southern and southeastern regions of the country, only 2 and 7% of the displaced population have settled, respectively. The data also show that over 60% of the displaced lived in the northern region of the country until 1989, 32% in the south, and 8% in the central and western region (INSTAT 2001 and 2011).

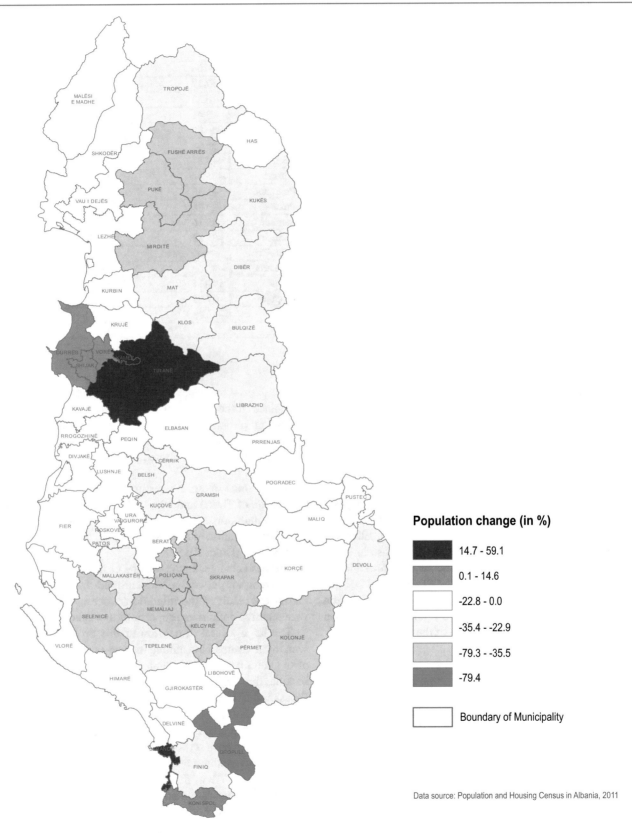

Fig. 11.13 Population change 2001–2011

Data source: Population and Housing Census in Albania, 2011

The flow of migratory arrivals has therefore been 44 times greater in the central and western part compared to the northern part and 13 times greater compared to the southern part. Similarly, the largest number of migrants in the central and western parts have come from the northern and northeastern regions of the country (INSTAT 2001 and 2011).

The reasons for this movement are related to the more difficult living conditions and the general economic situation in the mountainous regions. But there are also other reasons: Better educational opportunities, better health-care services, and better infrastructure in general in the central part of the country make some of the main motives that stimulate the population movement toward this region.

One of the main effects of this massive migration movement within the country has been the change in the population structure according to the city-village division.

Thus, from 1990 onward, there was a mass population movement from rural areas and small towns toward the large city centers. The best indicator of this movement has been the rapid increase of the urban population, which, according to the data of the last census (in 2011) for the first time in the history of the population in Albania, became the majority in the total population of the country, amounting to 53% versus 47% country population (INSTAT 2001 and 2011).

This large influx of people moving into cities has led to significant population increase for most cities in recent years in Albania. Among them, the city of Tirana should be singled out, but also the cities of Durrës, Elbasan, Shkodra, Fier, etc.

All these factors together show that internal migration in Albania has emerged with all its spatial dimensions, thus constituting a phenomenon with numerous socioeconomic and regional problems.

This phenomenon of massive internal migration in Albania is predicted to continue for many years to come, due to two main reasons:

Firstly, the more difficult living conditions in the villages force the population to flee toward the cities. Secondly, in the population structure by the place of residence, village population still constitutes a large percentage in comparison to the entire population of the country, and as in any other developing country, the general trend is the population moving into city centers. In this sense, migration is the main instrument of carrying out this process.

The problem lies in the fact that this internal migration in Albania after 1990 has mostly developed as a chaotic and disorganized phenomenon, without clear strategies (Heller et al. 2004). Therefore, it has been associated with several problems, both in the areas of settlement of the incoming population, as well as in the areas of its displacement. Thus, such problems related to lack of infrastructure in the areas of newly settled newcomer population, aggravation of the labor and housing market, difficulties in the social integration of the newcomer population with the local residents, etc., are more than present in the areas where the incoming population has settled.

Meanwhile, the areas of population displacement are currently facing a series of other problems, such as the risk of total abandonment of these areas, lack of prospects for economic and social development for the future, etc. The risk of total abandonment is higher in the rural peripheral areas. The effects of depopulation are present and obvious, for instance, in the abandonment of arable land and their settlements. This abandonment is happening on different levels.

Besides a general population decline, there is evidence of a process of interregional concentration or structuring within the internal migration system in Albania. Many peripheral villages are gradually being abandoned, and some of them have been depopulated over the last two decades. Medium-sized settlements are in the position of the first reception center for migrants from the more rural areas. Thus, in the prefecture of Kukës, cities like Bajram Curri, Kukësi, and Kruma serve as stations of these interregional movements (Fig. 11.14).

The reasons for emigration are more than the difficult economic situation. Agriculture provides a modest income, and it is not considered an opportunity for the future of the children. In this way, searching for a better future is the main reason for planning to migrate. The problems of low crop yield per hectare and the low level of technology used remain persistent, along with the risk of impossibility to trade the agricultural produce (Doka 2015).

One of the most obvious consequences of this process is the mentality of constant abandonment and departure, with no intention of turning back, that prevails in most of the population in the areas of displacement. It seems that people in these areas have lost their stability, their work, and their territory. Because these factors do not change in the short term, solutions to this problem must be sought in the "internal" circumstances within these areas themselves.

It is very important that the internal migration movement of the population in Albania should not continue to be spontaneous, but it should be organized based on clear migration policies and strategies harmonized to meet the requirements of both the abandoned and settled areas (Göler and Doka 2015).

From the analysis of the migration situation, inside and outside the country, the following can be concluded:

- The migration of Albanians, external or internal, has been a very important phenomenon since 1990 in the entire demographic development of Albania and at the same time one of the biggest problems that the country is facing since the beginning of the transition.
- Due to the difficult economic, political, and social situation of the country, compared to neighboring countries, especially Italy or Greece, the tendency of Albanians moving to other countries in search of better living conditions will continue for many years.

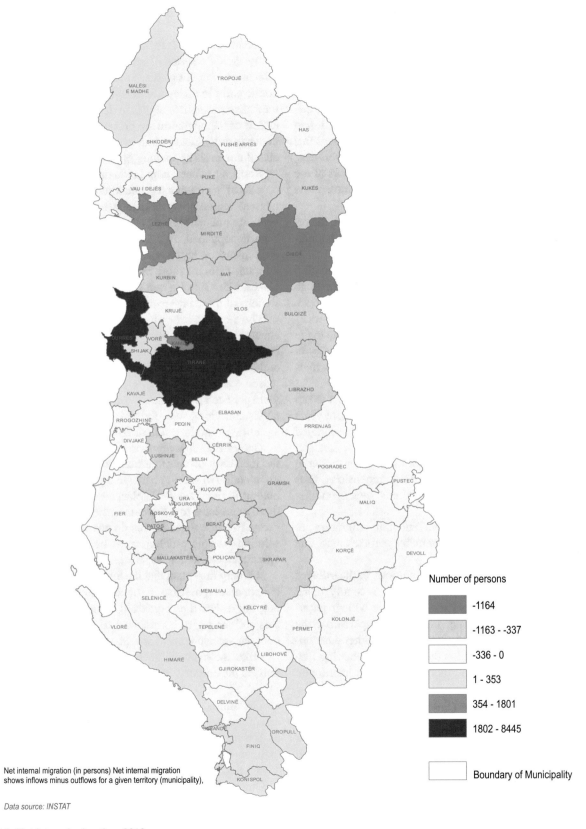

Net internal migration (in persons) Net internal migration
shows inflows minus outflows for a given territory (municipality),

Data source: INSTAT

Fig. 11.14 Net internal migration, 2019

A large portion of the population, almost half of it, still lives in the village. For this reason and due to the more difficult living conditions in the mountainous areas, the internal movement of the population between different areas of the country will continue in the upcoming years. The central and western region will continue to attract even more the displaced population from other areas, especially from the mountainous northern and northeastern part of the country.

- Since the migration of Albanians throughout the period from 1990 to 1991 onwards has been generally a chaotic phenomenon, it is important for the future to take various measures to regulate this phenomenon. Such important measures are the following:
  - Design and implementation of clear migration policies
  - Study of areas and regions most suitable for the concentration of the displaced population
  - Promoting economic development and infrastructure in the regions of population displacement to avoid their total depopulation
  - Political and economic cooperation with countries where the largest number of Albanian emigrants is concentrated to help them recognize the rights of emigrants

In conclusion, it can be said that migration in all its forms will continue to play a very important and decisive role for many demographic, economic, regional, and local developments of the country.

Under these conditions, it is necessary that all projects and strategies that are elaborated for the development of the country should consider the changes and effects that the migration process brings.

11.6   Structural Features of the Albanian Population

By the structural condition of the population, we mean its different groupings based on their specific features, such as age; gender; social, ethnic, and religious affiliation; place of residence, etc.

In this sense, the population of Albania is presented as diverse but also with its various specifics based on different structure categories.

11.6.1   Age and Gender Structure of the Population

In terms of age structure, the population of Albania has been first and foremost distinguished for its vitality. This means that in terms of population structure, young ages have historically dominated compared to the older ones.

The period from 1945 to 1990 is especially distinguished for its dominance of young ages. Thus, almost 35% of the population of Albania belonged to the nonworking age group (i.e., of 0–15 years), about 58% were of the working age group (i.e., 15–60 years), and only about 7% were of the overworking age group. Such an age structure, with predominant young population, has been entirely the result of the high rates of our own natural population increase. This was also reflected in the average age of the population, which at that time was 29 years old, thus being among the youngest populations in Europe (Misja et al. 1987).

This age structure of the population in Albania with the dominance of young age groups has been "endangered" since 1990 onward, and this is a result of both the decline of the natural population increase, and of the massive influx of external emigration, which has mainly been a feature involving young age groups. These factors have brought about changes in the age structure of the population of Albania in the last three decades. Although these changes are not immediately apparent, on the one hand, there is already a decreasing tendency in the percentage of young people in the entire population of the country, which means an increase in the percentage of older people, on the other hand. Thus, the latest census data (October 2011) showed a doubling of this indicator, from 7% in 1989 to about 15%. Another indicator has to do with the increase in the average age of the population, which is currently estimated at around 37 years old from 29 years old in 1990 (INSTAT 2011).

When considering the spatial dimension, it has been noticed that Tirana and its environs stand out in these age structure changes of the population, because this city has experienced rapid demographic developments in recent years which have been characterized by the significant increase of all age groups of the population. This is due to the large number of arrivals of different age groups from all other areas of the country, while many other regions and areas of the country have been characterized by losses in almost all age groups of their population, especially young age groups, because of mass departures of people in external or internal emigration.

In terms of the gender structure of the population, males have traditionally dominated compared to females. According to statistical data until 1991, for every 100 females, there were 105 males in the population (Misja et al. 1987).

However, even this gender structure has been shaken during the period 1990–2001, due to the phenomenon of mass emigration of the population, where the vast majority has been males. This change was reflected in the data of the general census of 2001, which revealed for the first time the dominance of the females in the gender structure of our population, with about 52% female versus 48% male. These data

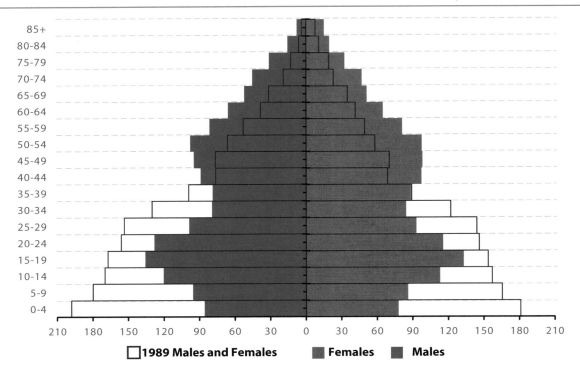

Source: Population and Housing Census 2011

Fig. 11.15   Population in 1989 and 2011 by age and gender

would also indicate the dominance of the females in older ages of over 70 years, which is related to the higher mortality rate of males of these age group (INSTAT 2001), while the data of the last census of Albania, in 2011, showed a small dominance of males, with 51% male versus 49% female. Among other reasons, this can be explained by the inclusion of females in external emigration, mainly through the process of reunification with families or relatives who had emigrated years ago (INSTAT 2011).

The graphical (see Fig. 11.15) representation of the age and gender structure of the population is illustrated through the age pyramid. The population of Albania has always been distinguished for progressive age pyramid because of the dominance of young age groups. However, comparing the databases of the different population censuses conducted, the changes that have taken place in terms of the weight by age groups and genders can be observed. (Graph.) (Fig. 11.16)

Considering the pyramid graph by age and gender, on the one hand, the dominance of the young ages of the population can be observed, which creates the broad base on which the pyramid stands and which promises the normal population development in the future. On the other hand, there are changes that have occurred in age groups and genders after 1990, where again the main factors that have influenced in this regard are the decline in natural population increase and the numerous population departures abroad.

Such differences are also observed by comparing different areas and regions of the country. This makes the appearance of the age pyramid of the population reflect great differences between the areas that have had significant population attractions both in terms of the total number but also for each age group and gender, in which the pyramid reflects its highly progressive form, whereas in the areas of population abandonment, the age pyramid appears to be stationary or, in many cases, even regressive (Doka 2015).

Thus, the situation is particularly problematic in the regions where mass exodus of the population has been dominant, especially among young ages. This has been accompanied by a decrease in the percentage of young people in the population, as for instance is the case of the southern and southeastern regions of the country, which is also reflected in the anomalies in the appearance of the age and gender structure of their population.

While the northern part is indeed the region with the largest population outflows, it nevertheless maintains high rates of natural population growth. During the last two decades, the birth rate in the districts of Hasi and Kukësi was twice as high as in the districts of Përmeti and Kolonja (INSTAT 2011). In some cases, it is this factor that has preserved the progressive or stationary appearance in their age pyramid.

The appearance of the age pyramid for the central and western part of the country is completely different, because the numerous arrivals of the population from all age groups,

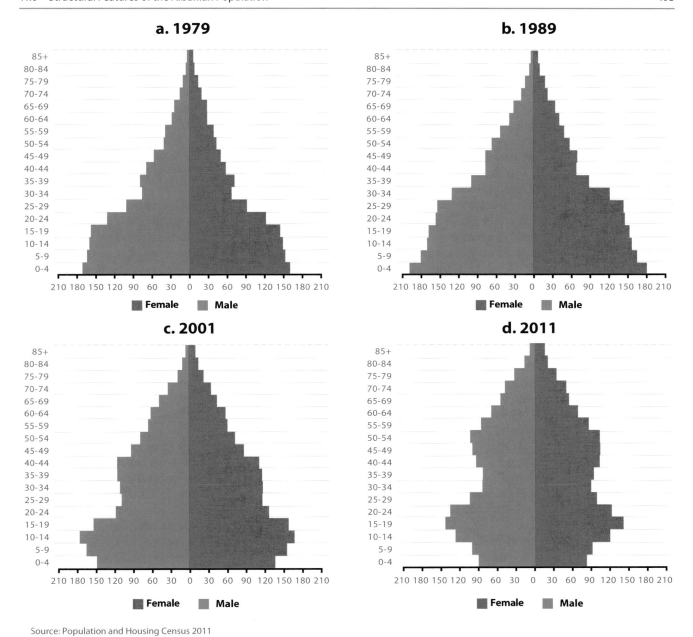

Source: Population and Housing Census 2011

Fig. 11.16 Age structure of the population in Albania 1979 and 2011

especially the young ones, determine the highly progressive form of the age pyramid.

All these changes in the age and gender structure of the population of Albania should be well taken into account in future projections of the population and socioeconomic and regional developments of the country, as the state of the age and gender structure and their spatial distribution will determine many indicators related to the labor market and the number of employees, the ratio between income contributors and their beneficiaries, the distribution of educational and health institutions, and so on.

11.6.2 Population Structure by Place of Residence

The main feature that has traditionally distinguished the structure of the population of Albania by place of residence is that it has been dominated by population residing in the village, compared to the city-dwelling population.

Although the Albanian territory has been inhabited very early and important urban settlements were built in Illyria, besides rural ones, the first statistical data about this structural category belong to the period 1917–1918, and the data reveal that the population living in cities at that time was calculated between 70,000 and 80,000 people. Out of 21 urban settlements registered in total at that time, Shkodra alone had about 23,000 inhabitants, Tirana and Elbasan had about 10,000, and all the other cities had less than 10,000 inhabitants (Berxholi et al. 2003).

The first general population census conducted in 1923 reveals important data in terms of rural versus urban population division. Thus, the data from that census show that 127,000 people, or 15.9%, lived in cities, and all the rest of the population lived in villages, whereas the data from the census in 1938 show a slight decrease of the urban population by 0.5% compared to 1923 (Berxholi 2000). Although the number of cities had already increased to 23, the population living in them was small. A particularly important role in this decrease was played by the huge influx of external emigration (*kurbeti*), which attracted a considerable portion of the population, especially in the cities of the southern and southeastern part of the country. Thus, while Tirana, Durrës, Lushnja, and Vlora resulted in population growth during the years 1923–1938, cities like Korça, Gjirokastra, or Përmeti resulted in its population loss (Misja et al. 1987).

Another feature of the population structure by residence for the period up to 1945 had to do with very large differences in the geographical distribution of the urban population between different regions of the country. Thus, about 55% of the population at that time was concentrated only in the western part of the country, where the largest cities of that time were located, such as Tirana, Durres, Elbasan, and Shkodra, while the number of cities in the other parts of the country, especially in the north and northeast, was very small (Berxholi 2000).

Such a distribution also affected the average distances between different urban settlements. Thus, while in the western part the average distance between urban settlements was 29 km, in the northeastern part, it was 59 km (Misja et al. 1987).

During the period 1945–1990, significant changes were observed in Albania in terms of population structure by place of residence.

Among the main factors that influenced this aspect was first of all the demographic policy of that period, which did not allow the free movement of the population from the countryside to towns or cities, aiming at what would be called "rural urbanization."

Under this demographic policy, the opportunity to move to the cities would be granted to that portion of rural population which was deemed as appropriate for city residence by the state. This policy was mainly based on the city needs for qualified labor force from the countryside.

In this regard, a major role has been primarily played by the process of industrialization of the country, especially in the 1960s and 1970s when the need for labor force in many cities and new industrial centers was high. This need stimu-lated a significant movement of the population from rural to urban areas, and this process involved not only the large cities in the central and western part of the country but also almost all the small towns of the northern and southern part of the country.

Along with the population increase in the existing cities during this period, the construction of many new urban centers would take place at the same time. Thus, 42 cities were built in Albania, scattered throughout the country, with the main motive being the construction of industrial facilities (Berxholi 2000).

The population movement toward cities and new industrial centers would decrease significantly in the late 1970s and especially in the 1980s. Fulfillment of the need for man-power; the rapid population increase of newly settled active population, along with the general population increase; the onset of economic slowdown; and the halt in industrial activity would be some of the main factors for the slump in the demand for new labor force, which, in turn, brought about the reduction in the population movement toward urban centers. Under these conditions, the communist state would undertake a series of administrative measures to prevent the free movement of people toward the cities and other areas of the country. The result of such a situation would be the fact that urban population during the period 1945–1990 increased by only 14% (Doka 2015).

The data in the table clearly show the much faster growth of the urban population in Albania during the years 1945–1969, by over 10%, compared to the following years (1970–1989), when this growth was only 4% (INSTAT 1989).

These data show (see Table 11.12) that Albania, in the years 1945–1990, compared to all other countries not only in Western but also in Eastern Europe, was characterized by lower rates of population growth in cities. This was the consequence of the demographic policy pursued during that period, and that was the demographic structure by residence

Table 11.12 Population by residence (1923–2011)

Years of census	The population according to residence (in %)	
	Urban	Country
1923	15,9	84,1
1930	18,2	81,8
1945	21,3	78,7
1950	20,5	79,5
1955	27,5	72,5
1960	30,9	69,1
1969	31,5	68,5
1979	33,5	66,5
1989	35,5	64,5
2001	42,1	57,9
2011	53.7	46.3

Source: 1. Statistical Yearbook of Albania 1991, p. 361
2. INSTAT: Census Results, 2001 and 2011

until the political change of the system in 1990–1991. One of the most important population developments in Albania during the period from 1990 onward is undoubtedly the change that has occurred in the ratio between the population residing in the rural and urban areas. The general tendency has been the rapid increase of the resident population in cities through the arrival of the population mainly from rural areas or the movement of the population from small towns to larger cities (Becker et al. 2005). The data from the latest census (2011) fully confirmed this trend as they reveal that, for the first time in our history, most of the Albanian population (53.7%) lives in cities, compared to 46.3% still living in rural areas.

These data show that, in a 30-year time span (1990–2020), there is an increase in the urban population by over 18%. For the previous 30 years (1960–1990), this increase was only about 5% (INSTAT 2019).

However, despite the rapid developments of the last three decades in favor of the urban population, in terms of Albanian population structure by residence, as compared to almost all other European countries, there is still a large percentage of rural population, which is proof that the tendency of the population moving away from rural toward urban areas will continue.

Similarly, large differences in terms of population structure by settlement can be observed by comparing indicators between different districts of the country. Thus, most of the urban population is concentrated in the districts of the central and western part of the country, especially in the largest cities of Tirana, Durresi, Elbasani, Fieri, Vlora, and Shkodra (see Fig. 11.17).

The districts with the lowest percentage of urban population are mainly in the northern and northeastern part of the country. These districts, on the one hand, have not traditionally been distinguished for any significant weight in terms of their urban population. On the other hand, they are the districts that have lost most of their (mainly urban) population after 1990 because of massive internal migration (Becker et al. 2005).

Under these conditions, most of the cities in the north and northeast, but also in the south and southeast of the country, have either maintained their population with difficulty as a result of the replacement of the leaving population with newcomers from rural areas, thus facing the phenomenon of "ruralization" of the cities, or these cities have suffered a decrease in their population as a result of the impossibility of replacing the leaving population with the incoming newcomers, as is the case of the city of Bajram Curri, Puka, Erseka, Përmet, etc. (Göler 2017).

This population structure by place of residence also determines the number and style of urban and rural settlements. Thus, out of the total number of settlements in Albania, only 65 of them are cities, while the other 2955 are rural settlements. However, even considering this figure, the city status for many small towns in Albania has been put into question,

as real city centers. Some of them have either already lost or risk losing this status.

Similarly, the size and weight of many cities remains small. Thus, only five cities in Albania have more than 100,000 inhabitants (Tirana, Durresi, Elbasani, Shkodra, and Vlora), and among these cities, only Tirana has over 500,000 inhabitants (Berxholi 2000).

The analysis of the history of the creation of urban settlements in Albania is of special importance and interest, which testifies to their existence and functioning since the Illyrian period, with cities such as Antipatrea, Antigonea, Bylis, Dimali, Lisus, etc. The period of antiquity is also of particular importance in terms of urban settlements, with such cities like Butrinti, Apolonia, Dyrrachium, etc., which are present with all their evidence even today.

A very important period in the history of Albanian settlements is the Middle Ages, which is evidenced today by a series of settlements in the form of citadels, castles, trading cities, and so on. Their role has been so important that a good part of them is habitable even today. Among them, it is worth mentioning such cities as Gjirokastra, Berati, Kruja, Elbasani, Korça, etc.

The reign of King Zog and the fascist occupation had their impact on the urban planning and construction style of our cities. In many urban settlements, especially in the construction of several major city centers, significant importance was given to the prevailing style of the time.

During the period of communist rule (1945–1990), significant changes took place in the structure and style of settlements in Albania.

The communist period in general disrupted the previously inherited style of the collective features of the settlements, replacing it with standardized socialist-style buildings, mainly in the form of four to five-story blocks of flats. There was also a significant increase in the number of new urban settlements (42 in total) during this period.

Major changes are observed in the structure, style, and functioning of settlements after the political and social change in 1990. Thus, there has been a major boom in construction during the last three decades, due to the great needs of the population for housing and increased living space, but also due to the significant growth of the trade network, services, and other facilities, which the market economy operating in the new conditions would greatly need.

The data show that the Albanian housing fund is relatively young. Thus, by analyzing this housing fund by specific periods, it turns out that only 8% of the buildings are constructions made until 1945. Most of the constructions in Albania, (Becker et al. 2005) about 53%, were built during the period 1960–1990. However, the last period, within a time span of only 20 years, especially after 2000, is really distinguished for its construction boom, as over 25% of all buildings in the country have been built during this period.

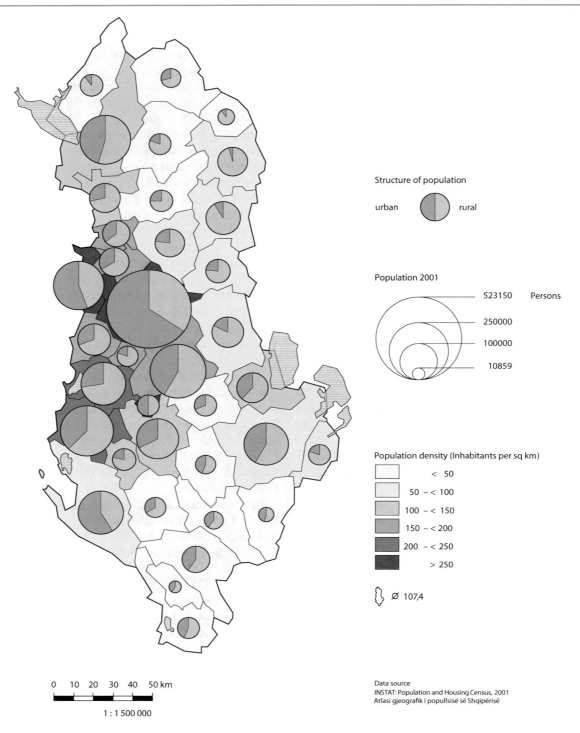

Structure of population

urban   rural

Population 2001

523150   Persons

250000

100000

10859

Population density (Inhabitants per sq km)

	< 50
	50 – < 100
	100 – < 150
	150 – < 200
	200 – < 250
	> 250

Ø 107,4

0   10   20   30   40   50 km

1 : 1 500 000

Data source
INSTAT: Population and Housing Census, 2001
Atlasi gjeografik i popullsisë së Shqipërisë

Fig. 11.17   Urban and rural population in Albania

The largest number of constructions is observed in the big cities and their suburbs, where most of the population that came through the process of internal migration is concentrated.

Along with this huge construction boom, the architecture and style of the new buildings has also changed. Today, these buildings are much more modern and display mixed style and architecture, starting from small houses or villas to multistory buildings. Many of these buildings have also adapted to suit the current demands for economic activity, especially in the sector of services, trade, tourism, etc.

There is now a variety of building forms and shapes, especially in big cities like Tirana, but also in Durrësi, Vlora, Fieri, and other cities where, besides small ones, there are also many new multistory buildings.

Moreover, with the increase in the number of dwelling buildings, a significant improvement in construction technology and functioning of these buildings has been witnessed, as well as an improvement in their equipment and furnishing, in the increase the living space per inhabitant, etc.

Similarly, with the increase in the number of settlements and residential buildings in Albania, the types of housing centers have evolved based on their geographical location, origin and manner of creation, the functions they perform, etc.

Thus, as far as the rural settlements are concerned, according to their geographical extent in the conditions of the very diverse relief of Albania, we distinguish three types: villages located on a plain, on a hillside, and a mountain slope. In terms of population residing in these types of villages, the lowland villages are the most populous ones, especially those located in the Western Lowlands, in the Korça Plain, and in some river valleys. The hilly villages are far more numerous, but they have smaller number of populations, compared to the lowland villages, while mountain villages do exceed 7% of the total number of rural settlements in Albania, and they are distinguished for a very small number of populations, as well as for their greater geographical distance between them (Doka 2015).

According to the origin, there are very old rural settlements in Albania, as well as many new ones, created especially during the period 1945–1990.

According to their function, the villages in Albania depend on their agricultural and livestock activities, but there are also villages with mixed economic activities. Besides agriculture, other economic activities are carried out such as agribusiness, rural tourism, trade, services, etc.

According to their geographical location, urban settlements in Albania are classified into plain, coastal, and lakeshore cities (like Tirana, Elbasani, Korca, Durresi, Vlora, Shkodra, Pogradeci, etc.), but Thera are also in cities that lie on steep terrain (like Gjirokastra, Berati, Kruja, etc.) (Berxholi 2000).

As far as the origin and construction method, in Albania, we distinguish cities of the Illyrian period and the antiquity (Bylis, Antigonea, Antipatrea, Dyrrahu, Apollonia, Butrinti, etc.), citadel and castle cities (like Kruja, Berati, Gjirokastra, etc.), commercial cities (Korça, Tirana, Shkodra, etc.), and new cities (Laçi, Patosi, Cerriku, Rrogozhina, etc.).

According to their function, in Albania, there are multi-functional cities (housing, industrial activity, administration, etc.), and this group includes the largest cities of the country (such as Tirana, Durresi, Shkodra, Korca, Elbasani, Vlora, etc.); cities with genuine economic function, where purely industrial or agricultural activity dominate (like Kuçova, Selenica, Roskoveci, Miloti, Kopliku, etc.); cities with primary administrative function, such as Bajram Curri, Puka, Erseka, and Çorovoda, etc. (Doka 2015).

At the end of this population structure analysis in Albania by place of residence, it should be emphasized that in the future, the tendency of movement and concentration of the population in cities will continue to be dominant. This is mainly explained by the fact that a large part of the Albanian population still lives in the countryside, and as in all other developed countries, this ratio will change in favor of the urban population. The largest spatial concentration of the urban population will continue to be dominant in the important urban centers of the central and western part of the country.

References

Becker H, Göler D, Doka D, Karaguni M (2005) Banimi në zonat ilegale të ish-objekteve Industriale të qytetit të Tiranës. Botuar në revistën shkencore gjeografike "EUROPA REGIONAL". Leipzig

Berxholi (2000) Regjistrimet e përgjithëshme të popullsisë së Shqipërisë. Vështrim historik. Tiranë

Berxholi A, Doka D, Asche H (2003) Geographic Atlas of the Albanian population

Doka D (2015) Gjeografia e Shqipërisë – Popullsia dhe Ekonomia. Tiranë

Gëdeshi I, King R (2018) Research Study into Brain Gain: Reversing Brain Drain with the Albanian Scientific Diaspora, Tirana

Göler D (2014) Rückkehr nach Albanien – Migration in Zeiten der Krise. In: Südosteuropamitteilungen 02/2014. München

Göler D (2017) From an isolated state to a migration society – transnationalism and multilocality as social practise in Contemporary Albania. In: Jordan P (ed) 10 years of EU enlargement. The geographical balance of a courageous step. Verlag der Österreichischen Akademie der Wissenschaften, Vienna, pp 137–158

Göler D, Doka D (2015) ReEmigration in Albania. From Emigration to Remigration and Vice-Versa? In: SUDOSTEUROPA Mitteilungen 01/2015. München

Heller W (1997) Migration und sozioökonomische transformation in Südosteuropa, Südosteuropa- Studien 59. Südosteuropa-Gesellschaft, München

Heller W, Doka D, Berxholi A (2004) Hoffnungsträger Tirana. In: Geographische Rundschau 56/2004 Heft 1. S. 50–58. Westermann, Deutschland

INSTAT (1989) Regjistrimi i Popullsisë dhe Banesave 1989

INSTAT (2001) Regjistrimi i Popullsisë dhe Banesave 2001

INSTAT (2011) Rezultatet kryesore të regjistrimit të popullsisë 2011, f.16

INSTAT (2014) Të dhëna demografike 2014

INSTAT (2019) Demographic Indicators 2019

King R, Piracha M, Vullnetari J (2010) Migration and development in transition economies of Southeastern Europe Albania and Kosovo. East Eur Econ 48:3–16. M.E. Sharpe

Misja V, Vejsiu Y, Berxholi A (1987) Popullsia e Shqipërisë. Naim Frashëri, Tiranë; Mijsa V. (1998) Emigracioni ndërkombëtar në Shqipëri gjatë periudhës së tranzicionit. Tiranë

Sociocultural Features

12

Dhimitër Doka

Abstract

Sociocultural features are a very important part in addressing the state of a country's population. They include the analysis of several indicators related to living standard, employment and unemployment situation, housing conditions, educational level, quality structures of population, etc. The population of Albania, although small in its number and geographical extent, is quite diverse in its sociocultural features. This is expressed in the specific ways of life, traditions, customs, its social stratification, etc. This can be explained by the specific territorial conditions, dominated by the mountainous and rugged terrain, as well as by the ancient history and tradition, numerous historical and cultural events, etc. Below, we address in more detail some of these features.

Keywords

Standard of living · Employment and unemployment · Education

12.1 Standard of Living

Determining the living standard of the population of each country is based on the analysis of various indicators, such as the amount of income per capita of the population, the expenditure of this income in accordance with the needs of the population, poverty level, housing standard, people's access to various services, etc.

During the communist period up until 1990, poverty was evenly distributed, reaching almost the level of starvation in the last years of communism. However, the population of Albania is distinguished for its great social stratification between the rich and the poor today. The average income per capita in Albania is estimated at around 5.000 Euros per year today. However, there is a big difference between the lowest

monthly salary that does not exceed 120 Euros and the highest monthly salary that goes to around 1.500 Euros. Thus, there is a difference of over 10 times, whereas the average monthly salary is estimated at 30,000 Albanian Lek, or around 250 Euros (INSTAT 2018). While the Albanian economy continues to be severely affected by informality, amounting to over 30%, it is difficult to accurately determine the amount of real income of the population. In this regard, a major role is played by emigration, through which Albania and Albanian families receive a large amount of income, from more than 1.5 million Albanians who have emigrated after 1990 and who send money to their country (Doka 2015).

Although accurate data are lacking, various estimates show that about 20% of the population of Albania can be considered rich, i.e., with significant income for the individual and the family, and even with significant investment opportunities. Another portion of the population, about 30%, is estimated to earn a satisfactory monthly payment with which it manages to meet its living needs and may even put aside savings (INSTAT 2018), while about half of the Albanian families live in conditions of financial difficulties to cover the daily costs of living; therefore, they are forced to live in permanent debts. The situation is even more difficult for that part of the population living below the poverty level, which means living with an income of less than 2 Euros per day. About 15% of the population of Albania lives in such a situation (INSTAT 2018).

According to INSTAT, the risk limit for being a poor person in Albania is estimated at 160,000 Albanian Lek per year, and there are about 670,000 individuals living at the risk limit of being poor (INSTAT 2020).

Risk of being poor or socially excluded refers to those individuals who are at risk of being poor or materially deprived or living in families with very low employment intensity. In Albania, this indicator is estimated to be around 50% (INSTAT 2020).

Family employment intensity refers to the number of months that all family individuals have worked versus the total number of months they should have theoretically worked during the year (12 months). Households that are considered to have very low employment intensity are estimated at about 20%. Individuals aged 18–59 who live in families with very low employment intensity make up about 14% (INSTAT 2019).

Social benefits, such as old-age and family pensions, included in the available family income, reduce the risk of being poor. If the income level did not include all types of social benefits, the risk of being poor would be greater. However, based on the relatively low income that individuals and families receive as old-age pensions or other social assistance, they seem to be merely a form of survival rather than have any real impact on poverty alleviation.

The risk of being poor by age group and gender has been analyzed for three age groups. The downward trend is observed in the 18–64 age group where the risk of being poor is somewhat lower. In the 0–17 age group, the risk remains greater, as their income is largely dependent on parents or other family members, while in the age group over 65 are included especially the persons who do not receive pension or other social assistance. The poverty rate by type of family, in 2018, in families without dependent children, is estimated to be lower compared to families with dependent children, respectively, 15.2% and 27.0% (Table 12.1) (INSTAT 2019).

A comparison of the poverty risk between Albania and other countries of the region and the European Union (see Fig. 12.1) shows that the highest value of relative poverty is recorded in Serbia (24.3%) and Romania (23.5%) followed by Albania (23.4%), while the lowest poverty rates are recorded in the Czech Republic (9.6%), Finland (12.0%), Denmark (12.7%), Hungary (12.8%), and Norway (12.9%). The average of European Union countries (28 countries) is 16.9% (Eurostat 2018).

Another indicator that clearly expresses the standard of living of a society is the structure of consumption expenditures. In the case of Albania, the average monthly expenditure for families consisting of an average of 3.7 persons is 76,000 Albanian Lek. A more detailed structure of consumption expenditures is presented in the table below (INSTAT 2019).

From the table data (see Table 12.2), over 60% of the average monthly expenditures of Albanian households go for basic needs such as food, energy, water, and transportation, while the possibilities of spending for other needs such as

health, education, communication, or entertainment remain very few, fluctuating at 3–4% each. This structure reflects the low standard of living for Albanian families, most of which simply aim to cover the basic living expenses required. This means that about 60% of them do not have the opportunity to go on vacation or tourism.

There is also a noticeable change in the structure of expenditures at the spatial level based on the large administrative units of the country (regions).

Thus, the region with the highest monthly consumption expenditures is Tirana (88,700 Albanian Lek per month), and this is explained first and foremost by the presence of capital, where, on the one hand, there are more employment opportunities and, on the other hand, there is more income for the population, which means more for expenditures and consumption. The regions of Durrësi, Korça, Fieri, and Elbasani come next, where the level of economic development is satisfactory, while the regions with the lowest incomes and expenditures are especially that of Kukës and Dibra, where along with the mass exodus of their population, the opportunities for income and expenditure have also decreased. Most families in these areas survive on low incomes from agriculture and livestock and on social assistance or remittances. These meager incomes can only suffice for the basic household expenses like food and housing (Doka 2015).

12.2 Employment and Unemployment

Another important indicator of the sociocultural structure of the population with direct or indirect reflection in many other indicators is the state of employment or unemployment. In contrast to the previous communist period where the main role was played by industry and agriculture, the main share of employment in Albania after 1990 is borne by the services sector, with about 54%, followed by agriculture with 26% and industry with about 20%. Most employees (about 55%) are salaried, and the rest are either self-employed or contributing workers in their family businesses (INSTAT 2020).

Meanwhile, from a country where unemployment was either an unknown or unacknowledged phenomenon until 1990 because it was not officially declared, from 1990 onward, unemployment is one of the biggest problems faced by Albanian society. It is difficult to judge based on statistical data on the performance of the unemployment rate in Albania. This is because these data have not been complete, due to both the great impact of informality and undeclared work characteristic of the economy in the last three decades, and the unclear criteria in determining the status of the unemployed.

However, based on the official existing data about unemployment rate in Albania for the period from the early 1990s until today, there is a large increase of about 30% of the unemployment rate in the first years of transition, especially

Table 12.1 Risk of being poor by age group and gender (in %)

	2017			2018		
Age group	Male	Female	Total	Male	Female	Total
0–17 years	27,7	31,8	29,6	28,7	30,6	29,6
18–64 years	24,0	23,4	23,7	23,2	23,3	23,2
65+ years	13,0	13,7	13,4	12,5	15,4	14,0
Total	**23,5**	**23,9**	**23,7**	**23,0**	**23,8**	**23,4**

Source: INSTAT Survey on Income and Living Standard 2017–2018

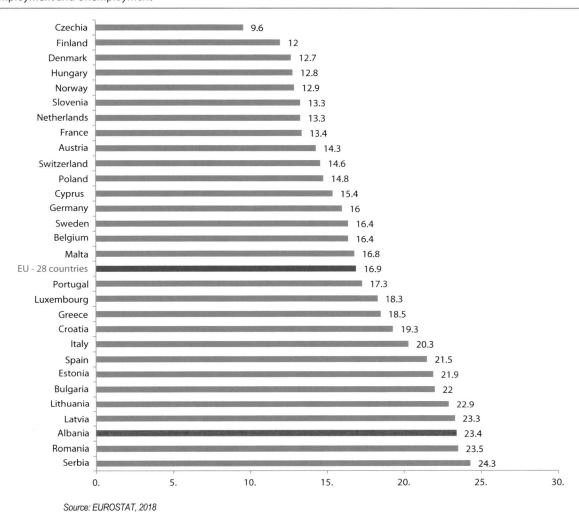

Fig. 12.1 Indicator of the poverty risk for Albania, other countries of the region and the union. (Cartography: Ledjo Seferkolli)

Table 12.2 Structure of average monthly household consumption expenditures

Main consumption groups	Albanian Lek	Percentage (%)
Food and nonalcoholic beverages	32.347	44.1
Alcoholic beverages and tobacco	2.558	3.5
Clothing and footwear	3.128	4.3
Household expenses (electricity, water, rent, etc.)	7.998	10.9
Household appliances and furniture	3.505	4.8
Health	2.879	3.9
Transport	5.294	7.2
Communication	2.479	3.4
Culture and entertainment	2.224	3.0
Education	2.868	3.9
Restaurants and hotels	3.638	5.0
Other goods and services	4.480	6.1
Total average expenses	**73.400**	**100.0**

Source: INSTAT 2019

because of the almost complete closure of previous industrial activity (INSTAT 2011a). In the following years, due to the opening of many new employment opportunities, especially in the services sector (trade, transport, tourism, etc.), there was a decrease in the unemployment rate (Doka 2015) to about 12.5% (in 1996).

The difficult situation created in Albania after the riots of 1997 and its negative effects would be reflected in the increase of the unemployment rate which went over 20% (INSTAT 2011b).

The stabilization that characterized the economy from the beginning of 2000 onward would also reduce the unemployment rate to about 13%. But based on new ways and criteria in determining the level of unemployment (also calculated for village population and the effects brought about by the financial crisis in 2008–2010, as well as the pandemic situation of 2020), the level of unemployment in Albania exceeds 20% (INSTAT 2020).

Changes in employment and unemployment rates can also be observed by age group and gender. Thus, the labor force participation rate for the population aged 15–64 years of age is 69.7%, while the labor force participation rate for

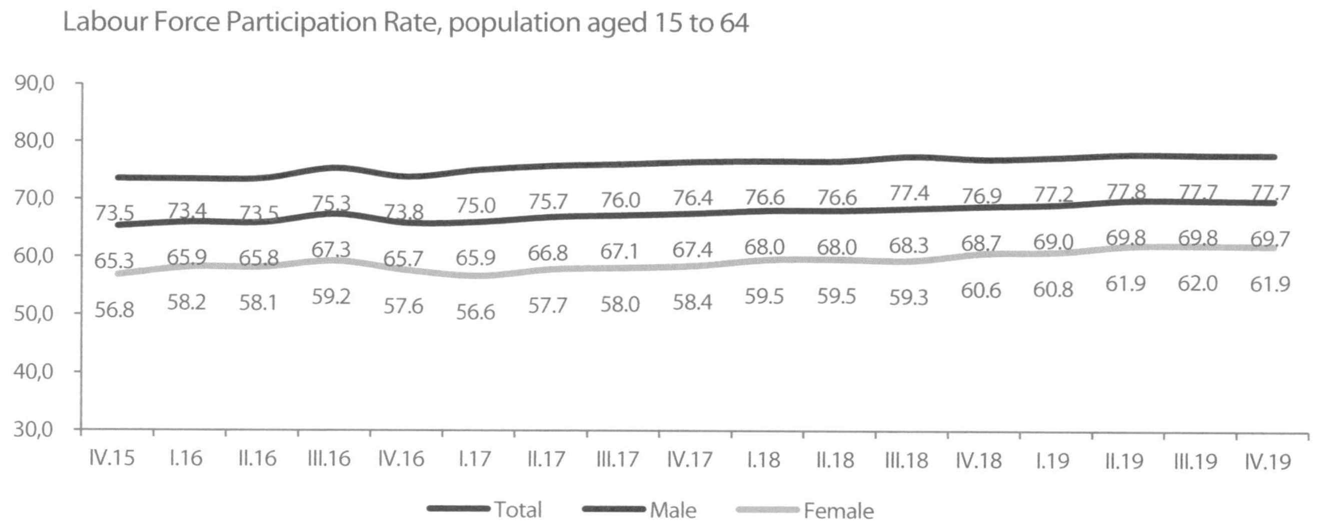

Labour Force Participation Rate, population aged 15 to 64

Fig. 12.2 Source: INSTAT 2019. Cartography: Ledjo Seferkolli

young people aged 15–29 is 51.7%, whereas the highest percentage is for the population aged 30–64 with 79.2% (Fig. 12.2) (INSTAT 2019).

For women, the labor force participation rate is 61.9%, while for men, this indicator is 77.7%, i.e., there is a difference of 15.8% more for males (INSTAT 2019). As far as the spatial distribution of employment and unemployment in Albania is concerned, large differences between different regions can be observed. The situation is better in the central and southern part of the country and more difficult in the northern and northeastern regions (Fig. 12.3).

Under these circumstances, unemployment remains one of the biggest problems and social difficulties of the Albanian population, the effects of which are reflected in many aspects of life, such as unabated emigration, uncertainty about the future, crime rate, etc.

12.3 Housing Conditions and Standard of Living

Other important indicators that reflect the sociocultural situation of the Albanian population have to do with the housing conditions and living standards. The 2011 census revealed the significant increase in housing stock and living space per capita.

In the meantime, about 30% of dwellings in Albania are vacant, and about 22% are not in use (empty). On average, each household has three rooms available, an indicator that has increased significantly compared to previous periods. Most of the buildings, especially those built in the last 20 years, are in the central part and coastal areas of the country (INSTAT 2011). As a result, in the southern part and along the coastline, each person has an average of one room and

20 m² of available space, thus undergoing a radical change compared to the period until 1990, when families with many members were forced to live in apartments with very small living space (Doka 2015).

Simultaneously with the increase in the number of dwellings and living space, the living conditions and the furnishing have improved, and so have the provision of running water, the expansion of heating possibilities, and other facilities. Despite these changes, the improvement of the social situation and its further expansion in line with European standards remains one of the most important challenges for the future development of the country.

12.4 Relationship with Property Ownership

The relationship of the Albanian population with property ownership is diverse and rather complex. This relationship has greatly changed depending on the socioeconomic systems the country has gone through. Thus, until 1945, the structure was diverse with such social groups as farmers and landowners in the village, craftsmen and city working class, industrialists, merchants, etc. It was precisely the diverse forms of property ownership that also defined a broad social structure, while other social structures based on well-being status, or educational, health, or professional level, were limited and also conditioned by the general difficult situation of socioeconomic development (Doka 2015).

During the period 1945–1990, the restriction of property, which was only in state ownership form, along with many other socioeconomic changes taking place in that period, would be reflected in the structure and social status of the

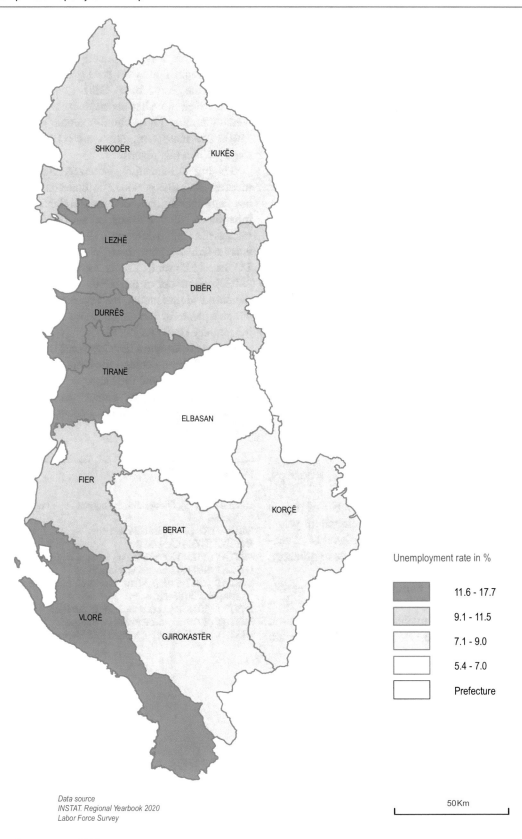

Fig. 12.3 Unemployment rate 2019

population. Thus, in relation to property, the social structure of the population would be limited to the existence of two classes: the peasantry, working in the agricultural cooperative, and the working class, but also the intelligentsia, which represented the most educated part of the two groups.

The comprehensive political, economic, and social changes that took place in Albania after 1990 would be reflected, among other things, in the situation and social structure of the population. Overall, there is a previously unprecedented expansion of different social groups based on the numerous criteria on which the social structuring of the population is based (Doka 2015).

Thus, based on the criterion of the relationship with the property ownership, with its expansion in various forms after 1990, that would also expand the social stratification, such as farmers and landowners in the village, traders and city entrepreneurs, etc.

12.5   Education

Education is a very important indicator that reflects the social status of the population. This is related to the weight of the number of pupils and students within the total number of the population, the absolute number of pupils and students according to the level of education, as well as the number of schools according to their type.

On this basis, we can say that the education system in Albania is complete, starting with the primary, secondary, and high schools up to the universities.

INSTAT data show an increase in the number of people over the age of 15 who have completed their school years; consequently, they have an increase in their level of education. On the other hand, there is also an increase in illiteracy, amounting to 2.8% (INSTAT 2019).

Significant changes in the number and size of schools have also been observed, especially starting from 1990 onward, and this is primarily related to the mass migration movement of the population, domestically and abroad.

Similarly, the number of vocational schools has decreased, from 513 in 1990 to 375 in 2001 (INSTAT 2001). The main reason for this was the closure of many agricultural vocational schools that were set up according to the socialist model in the period 1945–1990.

Moreover, there has been a decrease in the number of students attending high schools compared to the period until 1990. The mass emigration after 1990 has also played a major role in this regard.

The higher education system has had a positive development in the same period. The liberalization of high schools has significantly increased the number of students per 1000 inhabitants. The number of university centers has also increased. Thus, while until 1990 the only university center was in Tirana, today, there are several university cities, like Shkodra, Elbasani, Korca, Gjirokastra, Vlora, etc. From 2002, the number of private universities in Albania has also increased significantly, by about 30 institutions, most of which operate in Tirana, but there are also in other cities of the country (INSTAT 2019).

Despite these latest developments in education in Albania, there is still high demand for more admissions in university studies programs and for improved education infrastructure. These demands are the greatest in the central part of the country, especially in Tirana, because of the large influx of population from other areas.

References

Doka D (2015) Gjeografia e Shqipërisë – Popullsia dhe Ekonomia. Tiranë
EUROSTAT (2018) Buletin 2018
INSTAT (2001) Censusi 2001
INSTAT (2011a) Rezultatet kryesore të regjistrimit të popullsisë 2011
INSTAT (2011b) Censusi në harta 2011
INSTAT (2017–2018) Matja e të ardhurave dhe nivelit të jetesës në Shqipëri 2017–2018
INSTAT (2018) Anketa e buxhetit të familjes
INSTAT (2019) Forcat e punës në Shqipëri
INSTAT (2020) Niveli i varfërisë në Shqipëri

The Economy

13

Dhimitër Doka

Abstract

Considering its abundant natural resources, Albania should be regarded as a country with favorable conditions for economic development. Moreover, should only these natural conditions be sufficient for the economic development of a country, with its natural resources, Albania should have been among the most developed countries in the region and in Europe. But despite its abundant natural resources and favorable geographical position, Albania is little developed. One reason for this is related to the influence of the human factor which is equally important in this development. How the natural potentials are used in favor of economic development is determined by man. There are also several other factors, such as the historical, political, and social conditions which play their part in this regard.

Keywords

Natural resources for economic development · Performance of economic development · Structure of the economic sectors · Future of economic development

13.1 Economic Assessment of the Geographical Environment of Albania

Albania's geographical environment provides abundant natural resources for economic development. First and foremost, these resources include the favorable geographical and geopolitical position of Albania in the Balkans, the Mediterranean, and Europe. This position has influenced and may further affect, among others, the economic development of the country. The country's wide access to the Adriatic and Ionian Seas makes that several important roads that connect the region with other parts of Europe pass through Albania, which attaches importance to the development of trade, transport, tourism, and many other economic activities. By the same token, due to its geographical position, Albania can quickly become part of a global development market and attractive to many foreign investors, who like its accessibility. It is precisely the current perspectives for a better economic development of the country and its integration in Europe that can further increase the importance of its geographical and geopolitical position (Doka 2015).

Furthermore, the favorable natural conditions of Albania for its economic development are evident in all elements of its geographical environment, such as geological composition with plentiful underground resources, relief with its landforms, climate, waters, flora, and fauna. Thus, due to its diverse geological construction, the Albanian underground is distinguished for its abundance of mineral resources, such as chromium, copper, iron-nickel, oil, gas, coal, bitumen, etc., which constitute an important natural basis for the development of various branches of the economy (King and Vullnetari 2016).

In terms of its external forms of relief, Albania is considered a hilly and mountainous country, dominated by altitudes of over 200 m above sea level. The mountainous relief occupies about 25% of the country's total surface (Qiriazi 2019). However, within these forms, there is high diversity and interplay between the mountains, hills, plains, and valleys, creating favorable conditions for their economic development within its geographical areas. Furthermore, the country's relief structure includes whole flat areas such as the Albanian Western Lowlands, but also plains and plateaus, and several river valleys, which constitute the nuclei of general economic development, especially in agriculture.

A special potential of the Albanian relief, which can be used more in providing economic development, is the 427-kilometer-long coastline which stretches along the Adriatic and Ionian coasts along the entire length of the country from north to south (Qiriazi 2019). This coastline with its numerous beaches constitutes the main potential for the development of tourism and many other economic activities related to this sector.

Climate is also an important natural resource for the economic development of the country. Albania is distinguished for its diverse climate. Although a small country, there are different types of climates, ranging from the typical Mediterranean climate along the Adriatic and Ionian coasts to the climate with continental influence in the eastern and northern parts. Similarly, because of different forms of relief and the presence of water sources, there are different types of microclimates, which create distinctive conditions in specific areas and have an impact on their economic and vital activity. All this climatic diversity is an important condition for the development of various branches of the economy such as agriculture, livestock, tourism, etc. (Qiriazi 2019).

Other natural resources of Albania with an impact on its economic development are its water resources. Besides the two seas, the country's water wealth includes several lakes, rivers, and other natural resources. All these constitute an important and valuable basis for the overall economic development of the country, and especially for the development of some branches of the economy that are directly related to the presence of water, such as fishing, water transport, tourism, and so on.

The wealth of the plant and animal world is seen in entire areas of forest and green spaces, which constitute a solid basis for the development of some branches of the economy, such as forestry, livestock, etc. These forests and green areas are home to different species of animals and birds, which also constitute an important basis for the development of hunting, leather industry, etc.

All these natural resources create the basis for the comprehensive economic development of the country, at present and in the future. The preservation, evaluation, and utilization of all these natural assets is crucial for the time being so that they remain a continuous natural potential for a sustainable development of the Albanian economy (Doka 2015).

All these factors combined corroborate the fact stated above that Albania contains abundant potential for its economic development. However, in the case of Albania, this development is also highly conditioned by historical, political, and social factors.

13.2 The Performance of Economic Development

Although the Albanian nature with its plentiful natural resources offers ample opportunities for the economic development, several historical, political, and social factors within the country have traditionally hampered the proper economic development of Albania. Thus, frequent wars, political instability, numerous and continuous population emigration, and

the like, have always been the main obstacles to its economic development. For almost 500 years, Albania was under Ottoman rule, which moved the country away from Western development. The continuous partitioning of Albanian territories and the eventual establishment of national borders in 1913 would not only unfairly divide a people with the same history and culture, but it would also directly affect its economic development (Doka 2005).

The numerous ensuing wars and subsequent historical events would further deteriorate the country's economic situation. Not only the two world wars but also many other wars in the Balkans would hinder and destroy even the scarce economic progress made with sacrifices by Albanians. In the history of Albanian economic development, three main periods should be distinguished:

– *Until 1945,* Albania was considered a typical backward agrarian country. This situation was conditioned by several factors, such as the backwardness inherited from almost all previous periods that the country had gone through, the centuries-long Ottoman rule and the numerous wars the country went through, lack of knowledge on the capitalist elements of economic development, lack of political and governance stability of the country, lack of clear and long-term strategies for economic development, etc.

This situation was also reflected in all the economic indicators, dominated by agricultural and livestock activity. Therefore, over 85% of the active population of the country was engaged in agriculture, which at that time provided about 92% of the country's total production (INSTAT 1989). However, although being the dominant branches of the economy, agriculture and livestock were underdeveloped and provided inadequate income in relation to the country's overall demands. Mostly, these two sectors operated in almost natural conditions of an extensive economy and provided very low product yields.

The role of industry was almost negligible. It accounted for no more than about 8% of the country's gross domestic product. Industry remained in its infancy stage of development and was represented mainly by few factories and workshops, mostly operating in the processing of various agricultural and livestock products. Only in some areas and centers around the country, mainly during the reign of King Zog and the Italian capital flow into the country, several mines for the extraction of chromium and bitumen or the exploitation of oil and gas were established (Doka 2015).

Transport, trade, services, and the like were in an even more dire situation. Therefore, the road infrastructure was underdeveloped and mainly with unpaved roads. Commercial activity was limited and was carried out only in the most important civic centers of the country. The service sector network was represented by only a few hotels, bars, and cafes.

The trade balance was negative because imported goods clearly dominated and foreign trade was conducted mainly with the neighboring countries.

This situation was also reflected in the quite limited geographical distribution of the economic activities in relation to different areas and regions of the country. This affected the economic map of that period which was dominated by only such areas as the western lowland of the country with its most important urban centers (Durresi, Vlora, Shkodra, Tirana, Elbasani) and the southeastern region with its main center Korca. These cities were distinguished for their concentration of the main economic activities of the country, while in most other areas of the country, there was a significant lack of other economic activities, but for husbandry (Doka 2015).

This difficult situation of the country's economy would be further exacerbated by the damage of the Second World War, which destroyed even those few industries and the agriculture that Albanians had managed to develop.

– *The end of the Second World War* marks a new period for Albania and its political, economic, and social development. Between the two political and economic models of development that would dominate the world for over 40 years, the capitalist and the socialist one, Albania would choose the latter. It was precisely that political and economic model which would determine all other socioeconomic developments within the country until 1990 (INSTAT 2000).

Consequently, based on the socialist model, a series of reforms were undertaken which aimed at converting the country's economy from a private one to a centralized economy entirely controlled by the state. This would lead to the gradual and, later, total abolition of private property and its conversion into state property, which significantly narrowed economic maneuverability due to total state control and management. By 1960, the result of all these comprehensive and combined reforms would be the return of the entire economy to the ownership of the centralized state, according to the Stalinist communist model, which went unchallenged for the entire communist period until 1990.

This transformation process was accompanied by major changes both in the structure of the economy and in its geographical distribution. Consequently, the new economic structure prioritized the socialist industrialization of the country, making heavy industry a key priority, following the Soviet model. This expanded the country's industrial sector and production, which manage to provide over 40% of national income and offer employment for over 250,000 workers (INSTAT 1989).

However, such an industrial development, not aligned with the country's conditions and possibilities and dominated mainly by the import of a backward technology, would turn out to be unprofitable and costly to the poor Albanian economy. This industry would also devour the income provided by agriculture and other sectors of the economy. During the period 1945–1990, agricultural economy would play a secondary role compared to industry. However, this sector underwent significant transformations during this period, such as the general increase of arable land, mechanization of agriculture, establishment of systems for land fertilization, irrigation, and drainage (King and Vullnetari 2016).

All these measures would result in the increase of the production volume and yield, thus expanding the types of products, but also in the wide geographical distribution of various agricultural and livestock products. Compared to the previous period, until 1945, it can be said without hesitation that the agricultural economy in Albania experienced unprecedented changes from 1945 to 1990. Under these conditions, even during this period, the agricultural economy continued to remain the main branch of the country's economy, providing about 60% of total national income, despite the prioritization of industry. However, most of the income generated by this sector was often spent on unprofitable investments in industry, and very little of that income was used to increase the welfare of the population (INSTAT 2000).

In contrast to industry and agriculture, other branches of the economy (transport, trade, services, etc.) remained in the minimum developmental conditions, providing not more than 5% to the total national revenue. Although improvements were made, compared to the earlier period until 1945, such as the expansion of road network and types of transport, trade network and services, etc., in terms of the economic policy of that period, these activities would not prioritize development.

During the same period, 1945–1990, Albania's economic ties with the outside world would be totally determined by the course of politics. These ties were limited to within the countries of the communist bloc, swaying unexpectedly depending on the political relations between Yugoslavia, the Soviet Union, and China. Under these conditions, they would not have major effects on the Albanian economy.

All these factors together, coupled with the situation that was created following the collapse of the communist system in many countries of the former socialist bloc, forced the Albanian economy in the late 1980s and early 1990s to be subjected to bankruptcy, thus paving the way for the democratic reforms and the market economy (Doka 2015).

– *Since 1990*, the Albanian economy has entered a difficult, but also necessary, period of transformation from a centralized and planned economy to a free market economy. This transformation has required a total overhaul of the previous structures of the functioning of our economy.

The transition from a country where private property had been completely abolished until 1990, the permitting and strengthening of the use of private property in all sectors of the economy, is now at the core of the current economic transformation. A whole process of comprehensive reforms was needed for this, such as the privatization of agriculture and industry and especially the establishment and operation of many new private enterprises.

In the villages, the previous agricultural cooperatives were dismantled, and the land was distributed to the villagers. By 1995, about 98% of the agricultural land area in the villages around the country had been allocated to the respective village dwellers. Today, there are about 440,000 small agricultural holdings with an average arable land area of 1.4 hectares (INSTAT 2000), while in the cities until 1995, small and medium enterprises were privatized with financial turnover up to 500,000 dollars and employing up to 300 workers. At the same time, since 1995 onward in the privatization process have been included large enterprises with financial turnover over 500,000 dollars and employing over 300 workers (Becker et al. 2005).

This has led to over 80% of the Albanian economy being privatized today, and the private sector currently provides about 85% of the country's gross domestic product. Another significant indicator is that the private sector currently has the largest number of employees in the country (INSTAT 2019).

However, privatization in Albania has not been an easy process. It has been accompanied by many debates and social problems. As a result of the lack of a clear and long-term model for the country's development in terms of free market economy, there has been a lack of clear privatization strategies based on the respect for social criteria in this very important process.

The entire process of privatization and increase of economic activities has been accompanied by major changes in the weight and role of different sectors of the economy in country's overall production and revenues.

Table 13.1 Role of economic sectors in the gross domestic product (in %)

Sectors of economy	Contributions to GDP	Contributions in employment
Agriculture	21.4	45.8
Industry	19.4	15.0
Service	59.2	39.2

Source: INSTAT, Bulletin of Statistics (2019)

13.3 The Structure of the Economic Sectors

Since 1990 onward, the general trend in this direction has been the decline of the role of industry and the return to agriculture and livestock but also the increase of the role of the tertiary sector of the economy. This tendency is also expressed by the following table which shows the weight of different branches of the economy in the country's gross domestic product (Table 13.1).

Although *agricultural economy* continues to be an important sector for economic development and almost 46% of the country's population is directly related to it, this sector is less productive due to the difficulties and problems that rural areas are facing today.

Because of the excessive fragmentation of agricultural land, lack of adequate equipment, irrigation, fertilization, and the like, agriculture cannot provide the necessary amount of production to satisfy the market needs. In addition, even the product that is provided often does not meet the conditions and standards to compete with the many products that come through imports (Doka 2015).

One characteristic in today's Albanian agricultural economy is the fact that it is no longer dominated by the production of food grains and industrial plants but by a larger variety of agricultural and livestock products.

Consequently, the importance of livestock, vegetables, orchards, and the like has increased. Likewise, the general view of the agricultural landscape of Albania is no longer dominated by large plots areas of wheat or corn but by numerous small plots, where Albanian farmers cultivate almost everything the daily domestic market needs in accordance with the natural seasons.

This has made the agricultural sector of the economy, with all its respective branches and subbranches, extend almost everywhere in Albania. However, there are several agricultural regions, which bear the brunt of all the agricultural development, such as the Western Agricultural Region, the Southeastern Agricultural Region, the River Valley Agricultural Regions, the Albanian Coastal Agricultural Region, and hilly and mountainous agricultural regions (Fig. 13.1) (Doka 2015).

Each of these regions is presented with its own structure of agricultural and livestock products. Among them, it is worth mentioning the western region, which carries the main weight in the entire agricultural development of the country through the production of almost all agricultural crops. It is followed by the Southeast Region, which is especially distinguished for its production of fruits (like apples, plums, cherries, etc.) and its livestock products. The coastal region is also distinguished for the abundant production of olives, citrus fruits, viticulture, etc.

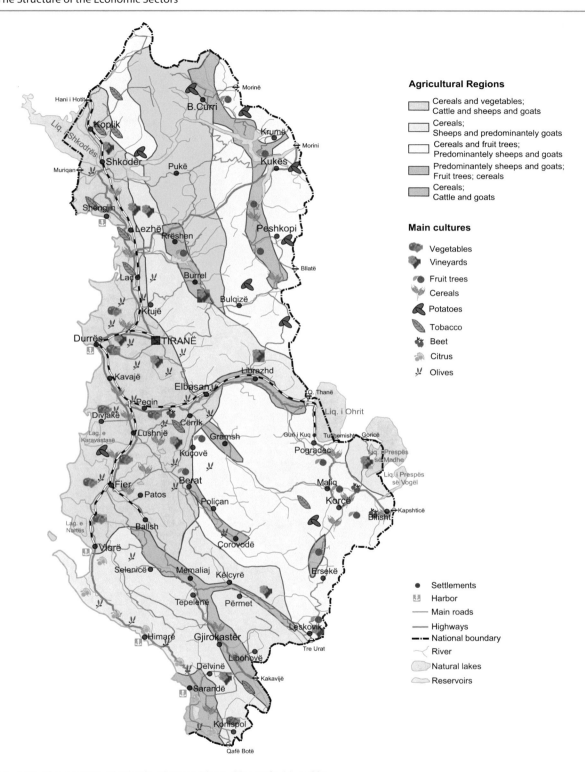

Fig. 13.1 Map of agriculture of Albania. (Source: Ideart. Gjeografia 11, p. 89)

Industry, following a difficult transformation phase after 1990 and the drastic decline of industrial production from 41.3% in 1989 to less than 20%, has currently undergone major changes in both structure of production and its spatial distribution (INSTAT 2019). Unlike in the past when heavy industry dominated in Albania, today, it is the light and food industry that prevails. Moreover, this industry is no longer scattered around the whole country (see Fig. 13.2), but it is

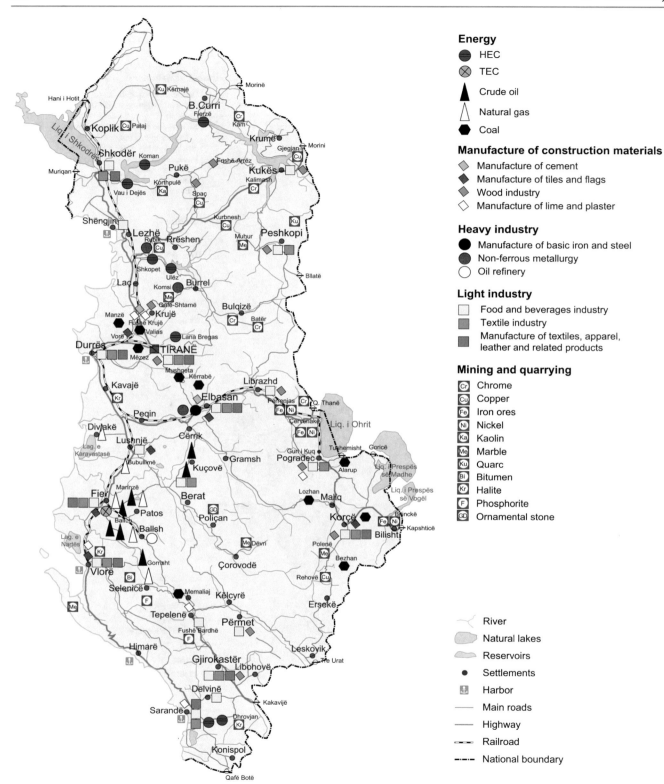

Fig. 13.2 Map of Albanian industry. (Source: Ideart. Gjeografia 11, p. 90)

mainly located in the central (Tirana-Durrësi area) and western part.

Despite the difficulties in the economic development of the country and the financial impossibilities to support large production enterprises, economic activities such as energy, with its numerous hydropower resources (which provide almost 100% of electricity), oil and natural gas remain of special importance for the country.

Albania is also one of the most important countries in the world when it comes to the extraction and export of **chromium**.

Albania, with about 37 million tons of chromium reserves, is estimated as the fourth country in the world for these reserves, after South Africa, Kazakhstan, and India (Göler et al. 2015).

The main chromium-bearing mineral areas are mainly located in the northeastern part of Albania, such as Bulqiza area with spring of the same name; Martanesh area with springs Batra, Krasta, etc.; Kukës-Tropoja area with springs Kalimashi of Kukës, Kam of Tropoja, etc.

In these chrome mineral-bearing areas in Albania, their best development has been around 80s of the XX century. Yet they still contain a lot of potential important to develop even better in the future (Fig. 13.3).

In addition, the apparel manufacturing sector plays a part in today's Albanian industry, where many foreign companies (Italian, Greek, German, Austrian, etc.) use the low-cost Albanian labor force to manufacture their products.

In recent years, agroindustry has become important. This sector carries out the processing of many agricultural and livestock products produced by the different areas of the country.

The tertiary sector (services) has become particularly important in the Albanian economy in the last three decades. This sector with its main activities such as trade, services, transport, etc., has managed to provide about 60% of the gross domestic product today, and it employs about 40% of all the employees. This represents an almost total reversal of the Albanian economic structure during the communist period, when this sector provided only about 5% of the GDP (Göler et al. 2015).

In addition, the tertiary sector with its branches is currently spread almost all over the country. The perspectives, not only of the main cities like Tirana, Durres, Elbasan, etc., but also of all the other inhabited centers around the country are today dominated by the economic activity of the tertiary sector, such as trade, services, transport, and especially tourism.

Tourism is one of the most important economic activities for Albania today and for the future perspective of its general economic development. According to 2019 data from the INSTAT and the Bank of Albania, the direct contribution of tourism to the gross domestic product of the country amounts to about 8%, and calculating its indirect impact on the development of many other related activities, it reaches at over 20% of the GDP (Göler and Doka 2018).

Although the global COVID-19 pandemic hit tourism in Albania hard, the chances for its rapid recovery are great. This prediction is based on the many natural and cultural resources that the country has in the field of tourism but also

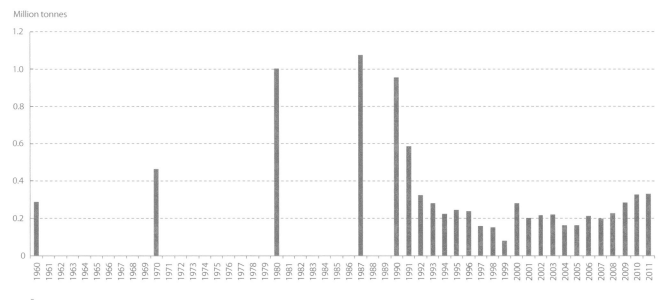

Source:
Göler, D., Bickert, M., Doka, Dh. 2015 from Schappelwein 1991: 148; Steblez 1994: 18; 1995: 4; 2000: 2.7; 2004: 4.9;

Fig. 13.3 Chrome production in million ton

on the fact that, unlike other Mediterranean countries, the potential of Albania has not been fully used for the benefit of tourism.

Albania can become an important tourist destination in the Balkans and in Europe considering its favorable geo-tourist position in the Mediterranean as the most important region of balneary tourism in Europe and the world, its diverse forms of relief and picturesque natural landscapes, its typical Mediterranean climate suitable for the development of different types of tourism, its numerous water resources of special tourist importance, its rich flora and fauna, and its numerous historical and cultural sites (Doka 2019).

In the meantime, for better tourism development in the future and proper use of the country's tourism potentials, it is important to overcome and solve a number of problems, such as improving the country's tourism image, avoiding environmental pollution, further development of general and tourist infrastructure, promotion of domestic and foreign investments in this sector, and the like.

The swift and adequate solutions to these problems improve the chances for the development of tourism, by strengthening of the country's position in the current regional tourist market, enhancing also the opportunities for wider recognition and European integration.

The areas with the best prospects for the development of tourism are all coastal areas from Velipoja in the North down to the border with Greece in the South, the inland mountainous areas (especially the Alps, but also the mountains of Korabi and Lura, as well as the mountain ranges in the Southeast and South, etc.), lake areas, museum cities, rural areas with agrotourism resources, etc. (Doka 2019).

In the impossibility to treat in detail all the economic activities of Albania, it is a fact to emphasize that the structure of the economy (see Fig. 13.4) is presented in recent time more and more expanded, where the main weight along with the agricultural economy is becoming more and more activities such as trade, transport, tourism, etc. There has also been a significant increase in recent years in such activities as hotel and services, information and communication, etc. The diagram below clearly presents such a situation.

The rapid developments that characterized the transformation of Albania after 1990 have been accompanied, among other things, by the deepening of inequalities in the state of economic development between different areas and regions of the country (Fig. 13.5).

The view of the map above clearly shows the particularly large number of economic enterprises concentrated in the western region, especially in the area of the capital, which indicates better economic and regional development compared to other regions.

In the development of **foreign trade**, the Albanian economy is currently characterized by a negative balance in the ratio between exports and imports.

This is related to the opening of the country after a long isolation and the need of the Albanian market for many goods unknown to it before, as well as the difficult situation of the Albanian economy, which is not able to provide in the country many of products (Fig. 13.6) (INSTAT 2017).

The large trade deficit would aggravate the Albanian economy even more, if it were not compensated to a considerable extent by the income that comes to the country through the foreign emigration of many Albanians working abroad.

Traditional goods exported from Albania are agricultural products and mineral resources, electricity, etc. Albania is also a very important country in the world in the export of medicinal plants collected in nature, especially sage. The import structure for Albania is wide and includes various goods in which, first of all, machinery, food, beverages and tobacco, textiles, shoes, etc., dominate.

Also from the statistical data, it results that over 90% of the current foreign trade of Albania is realized with the neighboring member states of the European Union, such as Italy and Greece. After them come other neighboring countries, such as Kosovo, Macedonia, and Montenegro, but increasingly with other countries in the region (Serbia, Bulgaria, Turkey) and beyond, such as Austria, Germany, France, etc. (Fig. 13.7).

In this way, there is a significant change in the structure of foreign trade according to the countries with which it is conducted, if until 1990 Albania's foreign trade was very limited and oriented to those few countries with which there were also good political relations, mainly with the countries of the former socialist camp of Eastern Europe. Today, the foreign trade space has expanded a lot, including in addition to the countries of the region many others in Europe and abroad (INSTAT 2017).

13.4 The Future of Albania's Economic Development

Despite the difficulties and problems accompanying this stage of Albania's economic transformation, the future is not without hope. There are many perspectives, alternatives, and opportunities for a better development of the country in the future.

The first opportunity in this regard is the important natural resources the country has for its economic development. Albania, with its abundant natural resources, offers more opportunities for development alternatives in the future (UNICEF 2000).

Therefore, if Albania were to use its transit position in Southeast Europe by further strengthening of the corridors passing through the country better than it has done until now, the impact on the economic development of the country would be significant. The current perspective for the integra-

Active enterprises by economic activity in 2019

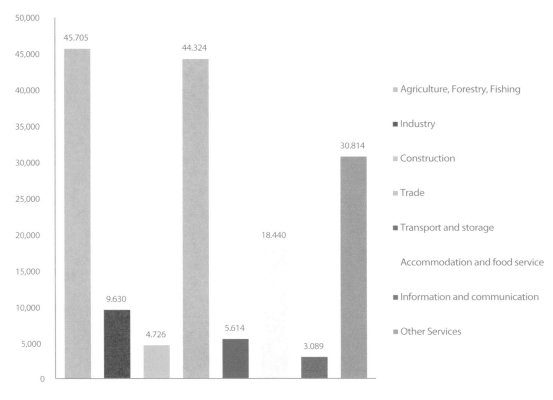

Source: INSTAT, Statistical Business Register

Fig. 13.4 Active enterprises by economic activity in 2019. (Cartography: Ledjo Seferkolli)

tion of this region in Europe would further increase the importance of the geographical and geopolitical position of Albania and the impact on its economic development.

By the same token, the use and reevaluation of the underground mineral resources (like chromium, copper, oil, gas, bitumen, etc.) would bring new and important impulses to the economic development of the country. Their impact would enhance not only the increase of the volume of exports but also the reactivation of various branches of the mineral extraction and processing industry, thus providing jobs and increasing income for the population, especially in those specific regions and areas, where economic development is more difficult to achieve (Doka 2015).

In addition, the use of the plentiful natural overground resources related to the warm climate, vast amount of water sources, and the rich flora and fauna remain great opportunities for development. It is important to preserve, evaluate, and make the best use of these assets so that they remain a replenished potential for sustainable development of the economy in the future. Based on these resource potentials and opportunities, several priorities of Albania's economic

development in the future can be determined, and among them are:

- **Agricultural economy** through the provision of even more agricultural produce and livestock production. This sector is favored both by the suitable natural conditions that the country enjoys for the development of these products and by the well-established Albanian tradition in this field. If this sector develops to its proper capacity, Albania can not only meet its own needs for such products but can become an exporter in the region and beyond. In this regard, Albania should make better use of its natural advantage in providing these agricultural products earlier compared to countries in the north and east of the region, thus enjoying a competitive advantage in the regional market (King and Vullnetari 2016).

- **Energy** is also a sector with bright prospects in the future economic development of the country. Albania has important natural potentials, which, if better used, would bring greater development in the field of energy. The numerous hydropower resources, fuels (oil and gas), great solar and wind power, etc., are important resources for increasing

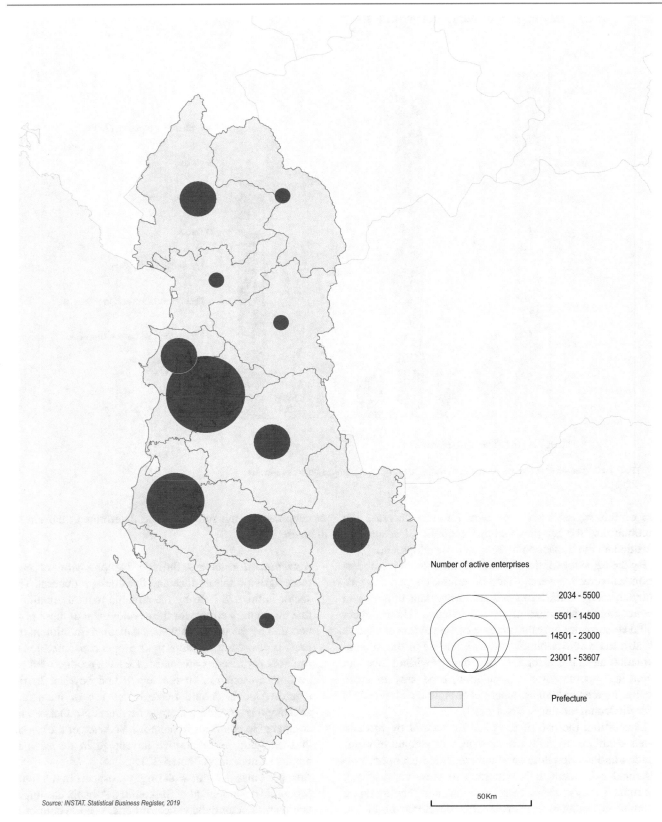

Number of active enterprises

2034 - 5500

5501 - 14500

14501 - 23000

23001 - 53607

Prefecture

Source: INSTAT. Statistical Business Register, 2019

50 Km

Fig. 13.5 Distribution of active enterprises, 2019. (Cartography: Ledjo Seferkolli)

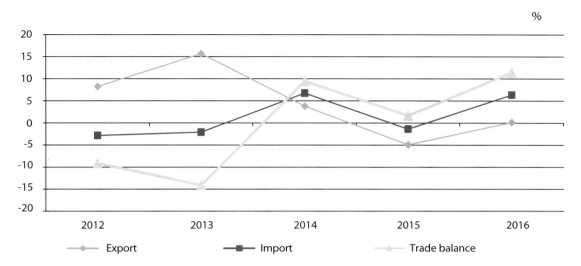

Fig. 13.6 Annual change in trade of goods, 2012–2016. (Source: INSTAT: Yearbook of 2017)

energy production, which would be sufficient not only to meet the needs of country but also to boost energy export. In this way, Albania could play the role of an important energy hub in the region (UNICEF 2000).

- *The services sector* is expected to further increase its weight and role in the overall economic development of the country. Only after 1990, Albania witnessed the increasing importance of this sector in the economy, especially when the country was finally opened to free market. Within the tertiary sector, tourism is expected to be of special and ever-increasing importance. Albania is in a favorable geotourism position and has just opened toward the regional and European tourist market. Under these conditions, tourism is a promising sector for important national and local developments, and it can serve as a promoting factor for other sectors. The development of tourism can directly promote and support many other activities in the service sector, such as trade, transport, infrastructure, but also indirectly increase agricultural produce and livestock production for tourism consumption. Tourism can also have a positive impact on the preparation and marketing of handicraft products, etc. (Göler and Doka 2018).

- *The light and food industry* can be developed and expanded significantly based on the potentials and opportunities that the country has in this field. Consequently, based on the natural resources and the large workforce, Albania can turn from a country where foreign companies simply use the cheap labor force in manufacturing their products into a producing and exporting country for these products, such as textiles, leather products, shoes, etc. Similarly, agroindustry represents much greater development opportunities for Albania. Due to the suitable natural conditions, the country provides numerous agricultural products and livestock production, which in turn serve as an important basis for the development of the processing industry in this sector. Recent developments have been encouraging in this sector with the establishment and operation of many industrial facilities specializing in the processing of milk, meat, fruit, and egg production throughout the country (Doka 2015).

- *The banking, insurance, and telecommunications sectors* are still in the early stages of their development in Albania. Therefore, there is significantly more space in promoting and expanding these activities in the future. Albania's increasing exchanges and connections with other countries, the large presence of Albanians living abroad, etc., are important factors that stimulate an even greater development of these economic activities now and in the future.

Taking further steps toward the country's strategic goal of regional and European integration will also play a particularly important role also in making use of all these resources and opportunities for the country's future economic development.

The opening of the country has been accompanied by Albania's membership in many international organizations, which enable comprehensive economic cooperation. Since September 2000, one such membership of particular importance for Albania has been the World Trade Organization (WTO). Albania gets significant benefits in the exchange of various products between member countries, which also affects stability and development and contributes to the further economic development of the country and the region (UNICEF 2000).

The signing of the Stabilization and Association Agreement with the European Union and the accession of

Exports by group of countries

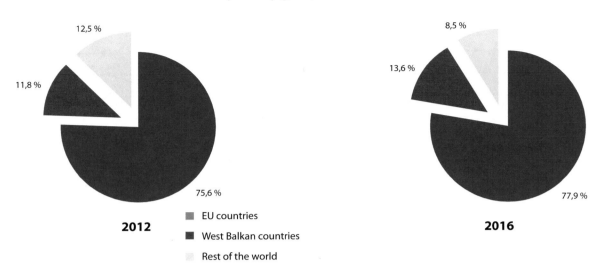

12,5 %

11,8 %

75,6 %

2012

■ EU countries
■ West Balkan countries
▨ Rest of the world

8,5 %

13,6 %

77,9 %

2016

Exports to EU countries, 2016

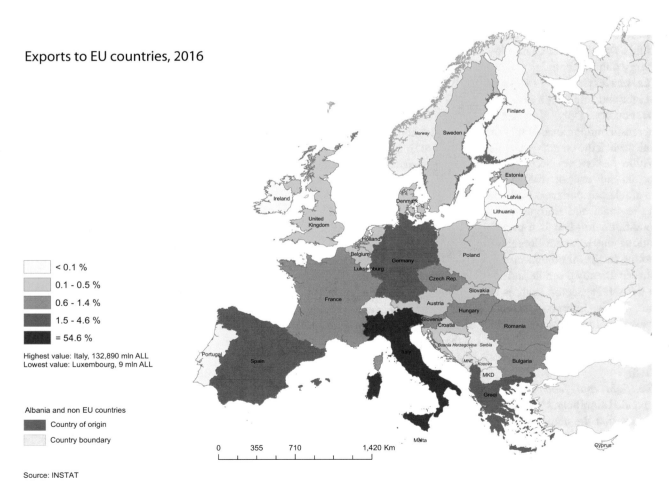

< 0.1 %
0.1 - 0.5 %
0.6 - 1.4 %
1.5 - 4.6 %
= 54.6 %

Highest value: Italy, 132,890 mln ALL
Lowest value: Luxembourg, 9 mln ALL

Albania and non EU countries
Country of origin
Country boundary

0   355   710   1,420 Km

Source: INSTAT

Fig. 13.7 Export by group of countries. (Source: INSTAT: Yearbook of 2017. Cartography: Ledjo Seferkolli)

Albania as EU candidate country in 2014 is of even greater importance.

The final goal of all these steps is the full integration of Albania in the European Union, at the foundation of which is the economic development of the country, as a necessary condition and basis of this whole integration process.

However, in achieving these strategic goals, it is necessary to overcome the difficulties and solve the problems the country still faces during its development, such as the aggravations in the country's political situation, corruption and crime, the difficult condition of infrastructure, the reduction of the informal economy, etc. The progress of many economic indicators depends on the solutions to these problems and the realization of goals which are related to:

- Economic and financial stability
- Proper functioning of various institutions promoting economic development of the country as a whole and different region in particular
- Mitigation of inequalities observed in the economic development of different areas of the country
- Improving the economic situation and increasing incomes for most of the population, especially in the peripheral and rural areas
- Increasing the Albanian exports and enhancing the competitive advantage in the region and Europe, etc.

To achieve all these goals and lay the foundations for a sustainable development in the future, it is necessary to design and implement various development strategies, both at national and regional level.

References

Becker H, Göler D, Doka D, Karaguni M (2005) Banimi në zonat ilegale të ish-objekteve Industriale të qytetit të Tiranës. Botuar në revistën shkencore gjeografike "EUROPA REGIONAL". Leipzig 2005

Doka Dh (2005) Regionale und lokale Entwicklungen in Albanien – ausgewählte Beispiele. Potsdam

Doka Dh (2015) Geografia e Shqipërisë – Populllsia dhe Ekonomia. Tiranë

Doka Dh (2019) Gjeografia e Turizmit. Tiranë

Göler D, Bickert M, Doka D (2015) Albanian chromium mining revisited. Die Erde 146/4

Göler D, Doka Dh. Tourism and Transition in the Western Ballkans – Albania as a laboratory of Tourism Development. In: Südosteuropamitteilungen 04/2018

INSTAT. Demographic yearbook of 1989, 2000, 2019

INSTAT. Treguesit e Tregtise se Jashteme 2017

King R, Vullnetari J (2016) From shortage economy to second economy: an historical ethnography of rural life in communist Albania. J Rural Stud 44

Qiriazi (2019) Perikli : Gjeografia Fizike e Shqiperise. Tirane

UNICEF – Tirana (2000) Evaluation of the social and economic situation in the regions of Albania

Part V

Geographic Regions

Geographic Regions

Perikli Qiriazi and Dhimitër Doka

Abstract

In view of the necessity for recognition and assessment of the region and its developments as the best way to smooth out regional inequalities, but also as an instrument for integration within the country, in the region, and beyond, after 1990, attention is being paid in Albania to the study of the geographical or integrated region and the geographic regionalization of the country.

Assessing the specific natural, social, and economic factors and regional consciousness, as well as relying on the current concepts and trends of the EU for the region and regional and local developments, efforts have been made for the geographical regionalization of the country. The most accepted opinion divides the country into north and northeast, southeast, south, and west.

Keywords

Geographical regionalization · Regional developments · Integration in European · Western region · The Northern and Northeast Region · The Southeast Region · The Southern Region

14.1 Geographical Regionalization of Albania

In its physical and human aspect, Albania constitutes a special and well-defined unity in the geographical space of the Balkan Peninsula. This unity has been identified by many scholars: *Ami Boué* (1840), who compares the Albanian highlanders with the European highlanders, but also distinguishes their special features; *Herbert Luis* (German geographer, 1927), who emphasized that Albania forms a separate physical unit in the Balkans, because it lies where the greatest push of the ophiolitic zone toward the west takes place; *Eqerem Çabej* (Albanian linguist, 2006), who recognizes Albania as a single and distinct unity from other neighboring countries and underscores that Albania has physical-geographical and ethnic-cultural cohesion, etc.

Within the unity of the Albanian geographical space, there are obvious natural and human differentiations related to the regions, which have been observed by different regionalization processes. The geographical regionalization identifies integrated regions. Natural, social, and economic factors and regional consciousness play a part in their differentiation. The interaction of these factors depends on natural uniformity or heterogeneity; forms of economic, social, and political organization; developmental stage of the country, and so on.

Unlike many developed countries, some other conditions apply in Albania: the marked natural diversity; the large weight of natural factors in the life and productive activity of society; the large share of rural population (46.3% in 2011); the country is still far from the super-technological world, where the homogenization of the social, cultural, economic, and political space is relentlessly carried out; the limited economic development and transport; major economic, social, and cultural differences between areas of the country; highly arbitrary and political administrative divisions; pronounced centralization; preservation of agricultural features of society and survival of the traditional culture, folk art forms, traditional family types, and social relations in the rural areas; the strong spiritual connection of the people with their region, etc.

These features stand at the foundation of the geographical regionalization of the country. But at the same time, this regionalization must be based on the EU concept of the region, as well as the regional and local developments, which are EU objectives for all countries seeking EU membership, including Albania. When great regional, social, and economic differences are deepening, regional development is the best way to smooth out regional inequalities, an instrument for domestic and regional integration. Therefore, the implementation of regional-based development strategies and policies is required, like administrative and territorial reorganization of the country based on the EU criteria, but also decentralization and deconcentration. These policies aim to create equal living conditions for people in all regions. The success of regional developments is closely related to their ability to function as administrative and development units. The functional integration of the region and regional developments is now emphasized, based on the connection, coordination, and cooperation between the regions (Fig. 14.1).

In this context, significant importance is attached to the Euro regions, as effective instruments for mitigating polarization between different areas of the country and within the EU, thus eliminating the negative impact of the size of member states, which can bring polarization, antagonism, and conflict. Europe aims to move from a community of small and large countries, poor and rich countries, into a community of integrated, developed, and equal regions. This trend of regional developments in the EU restricts the role of state borders, but it also brings economic and social development, stability, and prosperity. At the same time, it protects and develops the natural, material, and spiritual values of the regions.

It is also important for Albania to be reoriented toward regional developments, its adaptation and integration in European regional developments, in accordance with the EU current criteria and trends. To achieve this aim, the integrated regional geographical studies which started after 1990, when Albanian geographical studies entered the orbit of western development, should be deepened further.

Based on these concepts for the region, but also on the criteria and directions of regional, national, European, and wider developments, efforts have been made for the geographical regionalization of the country. The widely accepted view divides the country into north and northeast, southeast, south, and west. This division is like the physical-geographical zoning scheme, which is related to the great weight of the natural factor in the differentiation of geographical space and to the more in-depth study of problems and physical-geographical units.

14.2 The Western Region

14.2.1 Geographical Position

The western region occupies the entire western part of the country from the Bay of Hoti of Lake Shkodra (in the north) to Vlora (in the south). In the west, it is washed by the Adriatic Sea, while in the east, it reaches from the foot of the Western Alps in the north to the Kruja-Dajti mountain range in the south, including all the hills of Kërraba, Dumrea, Sulova, and Mallakastra and up to the hills of Vlora at its southern extreme (Qiriazi 2019).

The length of this region from north to south reaches about 200 km, while the width varies from several kilometers to about 50 km in the direction of the city of Elbasani and Berati (Qiriazi and Doka 2020).

This geographical position is suitable for the comprehensive development of the region because it is connected by road and rail with other regions of the country and the neighboring countries. Extensive access to the Adriatic Sea and, through it, to the Mediterranean provides maritime connections with other countries of the world and other activities: sailing, trade, tourism, fishing, etc.

14.2.2 Natural Conditions

In terms of geological formation, it is considered the youngest region as compared to the other regions, and this is related to the Plio-Quaternary folding movements, which created the anticline structures, which then came out of the water and the hills were formed and the syncline structures, which were covered by shallow sea bays, which were filled with alluvium brought by rivers. Thus, the great fields of this region were created. Tectonic subsidence occurred in some sectors (like Shkodra, Elbasani). As a result of this evolution, this region consists almost entirely of terrigenous formation (the plains and most of the hills) and very little of limestone (some hills) and gypsum (in Dumrea). These deposits contain fuels (coal, oil, and natural gas) and construction materials (clay, sand, etc.) (Qiriazi 2019).

Due to this evolution, the relief consists of plains and hills, which form large groups of hills (in Kërraba, Mallakastra, Dumrea, etc.) in the east and in the form of isolated ranges and hills inside the plains (Fig. 14.2).

The fields are flat and have very low altitude above sea level (up to 20 m). There are also some field sections located at the sea level and some others even below the sea level (for instance, Tërbufi -8 m, etc.) (Qiriazi 2019). Therefore, the downstream section of the riverbeds which cross these plains have very small steepness. Consequently, large amounts of solid materials have been accumulated in them, so they are shallow, too meandering, even above the level of the field, thus creating lower sectors between them. The constant change of the riverbeds and large floods are related to these phenomena, creating swamps in the lower sectors. These swamps covered large inland areas with sweet water, such as the swamps of Tërbufi, Roskoveci, Thumana, etc.

There were numerous swamps even in the coastal belt (in Velipoja, Lezha, Durresi, Hoxhara, etc.). Their formation

Fig. 14.1 Geographical regions of Albania Source: Ideart. Gjeografia 11, 2017, p. 98

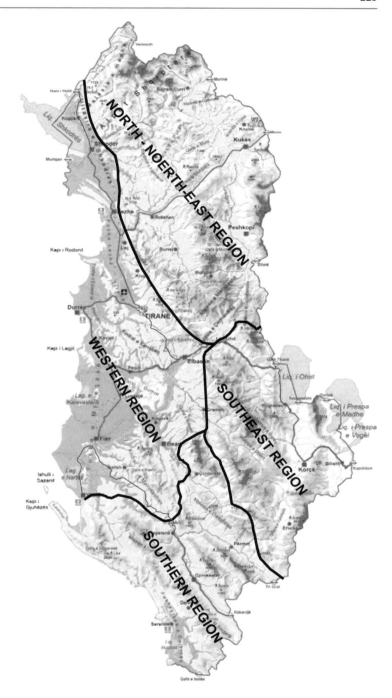

took place during the land's encroachment toward the sea, resulting in the isolation of the small marine inlets from river and marine accumulations. At first, lagoons were formed from them and later they evolved into salty marshlands (Fig. 14.3).

Until the middle of the twentieth century, the landscape of these fields was dominated by wetlands (either permanent or temporary swamps and marshes), by natural hygrophilous vegetation and forests alive with a variety of plants and animals, and by the frequent river floods. Due to these condi-

tions, the population was sparce and the inhabited centers were few. Similarly, arable land plots were small, and the economic development of the region was scanty.

In the 1960s and 1970s, great works were undertaken by society in the fields of this region. The riverbeds were deepened and systematized, and some were even changed (the Drini of Lezha was diverted into the Buna), and embankments on both river sides were erected, thus reducing the possibility of floods, but also the drainage system was put in place. Consequently, many swamps were dried up and

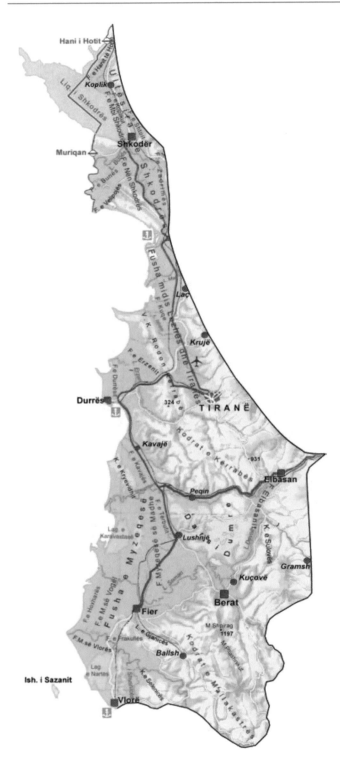

Fig. 14.2 Physical map of the western region
Source: Ideart. Gjeografia 11, 2017, p. 100

swampy lands were reclaimed. The swamps that were at or below the sea level were dried through water-scooping plants (Tërbufi, Hoxhara, Durrësi, Lezha, etc.) (Photo 14.1) (Qiriazi and Doka 2020).

All this work changed the entire appearance of the geographical landscape of these fields, which were arranged in regular plots of arable land, crisscrossed by drainage and irrigation canals. The residential centers increased, roads and railways were built, and before long, the region became the most populous and most developed in the country.

The climate is warm because the region has lowland-hilly relief, is washed by the sea, and is also protected from cold winds in the east. The region has hot and dry summers, while winters are mild and humid. The average annual temperature reaches 15–16 °C. January temperature varies from 5 °C (in Shkodra) to 9 °C (in the southern part), and in July, the temperature is around 24–25 °C. The average annual rainfall ranges from 900 mm (in Myzeqe) to 1700 mm (in Koplik). The rainfall is mainly concentrated in the cold half of the year, especially in the winter season. These climatic conditions are suitable for many crops and allow two to three crops to be harvested from the same land per year (Qiriazi 2019).

The hydrography of the region is very rich. The largest rivers of the country pass through this region, and it has many lakes: freshwater lakes (in Shkodra, Dumrea, etc.) and saltwater coastal lagoons. It is also rich in groundwater, especially the plains of Lezha, Fushëkuqja, etc. This great water wealth is of economic importance, especially for irrigation, supply of residential centers, development of tourism, etc. For this purpose, large irrigation canals have been built (Vjosa-Levani-Fieri, Naum-Panxhi, Miloti-Thumana, Mati-Lezha, etc.) but also water reservoirs and water pumping stations that take the groundwater of Fushëkuqja to supply Tirana, Durrës, Kavaja, and many villages (Qiriazi and Doka 2020).

These fields have gray-brown soils which are distinguished for their full profile and good productivity. There are also some soil subtypes, such as alluvial near rivers and saline near the coastline and former salty marshes, which, due to the salty composition, have not been put fully into agricultural use. In the high hills, there are brown soils, often degraded by erosion.

Natural vegetation has limited reach because it has been replaced by cultivated plants. It is found only in special sectors along the coast (the Divjaka Pine Forest), in the wet sectors (the wetlands with hygrophilous vegetation), and on some hill slopes covered with shrubs and Mediterranean oaks but quite degraded now (Qiriazi 2019).

14.2.3 Features of Demographic Development

The western region of Albania is the most populous region of the country, thanks to its better natural conditions, dominated by the flat terrain with good soils and warm climate, which provide also greater economic development compared to the other regions. These suitable conditions have made the region attract more and more population, which currently is

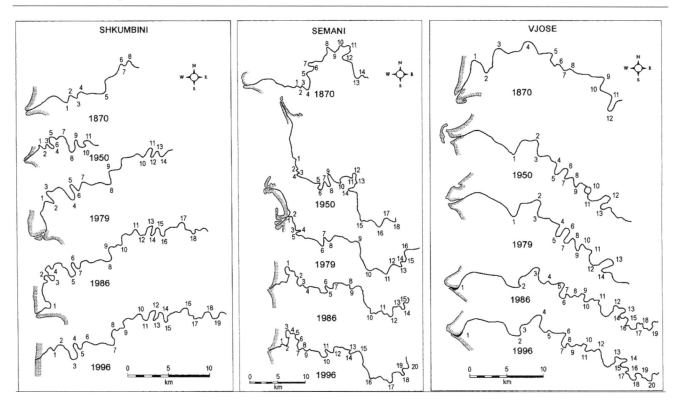

Fig. 14.3 Change of the downstream bed of Shkumbin, Seman, and Vjosa (Ciavola 1999)
Source: Mediaprint. Gjeografia Fizike e Shqiperise 2019, p. 520

over two million inhabitants, i.e., more than half of the entire population of Albania.

One indicator of the large concentration of population is the high population density, which in this region ranges between 300 and 400 inhabitants/km². Some special parts of this region have a higher indicator of population density per unit area, such as the Tirana-Durrësi area (Doka 2015).

The rapid population of this region took place particularly after 1945, until the end of the 1970s, and was the result of the drying up of the swamps and the development of various branches of the economy. However, the highest rates of total population increase in this region occurred after 1990, when the free movement of people was reinstituted, leading to a very large influx of people from other parts of the country settling in this region (INSTAT 1989).

The huge influx of population from the other regions of the country, especially from the north and northeast moving into the western region, is related to such factors as:

- The difficult living conditions in the highlands and mountainous areas, which force the residents of these areas to always be in search of better living and working conditions. The western region currently meets their demands for better conditions.
- The banning of the free movement of people until 1990 and "overcrowding" of higher areas of the country beyond

their sustenance capabilities. That was also related to the demographic policy, which aimed at keeping the population rooted in its birth areas and at increasing the overall population as much as possible in each area of the country. Such "overcrowding" is now decreasing more and more because of population departures from these areas.
- The large village population, which, as in all the other countries, tends to move toward the cities. The western region has been affected by this phenomenon more than any other region of the country. The urban population of this region in the last few years has increased significantly because of the arrival of the population from rural areas.

The data on the domestic migratory movements in Albania reveal clearly the fact that internal migration has been dominated by the tendency to move from the peripheral areas toward the center or the western part of the country, especially toward the Tirana-Durresi area. Thus, 91% of the people involved in the internal migratory population movement after 1990 have followed this general trend (Doka 2015).

Considering the administrative units, the largest concentration of settlement of incoming population occurred in the district of Tirana, where about 55% of the entire population departing from the other areas of the country is currently concentrated. Durrësi follows Tirana with about 18%. Thus, about 73% of the entire internal migratory movement is con-

Photo 14.1 Current plain landscape in the western region
Photo by Dhimiter Doka, June 2017

centrated in these two regions of the country. It is estimated that about 23% of the population of the Tirana region and 18% of the Durrësi region is made up of the population that came to these regions after 1991. This clearly proves the large concentration of the displaced population in the western region, especially in the Tirana-Durrësi area (INSTAT 2019).

This large concentration of population in the western region is predicted to increase further in the future, although the rates of population arrival may not be as high as they have been so far.

This large population influx toward the western region through the massive migration process after 1990 has been accompanied by a series of demographic and social transformations. This has affected the region's population structure by age, which is now characterized by the dominance of young people. This has also affected the high natural increase and the largest percentage of urban population in this region as compared to all other regions of the country.

One of the distinctive features of the population in the western region of Albania is its large portion of urban population. Thus, more than half of the population of this region resides in its city centers. The most important cities of Albania, starting from the capital city of Tirana (estimated at about one million inhabitants), Durresi, Elbasani, Shkodra, Vlora (with over 100,000 inhabitants), Fieri, Berati, and Lushnja (with over 50,000 residents), are found in this region (INSTAT 2019).

The number of cities and their population increased significantly in the period 1945–1970, when different branches of the economy developed bringing about the need for new labor force, but also after 1990, when the free movement of people in the cities was allowed causing many people from the suburban areas and other regions of the country to settle in this region.

However, besides the cities and their urban population, rural settlements with their respective populations play a very important role for the region, which has both the largest number of villages and the largest number of rural populations. The movement of the population from the villages to the cities of this region has not led to the abandonment of villages, as their population has been replaced, often even increased, by the population coming from other villages of mountainous areas or other regions of the country.

Photo 14.2   Photo of Tirana, the capital of Albania
Photo by Dhmiter Doka, May 2019

The most numerous village centers in this region are in its hilly areas, but there are also villages in the plains. The groups of hills in the region, such as the hills of Kërraba, Vora, Dumrea, Sulova, Mallakastra, etc., are typical rural areas, where the large number of rural settlements with their considerable population number is evident. However, many villages are in the plain area of this region, where their rural population makes their living. The number of rural settlements and their respective population increased significantly following the agricultural systemization that took place in this region (Heller et al. 2004).

In general, the villages of this region have good natural conditions for agricultural development and animal husbandry; therefore, in the recent years, following the land privatization reform process, farmers have had a significant increase in their income. As this is the most developed region in the country, the villages and their rural population have the best opportunities not only to produce various agricultural and livestock products but also easier access to markets for their sale within this region, where most of the population of the country resides.

Such a situation in terms of the demographic development of the region makes it stand out not only for its large number of population but also for its comprehensive distribution throughout the territory of the region. Besides the urban areas as the most populous in the region, all the other areas are also populated. The urban centers such as Tirana, Durresi, Elbasani, Fieri, Shkodra, Vlora, etc., are primarily distinguished for their highest population density (Photos 14.2 and 14.3).

However, many rural centers and areas in all parts of the region are also densely populated, whereas the hilly areas of this region, such as Kërraba, Dumrea, Mallakastra, etc., have lower population density (Berxholi et al. 2003).

14.2.4   Economy in the Western Region

The western region is the most economically developed region of the country. Due to the favorable conditions provided by its geographical position, lowland relief, mild climate, and great concentration of population, almost all the branches of the economy are developed in this region.

The present-day development of this region has gone through several periods, each one distinguished for its own specific developmental features.

Photo 14.3 Part of the city of Berat
Photo by Dhmiter Doka, June 2019

- *Until 1945*, the region was underdeveloped, and mostly most of it was occupied by swamps and marshes. Agriculture was the main branch of the economy, but it was unproductive and mainly focused mainly on livestock (Louis 1925).
- *From 1945 until 1990*, the economic development of the region was dominated primarily by industry. The main cities of Tirana, Elbasani, Durresi, Fieri, etc., became the most important industrial centers not only of this region but also of all Albania. This new development led to the rapid growth in industrial production, especially in the 1950s, in the 1960s, and onwards. Heavy industry, mainly the mechanical and chemical industry, dominated the region until 1990. However, light industry played an important role for this region and the whole country at that time (Qiriazi and Doka 2020).

As far as agriculture is concerned, this region underwent the most significant transformations, turning it into the most developed agricultural region, the breadbasket of the country. The various agricultural products of this region, such as cereals, grapes, vegetables, fruits, etc., were important not only to meet the domestic needs but were also used for export.

- *After 1990*, the structure of the economic branches of this region is diverse and complete, with all its possible activities (Fig. 14.4).

Agriculture is one of the main economic activities in this region. The agricultural products are diverse, including almost all types of crops such as cereals, industrial plants, horticulture, vegetables, etc. Livestock farmers are mostly (Doka 2015) engaged in raising cattle, small livestock, but also pigs, poultry, etc. Although the geographical distribution of this agricultural economy covers all parts of the western region, there are some special areas, such as the fields of Myzeqe, Shkodër-Lezha, Tirana, Durrësi, Elbasani, etc., which provide the bulk of the various agricultural produce, while the hilly areas of the region are distinguished for the cultivation of various fruit trees, such as olives, citrus, orchards, vineyards, etc. Similarly, these areas are also important for cattle breeding, mostly small livestock, sheep, and goats.

Fig. 14.4 Economic map of the western region
Source: Ideart. Gjeografia 11, 2017, p. 109

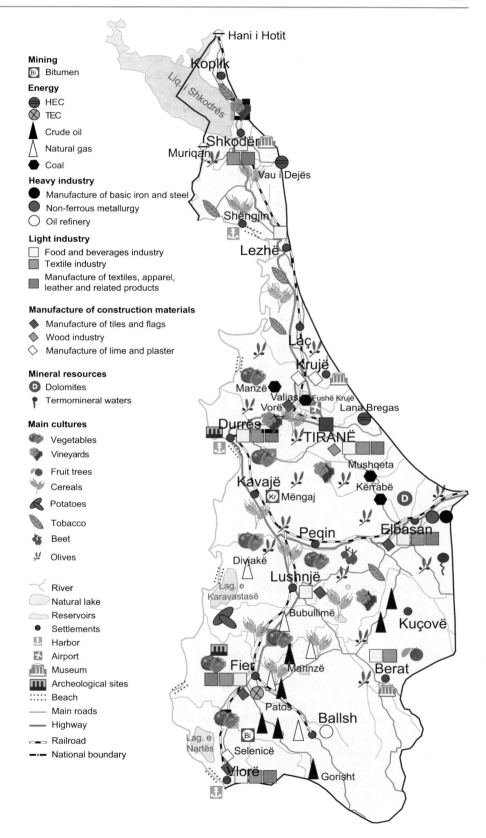

14.2.5 The Future of the Region

This region will continue to be the most important agricultural region of the country even in the future, which is favored both by its favorable geographical position, good quality soil and mild climate, as well as the increasing concentration of population, which needs its various agricultural and livestock products for consumption.

Industry in this region has undergone great changes compared to the previous period until 1990, much the same as it has all over Albania. Thus, heavy industry has been almost entirely replaced today by light and food industry, in line with market demands. The energy sector, based on large oil and natural gas reserves, remains important for the region. Likewise, the building material industry has been greatly stimulated by the boom in construction sector (Doka 2005).

Besides the two main sectors of the economy, agriculture and industry, the tertiary sector of the economy has become increasingly important in the region, particularly after 1990, when the economy began to move toward the free market model. The economic activities of the tertiary sector, such as transport, trade, services, tourism, etc., have become the main activities for a large part of the population in this region, with tourism playing a crucial role within this new economic framework. This region contains not only important natural and cultural potentials but also better infrastructure and greater human, economic, and financial resources (Qiriazi and Doka 2020).

Based on the general socioeconomic situation of the western region of Albania, we can say that it is the most dominant one compared to all the other regions, and it will continue to carry the main weight in the development of the entire country in the future.

14.3 The Northern and Northeast Region

14.3.1 Geographical Position

This is the largest region in Albania. It includes the mountainous territories north of the Shkumbini River. Its proximity and connection with the central part of the Balkans through gorges, river valleys, passes, hills, and low mountains make for the continental influence on its mainly Mediterranean landscape of this region.

This region has a transit geographical position between the Adriatic coast and the Balkan hinterland. The value of this position has been exploited since antiquity, then in the Middle Ages, and later in history. But the valuable position was disrupted in 1913 by the establishment of the state border, which separated the region from its natural urban centers (Dibra e Madhe, Prizreni, Gjakova, Peja, Plava, Gucia, etc.) This was deepened further during the communist period when the region became highly isolated.

This isolation was particularly intense in the Albanian Alps. Besides political reasons, the rough and rugged terrain of the Alps coupled with their long distance from the country's urban centers and lack of adequate roads play their part and increased the difficulty of communication with the other regions of the country. This highly isolated position of the Alps has strongly influenced the scanty socioeconomic development and have played a part in the mostly closed character of the economy, in the insufficient level of emancipation of the population, but also in the preservation of the rich material and spiritual heritage.

The region began to emerge from its intense isolation after 1990, when the political changes enabled the resumption of communication with neighboring and other Southeast European countries. Another positive development was the construction of the Durrësi-Prishtina Highway and the building of the roads in the interior part of the region and the Alps. These developments have created the conditions for the assessment and exploitation of the numerous natural resources of the region and its integration into the domestic and regional development framework.

14.3.2 Natural Conditions

This region has huge natural resources, but it has the lowest level of socioeconomic development. Its diverse and complex geological construction provides vast resources of metallic and nonmetallic mineral resources. The mainly mountainous relief of this region is distinguished for its sharp contrasts, its high degree of fragmentation, and its limited size of arable land. But this region offers great opportunities for the development of winter and green tourism, animal husbandry, forestry, etc. (Figs. 14.5 and 14.6; Photo 14.4) (Xhomo et al. 1991; Qiriazi 2019)

The region has a diverse climate which provides good opportunities for the development of some branches of agriculture and especially tourism. The region also has large water and hydropower resources, diverse vegetation, the country's largest forests, and a rich animal world, which has great economic value and tourist attraction potential. The complex and diverse nature of this region is rich in natural heritage of national and international importance (such as the primeval and old beech forests in Gashi, the strictly protected reserve of Gashi, and the national parks of Thethi, Valbona, Lura-Mount Deja, Qafë-Shtama). The region has also many natural monuments, natural parks, protected landscapes, and managed resource areas.

Fig. 14.5 Physical map of
the Northern and Northeastern
Region
Source: Ideart. Gjeografia 11,
2017, p. 122

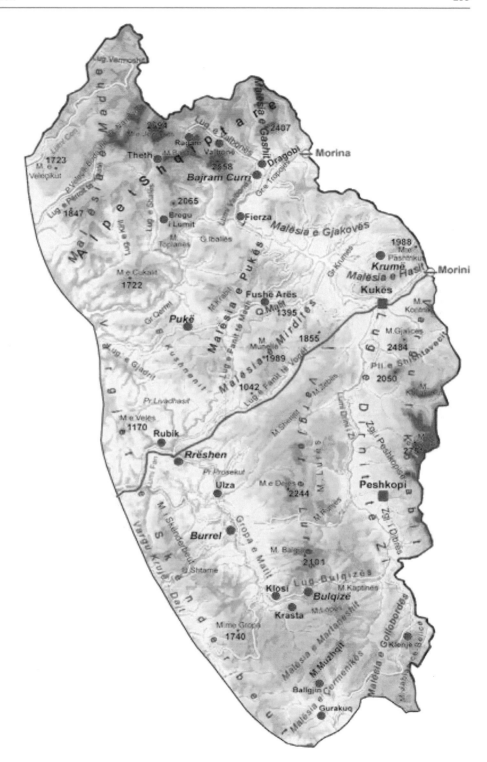

14.3.3 Features of Demographic Development

The early population of the region was determined by its generally difficult natural conditions but also by the political, economic, and social developments within the country and abroad. This conditioned the small population number, its increase and decrease during certain periods, its geographical distribution, and special features. The sparse population, mainly residing in the valleys and plateaus, has continued the decreasing trend in the last 30 years. In 2011, as compared to 1998, the decline in population ranged from about 20% (in Hasi) to 64% (in Tropoja). This exodus of the population

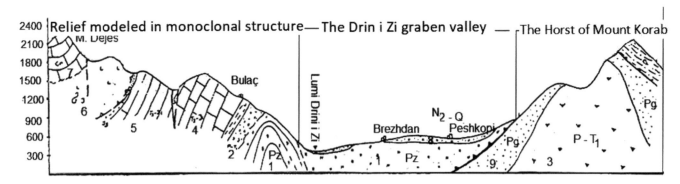

Fig. 14.6 Geomorphological schematic profile of Dejë-Korab (geology Xhomo et al. 1991, geomorphology Qiriazi 2019). 1. Schists of Pz. 2. Terrgeneof P-T. 3. Gypsums of P-T$_1$. 4. Limestone of T$_3$-J$_1$,5 Terrgene of Pz. 6. Ultrabasic. 7. Limestone of Cr$_2$. 8. Deposits of Plio-Quaternary

Photo 14.4 Alpine landscape in the Albanian Alps (web Albanian Alps)
Photo by Dhimiter Doka, July 2018

leaving the region is estimated to continue in the future, as the trend is toward complete depopulation of the poorest mountain areas. The negative consequences of this mass migration are numerous:

– Significant reduction of able-bodied human capital and reduction in the consumption of products and services, which leads to loss of interest for new investments in the region.
– Abandonment and ruin of many rural and even urban centers, which were established in the 1970s as hubs of the

chromium mining industry, which was closed, and the miners were left without work and livelihood. A notable example is the town of Kami, in the municipality of Tropoja.
– Abandonment of arable lands, which are scarce indeed. This has brought about the activation of erosion, which is impoverishing the land.

About 30–35% of the total population in this region is urban population. The small increase in urban percentage in recent years is mainly related to the much faster decline of its rural population (Qiriazi and Doka 2020).

Despite the great migration, the population structure by age still retains its vitality in the region. This is seen in the high percentage of young people, up to 14 years old (25–27%, which is above the national average) and in the small percentage of the elderly, over 65 years old (6–8%, which is below the national average) (INSTAT 2019). This is related to the still high level of births. This positive phenomenon distinguishes this region from the southeast and southern regions, where the population is aging (see these regions).

The population of this region has very strong spiritual ties with their homes, with the wider area and the rich material and spiritual heritage.

The rural inhabited centers in this region are small, sparse, far from one another, generally scattered, and adapted to the landscape. This is more obvious in some alpine villages (Thethi, Dragobia, Razma, etc.) in the Albanian Alps, where their urban and architectural features are more dictated by the rugged terrain rather than human organization. Due to the extremely rugged terrain, the villages are scattered, mostly in neighborhoods consisting of dwellings where blood-related people reside. These dwellings, especially the one on capes or near rock blocks, are surrounded by small plots of arable land and connected by narrow paths. These dwellings are built in tower-shaped architecture. They are made of stone and have very steep wooden roofs and small windows. They have been adopted for the traditional alpine landscape and its severe climate with heavy snowfall. But it has also been conditioned by the historical and political developments of the country and the isolated Alpine region, notorious for the Kanun of Lek Dukagjini, for its blood feuds, the isolated life within the dwelling, and strong blood connections. New modern elements were introduced in the urban structure and the architecture of the houses in these villages during the communist period but also after 1990 (Qiriazi and Doka 2020).

Even urban centers in this region are few and small. Their number of urban centers increased, especially during the communist period when the city status was given to some villages for administrative purposes. This status was also given to new industrial centers, some built on brand-new land. As mentioned, the demographic developments and the decline of industry after 1990 were accompanied by the abandonment of small administrative and industrial cities.

14.3.4 Economic Development

Although this region has vast natural resources, it is the least economically developed region in the country, and it has the lowest level of welfare. This contradiction between its great natural resources and the low level of development and well-being is deepening. This is related to political, historical, and social conditions and factors in the region, at home, and abroad.

This region has the largest reserves of metallic mineral resources like chromium, copper, bauxite, rare metals, and nonmetallic minerals: gypsum, kaolin, quartz, and construction materials like marble, decorative stones, clay, and so on. It has great water resources, with immense hydropower reserves. It is the region with the largest extension of beech and coniferous forests, with abundant reserves of timber. The region has many quality pastures and significant biodiversity. It boasts an extremely rich tourist offer, especially the Albanian Alps (well known for their natural diversity, clean air and water, healthy climate, breathtaking landscapes and natural beauty, great natural and cultural heritage). The region's geographical position serves as a connection between the western and eastern Albanian territories, between the Adriatic coast and the interior part of Southeastern Europe, through which regional and international roads have been important since antiquity. Following the political changes after 1990 and the construction of Durrësi-Prishtina National Highway, the centuries-old isolation of this region is coming to an end, thus enabling its integration with the other regions of within the country and beyond (Fig. 14.7) (Qiriazi and Doka 2020).

The vast natural resources represent a great potential for the economic development of the region. However, this development also requires sufficient human resources, urban traditions, cultural heritage, as well as socioeconomic management. But due to mass migration, this region is facing a huge reduction in able-bodied workers, in consumption, in trade, and in services. This remains the main obstacle in the development of this region. One way to overcome this obstacle is to find the means to encourage the population to stay in this region and incentives for those who have left to return. The creation of employment opportunities and the provision of quality services, including cultural and spiritual needs, are some of the ways to achieve this objective (Qiriazi and Doka 2020).

In many remote mountainous areas, roads are either missing or in poor condition, and this becomes a hindrance to the integration of these areas, but it also hinders the adequate use of the natural resources and their economic development. In terms of development of this region, we distinguish several periods:

– **Until 1945**, the region was characterized by significant backwardness. The main economic activity was agriculture, concentrated mainly in the small plains and valleys, while in the mountainous areas, the main activity was animal husbandry. However, the production was very low, only for household consumption. Industrial activity was in its infancy, mainly focused on limited extraction of

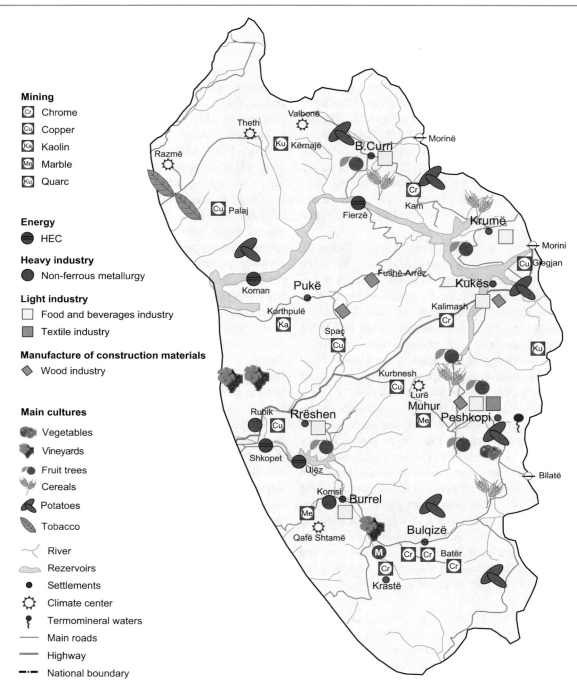

Mining
- Cr Chrome
- Cu Copper
- Ka Kaolin
- Me Marble
- Ku Quarc

Energy
- ⬤ HEC

Heavy industry
- ⬤ Non-ferrous metallurgy

Light industry
- ☐ Food and beverages industry
- ▢ Textile industry

Manufacture of construction materials
- ◆ Wood industry

Main cultures
- Vegetables
- Vineyards
- Fruit trees
- Cereals
- Potatoes
- Tobacco

- River
- Rezervoirs
- ● Settlements
- ✿ Climate center
- Termomineral waters
- — Main roads
- — Highway
- –·– National boundary

Fig. 14.7 Economic map of the Northern and Northeastern Region
Source: Ideart. Gjeografia 11, 2017, p. 136

chromium and copper ores, handicraft processing of agricultural and livestock products, etc.

- During the period from 1945 to 1990, the economic development of the region was dominated by industry, especially the mining of copper, chromium, decorative and construction materials, metallurgy, hydroelectric industry, timber, etc. The construction of large hydro-

power plants on the Drini and Mati rivers makes this region the largest producer and supplier of electricity to the country. In some areas (Puka, Mirdita, Bulqiza, Mati), industrial production surpassed agricultural production. But this industry polluted the environment. Despite efforts to increase the areas of arable land on the slopes of hills and mountains, they remain insufficient and underproductive.

– **After 1990**, due to outdated economic structures, obsolete technology, and lack of new investments, many mines, mineral enrichment plants, and copper and chromium metallurgy plants were closed. Some chromium and copper mines continue to operate, although not at their full capacity. The logging and wood industry also declined.

After 1990, the clearing of new lands was halted, while the existing land areas were privatized. Agriculture was limited to the river valleys while raising livestock in the mountains. Many farms struggled for survival. There are frequent forms of closed economy. Population departures and limited land production led to abandonment of many lands. Under these conditions, livestock became more lucrative and more developed, as the environment offers good conditions for this activity. However, poor market connections and long winters limit its further growth. Apiculture, mushrooms, cranberries, and medicinal plants are gaining importance, as are fruit trees, especially chestnuts (in Tropoja), plums, and apples. Bio products and traditional regional products are also gaining ground.

Tourism began to develop rapidly in some areas. The Alps are distinguished for several tourist attraction centers like Thethi, Valbona, Razma, Vermoshi, where private initiatives have set up tourist infrastructure. There is also fledgling tourism in some national parks. However, the development of tourism requires the fulfillment of all conditions. Overall, passive areas predominate in the region, while there are limited development impulses in several specific areas.

14.3.5 The Future of the Region

The North-Northeastern Region has all the potential to develop at a fast and stable pace. The change of political system after 1990 has created the political and social conditions for the integration and development of this region. One major problem for this region is insufficient human resources to exploit its vast natural resources. Development strategies for the region aim at applying a new way of assessing and managing the natural resources of the region, by building the necessary infrastructure and creating the right conditions for investment, employment, and quality services (in education, health, culture, trade, etc.). This will ensure sustainable development and general increase of the well-being of the population, motivating people not to leave and encouraging others to return to it.

Based on the current concepts of assessing the natural and human potentials, strategic priorities have been defined, as well as the vision for sustainable development of this region in line with environmental protection (Qiriazi and Doka 2020). Some of these priorities are:

– **Development of industry**, based on natural resources and in support of other economic sectors, especially tourism. Using modern technology, industry should provide income, without damaging the environment and the competitive advantage in tourism in this region. The following branches can be developed: energy industry, especially in electricity and other clean alternative sources (like solar and wind energy); mining and metallurgical industry (chromium, copper, bauxite, etc.); kaolin mining; decorative and construction materials (serpentinite, marble, etc.); wood industry based on developed and controlled logging; agro-food industry for processing agricultural and livestock products, mainly for tourist consumption; and crafts which have good tradition in the processing of stone, wood, metals, wool products, as well as souvenir production for tourists.

– **The development of tourism** should become a top priority, and it must become the most developed branch of the third sector and the entire regional economy. The region has all the conditions to achieve this starting with its healthy climate, water resources, stunning landscapes, rich natural and cultural heritage, spiritual traditions of hospitality and cuisine, and integration with neighboring regions, all thanks to its geographical position. However, this priority must be in harmony with the other branches and related sectors like agriculture, livestock, and forestry. To extend tourism season throughout the year and to increase the competitive capabilities of the region, various forms of tourism should be encouraged like blue, green, winter, mountain, curative, and cultural tourism.

– **Agricultural activity** is limited because of the small area of arable land and its fragmentation into very small-size farms. Therefore, livestock provides greater opportunities for development as it is related to rich and quality pastures. Agriculture and livestock must specialize in certified organic products, which can be used for tourism through food processing and traditional cuisine of this region, while the logging industry, as an important resource of the region, must ensure the controlled use of forests in complete harmony with the advantages of tourism. Collection of medicinal plants, forest fruits, but also fishing and apiculture present great opportunities for development.

Priority should be given to the development of road, water, and air transport which will ensure the economic development and integration of this region with other regions within inside the country and beyond.

The development of these branches of the economy will eventually curb the depopulation of the region, enable the return of those who have left, and encourage investors in the region. At the same time, this development will have positive effects for the other regions in the country. This would lead to a respite in the large concentration of population in the western region and would create opportunities for investment and employment from other regions. This would make this region an attractive vacation destination for all Albanians.

14.4   The Southeast Region

14.4.1   Geographical Position

This region stretches from the middle Shkumbini valley, and it includes plains, valleys, mountains, mountain ranges, and highlands reaching as far as Leskoviku. It is connected to the other Albanian regions and other countries of Southeast Europe to its east by road and rail. The large Trans-Adriatic Pipeline and the implementation of the Corridor Eight Project and the "AMBO" oil pipeline will make this region a major link between the country and the East, promoting the development and integration of the country in the European and Asian economic regions.

14.4.2   Natural Conditions

This region has diverse nature, especially in the vertical direction, and it has continental influences. Its terrain consists of terrigenous (molas and flysch), limestone, and magmatic composition, which created wrinkle structures, especially (horst and graben) intertwining between them. It is permeated by active tectonic faults, which generate earthquakes, often strong and causing great damage. It is rich in minerals like iron-nickel, nickel-silicate, copper, asbestos, talc, coal serpentine, quartz sand, clay, dolomite, marble, etc.

The irregular fragmented and complex relief consists of depressions, plains, valleys, highlands, ridges, and mountain blocks or ranges, which intertwine and intersect, thus forming a real mosaic of shapes. Elevations up to 1000 m predominate, but they range from about 200 m to 2523 m (Mount Gramozi). Flat or slightly sloping surfaces in valleys, depressions, and especially in plains, which provide good conditions for the population and agricultural development, have a large extent in this region. Depressions, plains, and valleys are connected through passes and gorges, where motorways also run through.

The climate, with its cold winters and cool summers in most parts of the region, is suitable for cereals, sugar beets, vegetables, fodder, apples, plums, etc. The climate is also suitable for the development of some types of tourism, etc.

As far as hydrography is concerned, the region is poor in large rivers, but it is rich in large lakes (Ohrid and Prespa Lakes) and glacial lakes, with groundwater and artesian springs in the Korça Plain (Figs. 14.8 and 14.9).

Brown earths and brown forest soils predominate. The soils are fertile in the plains, depressions, or valleys but poor on mountains slopes subject to erosion.

Plant species of the northern regions mix here with the species of the Southern and Mediterranean regions. The vegetation in the region consists of oaks, Mediterranean shrubs, mainly deciduous, beech, conifer, and alpine pastures in higher areas. Large and rare mammals like brown bear, lynx, etc., are found here (Photo 14.5).

The rich natural heritage is related to the complex and diverse nature and the important position as an intersection of floristic and faunistic migration routes. The natural heritage here is global (Lake Ohrid, Primeval Beech Forests of Rajca, Ohrid-Prespa Biosphere Reserve, Ramsar Wetland of Prespa Lakes) and national (Shebeniku-Jablanica National Park, Prespa, Drenova Fir; nature monuments, nature parks, protected landscapes, and managed resources).

14.4.3   Features of Demographic Development

Archaeological discoveries show that this region has been populated since ancient times. At the end of the sixteenth century, it was part of the Albanian territories with the highest population density. In the twentieth century, its population increased at a slow pace, especially in the municipalities of Korça and Kolonja. After 1990, growth rates slowed down, and the population even shrank. This phenomenon was related to the lower number of births and increased migration figures. Migration in this region had started long ago, it stopped entirely during the communist period, and it increased rapidly again after 1990. The population migrated to the western region of the country but also abroad, in Greece, Italy, USA, Canada, etc., where the former migration pattern was resumed. The population in the municipalities of Korça and Kolonja decreased by 32% in 2011, as compared to the population in1989 (Qiriazi and Doka 2020).

The population density in this region is lower than the national average. It varies between 20 and 30 inhabitants/ km^2 (in the mountainous areas) to about 200 inhabitants/km^2 (in the Korça Plain and the upper Devolli valley). In the Korça Plain, the population increased greatly due to the arrivals from the mountainous areas of the region, after the drying up of the Maliqi Swamp, the systematization of the lands and rivers in 1960s. The same thing happened in the Përrenjasi valley and in the Pogradeci plain.

The rural population accounts for about 60% of the region's population (in 2016). But its percentage varies from 22% (in Librazhdi municipality) to about 46% (in Korça municipality). After 1990, the urban population growth is closely related to arrivals from rural areas (INSTAT 2019).

Fig. 14.8 Physical map of
the Southeastern Region
Source: Ideart. Gjeografia 11,
2017, p. 143

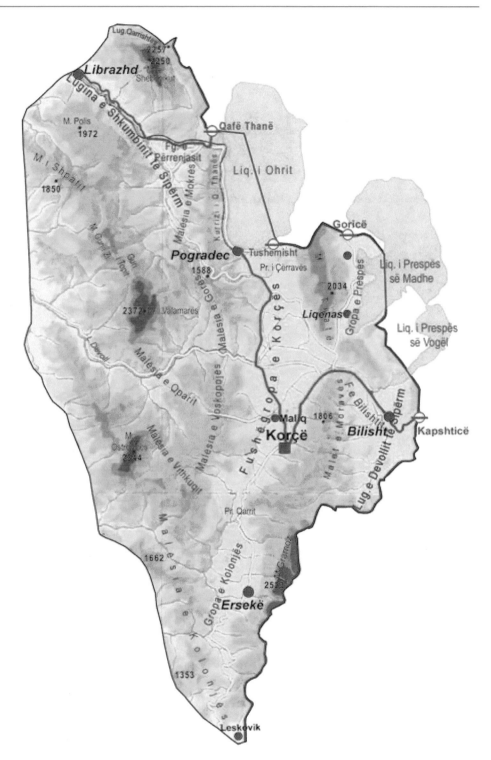

The age structure of the population, due to migration and declining births, shows signs of aging population: The young age group (up to 14 years old) amounts to 18.5%, the group between 15 and 64 years of age reaches 67.5%, and the group over 65 years old makes up 14% of the population (in 2011). The general trend is toward decreasing percentage of young people and increasing percentage of aging people (INSTAT 2019).

This population has strong cultural traditions that have contributed greatly to the development of the nationwide culture. The efforts for national education culminating with the

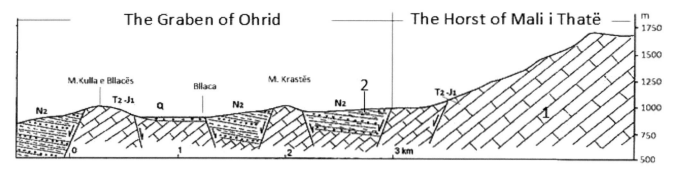

Fig. 14.9 Geological schematic profile of Kulla e Bllacës-Mali iThatë (Sh. Aliaj, 2012). 1. Limestone of T_3-J_1. 2. Terrgene of N_2

Photo 14.5 The city of Korça
Photo by Dhimiter Doka, May 2018

opening of the first official school in Albanian language (on March 7, 1887, in Korça) are well known.

The largest and most dense settlements are in the plain areas low-lying fields and the upper Devolli valley, while in the hill sides and especially in the mountainous areas, these settlements are rarer and smaller. Based on their geographical conditions, rural centers are divided into the following:

– Villages on the plains, the low-lying fields, and the upper Devolli valley are large and concentrated centers, and some have a regular urban planning.
– Villages on the hills are smaller and generally concentrated ones.

– Villages on mountain slopes, due to the more difficult conditions, are rare, small, and some of them even scattered (Librazhdi municipality). There are also villages with typical traditional dwelling architecture. Recently, their population has been greatly reduced, and some have even been completely depopulated.

This region has eight cities, with Korça and Pogradeci being the most important ones. The creation of these cities is related to their historical castles (like Korça, Pogradeci) or to their proclamation as administrative centers (like Erseka, Bilishti, Librazhdi) or to the industrial facilities they have like Përrenjasi (iron-nickel mining) and Maliqi (food).

According to their function, there are complex cities (like Korça, Pogradeci), industrial-agricultural towns (Përrenjasi, Maliqi), agricultural ones (like Leskoviku), and administrative-economic center (like Librazhdi, Erseka, Bilishti).

The region has a rich cultural heritage. There are very early settlements, such as the palafitte settlement of Maliqi and Lin (early Neolithic, 5000 BC); the Neolithic settlement of the Treni Cave, etc.; the Illyrian city of Pelion (in Lower Selca) mentioned for its monumental tombs (third century BC), etc. There are numerous archaeological and historical monuments: the Pogradeci Castle, built on an Illyrian settlement of the fifth century BC, probably Enkelana, which is evidence of Illyrian-Albanian continuity, etc. There are religious monuments: the Paleo-Christian church of Lin (fifth to sixth century), mentioned for its mosaics with great artistic values; the hermit churches of Prespa; the church of St. Mary (fourteenth century) on the island of Maligradi; the famous churches of Voskopoja, etc.; Mirahori mosque in Korça (fifteenth century), etc. There are architectural monuments (Golik bridge, traditional dwellings in the villages of Voskopoja, Dardha, Drenova, Boboshtica, Rehova, etc.), the environmental monuments (the Korça Bazaar, recently reconstructed, etc.), artistic monuments (early paintings on the Rock of Spille (Lesser Prespa)), mosaics, church paintings of great medieval Albanian painters (David Selenica, Onufri, Zografi brothers, Kostandin Shpataraku, etc.), Museum of Medieval Art in Korça (with rare religious icons and paintings, stone works of fifth to eighth century, metal works of eighteenth to nineteenth century, wood and wool handicrafts, etc.), ethnographic traditions (handicrafts, folk costumes), folklore, customs, etc.

14.4.4 Economic Development

Based on the level of economic development, the Southeastern Region occupies an intermediate position, between the western region as the most developed one and the North and Northeastern Region which has lowest level of development. However, this region's natural and human resources, its favorable geographical position, and its rich traditions offer great opportunities for development.

This region also has great natural resources. But its economic and social development, although higher than the North and Northeast Region, again for the same reasons, remains below the possibilities offered by its rich natural resources (Qiriazi and Doka 2020). The main branches of the region's economy are the following:

– **Agriculture** has its suitable development environment in plains, fields, and valleys, and it is associated with a large extension of flat or slightly sloping surfaces, as well as with climatic conditions, water resources, and soil quality. The great works like the drying up of the swamps, the systematization of lands and riverbeds, and the provision of irrigation had a great impact on agriculture. These works opened lands and made soil fertile. The region is distinguished for its good-quality agricultural production throughout the country. In the second half of the twentieth century, new lands were opened for agriculture use on hillsides and mountains slopes, which were later degraded by erosion and have been abandoned by farmers in recent years.

The percentage of agricultural land ranges from 14% (in Kolonja municipality) to 37% (in Devolli municipality). This region plays and important part in terms of agricultural produce in the country. The Korça Plain and the upper Devolli valley are distinguished in this respect. Although agriculture remains the main economic activity, its weight has diminished in recent years as livestock is gradually growing.

Forests occupy from 31% of the area (in Korça municipality) up to 51% (in Librazhdi municipality), while pastures are, respectively, 13% and 15%. Most of the arable lands are cultivated with cereals, vegetables, potatoes, and beans. The area used to cultivate beans has been constantly increasing, because of the regional and foreign market demands for this product. The area used for the cultivation of fodder for livestock has also increased. This region is well known for apples, plums, and pears (cultivated in the Korça Plain, in the fields of Pogradeci and Kolonja, in the Upper Devolli valley, etc.), where new orchards with these trees have been recently planted on the fields (Fig. 14.10) (Qiriazi and Doka 2020).

Livestock is gaining ground as market demands for these products have increased. Most of the livestock products are given by cattle in the fields and by sheep and goats in the hilly and mountainous areas. Fishing fish farms in the lakes are also gaining ground.

– The industry has good conditions for development in this region, because it has abundant natural resources. The region is connected both domestically and abroad, and it has a highly qualified labor force. Another asset is the rich cultural tradition.

Some noteworthy branches of industry are as follows:

The energy industry is represented by several small hydropower plants, and the most powerful one is built on the Devolli River. The vast reserves of lignite type were widely exploited before 1990, but its use was discontinued after that year.

The mining industry is related to the iron-nickel, nickel-silicate, chromium, and copper mines, no longer in use now but also related to the extraction of construction materials.

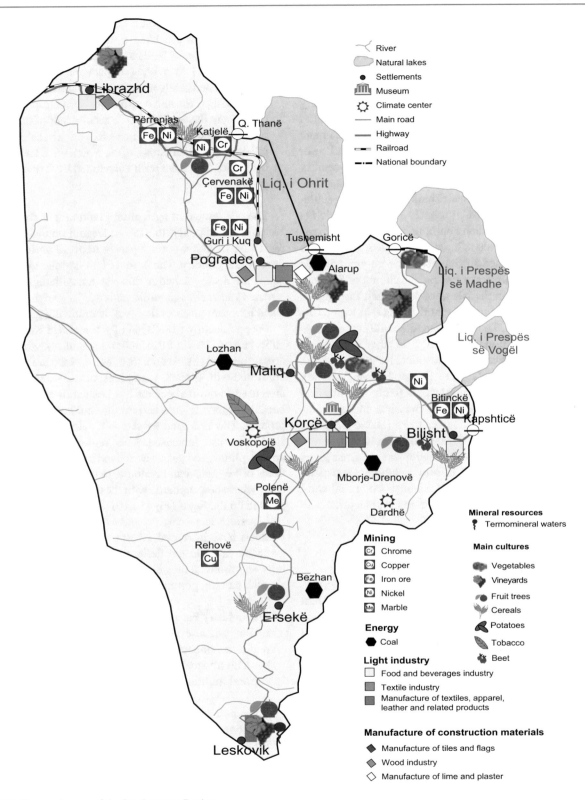

Fig. 14.10 Economic map of the Southeastern Region
Source: Ideart. Gjeografia 11, 2017, p. 152

The wood industry is related to forest exploitation and processing in factories and workshops.

The light and food industry has seen a development trend and is mainly represented by the apparel-manufacturing sector using design materials prepared by their foreign customers. There are also factories and workshops which process local agricultural products.

Craftsmen in this region have ancient traditions in the processing of wool, stone, wood, metal, etc.

Road transport is the main form of transportation, as the railway goes only as far as Pogradeci. The region is connected to neighboring countries and beyond.

Tourism has developed significantly in several centers like Pogradeci, Korça, Voskopoja, Dardha, etc. The development in tourism is related to the optimal natural conditions and cultural heritage but also to the high level of culture, hospitality, and emancipation of the population in the region. Blue tourism is focused on the shores of Lake Ohrid and Greater Prespa Lake, while civic and cultural tourism is focused on cities with special architectural traditions and archeological and historical objects. Green tourism is practiced in the rural areas with a tradition of hospitality and good cultural level, while mountain tourism is ideal in some mountains, and winter tourism is also gaining ground.

After 1990, the political and social conditions were conducive to the rapid integration and sustainable development of this region. The main problem even for this region remains insufficient human resources needed to exploit its natural resources. Another problem is the onset of the population aging, which is expected to get worse.

14.4.5 The Future of the Region

The right strategies for scientific evaluation, exploitation, protection, and preservation of natural and human resources and for their integration with other regions inside and outside the country help in determining the right directions for rapid and sustainable development (Qiriazi and Doka 2020).

The following main branches of the economy should be developed:

- *Agriculture* specialized in products demanded by the domestic and foreign market and tourism like cereals, fruits, vegetables, livestock products, honey, etc. It is important to increase the irrigated area; maintain the lands; deepen the trend of raising livestock, forestry, fisheries, apiculture, etc.
- *Tourism* in this region relies on the rich natural and cultural heritage, high cultural level, and the emancipation of its population, but also on the rich spiritual traditions, hospitality and cuisine, and the integration with neighboring regions and beyond, thanks to its nexus geographical

position. Tourism should be the most developed branch of the economy and practiced throughout the year.

- Industry, aiming to support other economic sectors, should be based primarily on the region's natural resources and its competitive advantages. Under these conditions, there is potential to develop the mining and energy industry, especially in generating electricity through hydropower and other alternative sources (like solar and wind energy, etc.); the agri-food industry for processing agricultural and livestock products; the light industry using its traditional branches (shoes, textiles, clothing, etc.); the industry of building materials and wood products; crafts with their established tradition in the processing of stone, wood, metals, wool, etc. The development of industry should not harm tourism and the ecological values of the region.

Road, rail, and air transport requires further improvements, such as the extension of the roads in the direction of Skrapari and Berati, the reactivation of the north-south road axis (Bitola-Korça-Ioannina), the construction of the anticipated Corridor Eight; the reconstruction of the Elbasani-Pogradeci railway and its extension into Greece and Macedonia, the construction of the Korça Airport, etc.

The deepening of the cross-border integration processes and the development of this region will create the proper conditions for the city of Korça to regain its former role as a developmental pole of a much wider area, including the entire Southeastern Region and the Albanian territories beyond the political border of the country.

14.5 The Southern Region

14.5.1 Geographical Position

This region, which has a north-south extension of about 120 km and east-west of about 115 km, includes the southern part of the country. Its wide access to the Ionian Sea conditions its most distinctive Mediterranean climate, hydrography, soils, and vegetation. This region has been connected through highways with the other regions of the country guaranteeing regular communication since the antiquity of the Hellenic world, with the Balkans and beyond. Now the region enjoys direct communication with the European Community through border crossing points with Greece and two seaports. It is crossed by the Hani iHotit-Kakavija Corridor, Tirana-Athens Corridor, and tourist tours around the Mediterranean, and it is part of the large Mediterranean tourist region. The region's geographical position has helped in its development since antiquity. However, this development was interrupted during the communist period when the country was isolated. Following the political changes in

1990, the right conditions were again created for the region's geographical position to gain once again its stimulating role in the region's domestic and regional development and integration.

14.5.2 Natural Conditions

The Southern Region, although mostly mountainous, has geological construction and relief simpler than other mountainous regions of the country. It consists of limestones, which builds large longitudinal anticline structures, sinking gradually toward the northwest, and flysch, which builds mainly syncline structures and molas that fills graben (Qiriazi 2019).

The dominant compression regime of neotectonic movements deformed the wrinkling structures, giving them the character of monoclines. Normal tectonic faults in the western wings of anticline structures are related to the tensile regime of neotectonic movements. Water sources emerge at the tectonic-lithological contacts of the western side of the mountain slopes. Hence, settlements and populations are denser on these slopes. The diving graben of Delvina, the graben of Dukati, the bay of Vlora, the separation of the Sazani island from the Karaburuni Peninsula is connected to this diverting structure. Frequent and strong earthquakes are associated with active tectonic faults. The region has underground resources like oil, coal, phosphorite, decorative stones, building materials, etc.

The relief extends from the sea level up to 2465 m (in Nemërçka). Heights up to 800 m predominate. This region consists of mountain ranges, highlands, valleys, and the plain of Delvina. The mountain ranges and valleys intersect regularly. The sparse highlands are fragmented by rivers in different directions. These forms of relief are connected by gorges and passes where motorways pass.

It has typical Mediterranean climate with milder, wetter winters and hotter, drier summers than the other mountainous regions of the country. It has great thermal and pluviometric properties. The rich hydrography includes the river Vjosa, with its branches the Drino and the Shushica, the Osumi river, etc. Hydropower reserves have been partially harnessed in Bistrica, etc. The Vjosa River is distinguished for its diversity of ecosystems and living organisms. Not used for generating hydropower, it has been described as "Europe's only wild river." Therefore, environmentalists demand the suspension of the construction of hydropower plants on this river. Other natural resources in the region include the Butrinti Lagoon, distinguished for the cultivation of mussels; the Blue Eye, a very large water spring; other mineral and thermo-mineral springs; artificial aquifers; and irrigation canals (Figs. 14.11 and 14.12) (Qiriazi 2019).

The soils are mainly brown and gray, which are fertile in the valleys and plains of Delvina but poor on the slopes of hills and mountains. The flora and fauna are dominated by Mediterranean species, while in the highlands, there are also species from the northern Balkans. The vegetation is represented mainly by shrubs and deciduous types. The most widespread trees are the Mediterranean oaks, damaged by man. Above the oak layer lie conifers, mainly fir and alpine pastures at higher elevation. The wildlife is represented by mammals like bear, roe deer, wild boar, jackal, etc.; rare birds (eagle, sea swallow), etc. There is also very rich marine and lagoon fauna.

The region's rich natural heritage is related to its natural diversity and southernmost position, as a crossroads of wildlife migratory pathways. The natural heritage is world (Ramsar Wetland Çuka-Butrinti-Cape of Stillo) and national (Strict Nature Reserve of Kardhiqi; Tomorri National Parks of Llogara, Hotova, Butrinti; the National Marine Park of Sazani-Karaburuni; natural monuments natural and municipal; Aos-Vjosa eco-museum) (Qiriazi 2020).

14.5.3 Features of Demographic Development

Archaeological discoveries in Xara, in the cave of Konispoli, and in other places indicate that the region has been populated since early times (Middle Paleolithic). As early as the seventh to sixth centuries BC, the first ancient cities began to be built. By the end of the sixteenth century, this region was part of the areas with the highest population density in the Albanian territories. Subsequently, the population increased at a slower rate (by a natural increase as much as half the national average). This rather small increase in population is related to the low number of births but especially to the constant migrations. Migration abroad taking place before 1945 ceased entirely in the communist period but increased again greatly after 1990. This migration was the main cause of the population decline in the period between the two world wars and especially the period after 1990. In 2011, the region had less than half of the population it had in 1989 (INSTAT 2019).

The average population density of the region (about 30 inhabitants/km² in 2011) is lower than the national average. Fewer than 20 inhabitants/km² are found in the hilly and mountainous areas, where the declining tendency continues, while higher figures (over 50 inhabitants/km²) are found in the Delvina plain and in the valleys of the region, which offer better living conditions.

Due to emigration and declining birth rates, the population is aging. The young population of age up to 14 years old occupies 17% (the lowest in the country), age 15–64 amounts to 67%, and age over 65 years occupies 16% of the population (in 2011), the highest in the country. This reveals an

Fig. 14.11 Physical map of the Southern Region
Source: Ideart. Gjeografia 11, 2017, p. 159

advanced rate of population aging, which is most striking in the Pogoni area, where the elderly makes up over half of the population. The population of Gjirokastra region is the oldest in the country (with an average age of 39.7 years in 2011) (Qiriazi and Doka 2020).

The urban population predominates accounting for over 53% of the region's population (2011). The natural increase of the urban population is low, and this is more related to the arrivals from the rural areas especially after 1990. The population has cultural traditions, which have national culture reverberation (Photo 14.6).

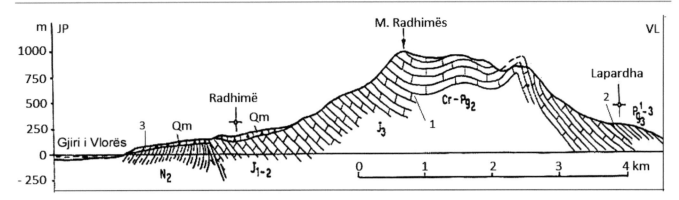

Fig. 14.12 Geological schematic profile of Radhimë-Lapardha (Sh. Aliaj, 2012).jpg. 1. Limestone of J_{1-2}, J_3, and Cr-Pg$_2$. 2 Terrigene of Pg$_3$. 3. Molasse of Qm

Photo 14.6   The city of Përmet in the Vjosa valley
Photo by Dhimiter Doka, May 2018

In this region, the largest and densest settlements are in the Delvina plain and in the valleys of the Vjosa, Drino, and Shushica rivers, while in the hills, especially in the mountainous areas, they diminish as the distance between them increases. Based on their geographical conditions, the rural centers are divided into lowland, hilly, and mountainous villages.

Most of the hilly and especially mountainous villages are concentrated, and they have traditional dwelling architecture, adapted to the Mediterranean geographical landscape.

Their traditional urban and architectural features are more related to their surrounding landscape rather than the result of human organization. This is quite clearly observed in the concentrated villages along the Ionian Riviera. They are excellent examples of human adaptation to the special qualities of the Mediterranean landscape.

Being built on certain forms of relief, these villages, depending on the time and their level of development, provided not a few advantages. Settling on the steepest relief with southern exposure saved their little agricultural land and provided light, sun, and protection from cold winds. The construction at the height of the relief gave the village the shape of an "acropolis," which provided ideal position for enemy observation while stretching away from the sea, which was both a source of livelihood but also of disasters, and increased their defensive ability from pirates and invaders, quite numerous in the long turbulent historical periods.

Based on these natural conditions, a special cultural landscape was created. It is distinguished by the grouping of roads and solid buildings, compacted one after the other and rising directly on the foot of an often-steep rock, and by the creation of interdependent natural and artificial environments, merging with the private and public space, where elements of private green areas abound. This creates a connection and a continuity of private yards on the public road while providing passersby with shade, comfort, security, and a sense of penetration into the privacy of the residents (Qiriazi and Doka 2020).

New elements, related to the modern tourist infrastructure, have been added to these villages recently. Some of these villages have been transformed into tourist attraction villages.

The region has 14 cities with rich cultural traditions, which provide great tourist values for them. Among these cities, Gjirokastra and Saranda are distinct. Some of these cities were created near their historical castles (like Gjirokastra, Tepelena, Libohova, Këlcyra). Çorovoda was declared a city with administrative function, while some others were created near their industrial facilities, such as Selenica (extraction of bitumen), Memaliaj (extraction of coal), and Poliçani (mechanical industry).

According to their functions, there are complex cities (like Gjirokastra), industrial-agricultural ones (like Përmeti, Këlcyra, Selenica, Memaliaj), agricultural ones (like Konispoli, Libohova), and administrative-economic ones (like Çorovoda, Delvina, Tepelena) but also seaports and tourist centers (like Saranda, Himara) and Orikumi, as a military base and recently as a tourist attraction.

The great wealth of cultural heritage is related to the ancient population of the region and its social, economic, and cultural development. There are world heritage sites here (like Butrinti and Gjirokastra): national heritage sites like the ancient Illyrian cities of Antigonea, third century BC;

Phenicia, third century; Amantia, c. fifth century BC; Oriku, sixth century BC; Onhezmi in Saranda, a city and ancient pier of sixth century BC; and many other monuments and cultural sites.

14.5.4 Economic Development

This region, like the southeastern one, occupies an intermediate position between the most developed western region and the less developed North and Northeastern Region. However, the natural resources and human capital enable a much higher level of socioeconomic development.

Not insignificant flat area or the not-so-steep areas in the Delvina plain and in some valleys, coupled with great climatic, especially thermal, resources, create good conditions for the development of agriculture and livestock. This works for the drying of the swamps in the Delvina plain, and the systematization of lands and rivers, as well as the provision of artificial irrigation during the second half of the last century, had a significant impact. These works made the lands fertile. New lands were also opened on the slopes of hills and mountains (large plantations of coastal terraces in Jonufëri, Lukova, Borshi, Shënvasili, Ksamili, etc.), where citrus, olives, and vineyards were cultivated. The post-1990 damage to plants grown on these lands has led to erosion, which is impoverishing the soil. Many degraded lands in recent years have been abandoned by farmers (Qiriazi and Doka 2020).

Agriculture plays the main part in the economy of this region. The main activity is occupied by livestock, which provides over half of the agricultural production. This branch will be further developed in the future because the demand for livestock products is increasing, mainly due to tourism.

The area of agricultural land ranges from about 15% in the municipality of Gjirokastra to about 30% in the municipality of Delvina. The largest area of forests, about 41%, is in the municipality of Përmeti, while the smallest area is in the municipality of Saranda, about 18% of the total area. The staple agricultural plants are cereals. The cultivation of vegetables, potatoes, and beans also plays a significant part as their cultivation area is constantly increasing, due to the increased demand for such products for the city markets and the growing tourist market. The region is distinguished for its production of olives, citrus, figs, grapes, and other fruits. The region has long tradition in livestock breeding, which is constantly increasing its economic weight. The largest production is given by cattle. The region is also distinguished for sea fishing and fish and mussel farming.

The suitable conditions for the development of industry are related to its natural resources, its connections inside and outside the country, its traditions, and the qualified labor force. There are the following branches of industry:

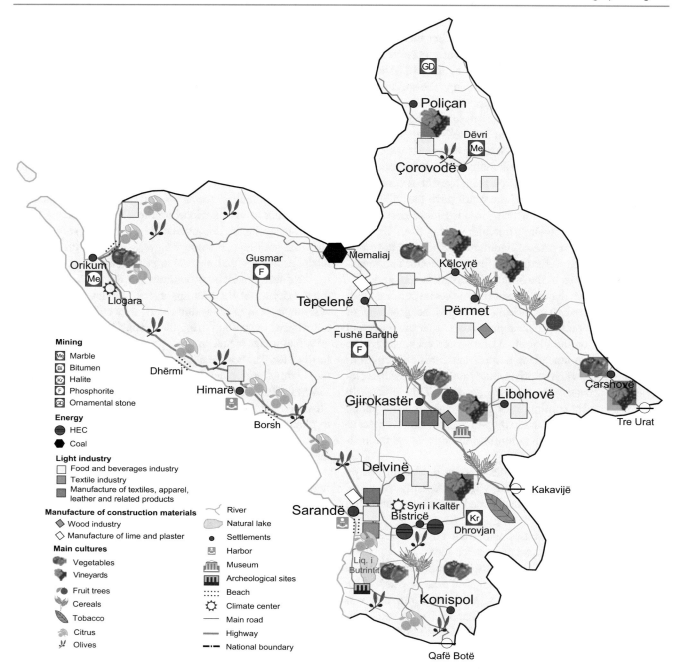

Fig. 14.13 Economic map of the Southern Region
Source: Ideart. 11, 2017, p. 169

Energy industry consists of two hydropower plants on the Bistrica and several smaller ones, the coal industry, which has long ceased to work, and the gas and oil reserves, etc.

As far as the mining industry is concerned, only the bitumen mine still works, while the extraction of phosphorites and mineral (rock) salt has stopped working. The building materials industry consists of the extraction of dolomite, limestone, decorative stone, and marble; the use of clay for bricks and the use of sand, etc.

The light and food industry are represented by the shoe and clothing factories, by factories processing agricultural and livestock products found in almost all major cities. The crafts have an established tradition in the production of wool, stone, wood, metal, and other products.

Transport consists mainly of road transport, with much greater weight and sea transport. The road transport is connected with Greece road system through the Kakavija, Tri Urat, and Qafëbota cross-border points. Sea transport plays a special role for goods, passengers, and especially for tourists.

Tourism is the most developed branch of the region's economy. This is related to the very good conditions for the development of several types of tourism, like blue, civic, green, mountain, and religious tourism. Several urban (Saranda, Gjirokastra) and rural centers are specialized on tourism and hospitality services, especially along the Riviera coast (Fig. 14.13) (Qiriazi and Doka 2020).

The economic structure of the region, especially the development of tourism and its road and sea connections with the other regions of the country and with Greece, helps the interregional integration, inside and outside the country. For the other regions of the country, this region offers tourism, and it also supplies their markets and foreign ones with the special agricultural products (citrus, especially tangerines, olives and quality olive oil, out-of-season vegetables, alcoholic beverages, etc.) and livestock products, especially quality cheese and meat. This region also supplies building materials, especially decorative stones, and marble, but also handicraft products. At the same time, this region receives from the other regions of the country agricultural products, especially during the tourist season; industrial products, such as electricity, fuels, and firewood; light industry products; construction materials; equipment; machinery, etc.

14.5.5 The Future of the Region

All prerequisites (natural resources and political and social conditions) are there not only for the integration but also for the rapid and sustainable development of this region. The main problem for this region remains insufficient human resources for the use of natural resources, especially population aging which is growing worse over time. Therefore, strategies for the evaluation and the scientific use of natural and human resources are needed, as well as strategies for its integration with other regions inside and outside the country (Qiriazi and Doka 2020). The following main branches of the economy should be developed:

Agriculture specialized mainly in agritourism for certified quality organic products such as citrus, olives and olive oil, fruits and vegetables, livestock products, fishing, and apiculture, especially products for the growing tourist market.

Tourism, the primary and most developed branch of the economy, should take advantage of the geographical position, the rich natural and cultural heritage, the high cultural level of the population, and its rich traditions. Many types of tourism can be developed here, like blue, civic and cultural, green, mountain, curative, or religious tourism. This will ensure the expansion of tourism season throughout the year, and it will also increase its competitiveness.

Industry, aiming to support other economic sectors, should rely on natural resources and its competitive opportunities. The energy industry (hydropower, solar and wind energy) should be developed; the natural gas and oil; the agri-food industry for local agricultural and livestock products; the light industry in its branches with established tradition (like footwear, clothing, etc.); the construction materials industry (limestone, dolomite, decorative, marble, etc.); wood industry, the handicrafts with long tradition in the processing of stone, wood, metals, wool, and other materials.

The well-developed transport system requires the improvement and extension of the highways, the construction of highways for the foreseen corridors, the expansion and modernization of seaports, the construction of tourist ports for yachts, the construction of Gjirokastra airport, etc.

References

Berxholi A, Doka D, Asche H (2003) Demographic Atlas of Albania. Tirana

Ciavola P (1999) Relation between river dynamics and coastal changes in Albania: an assessment integrating satellite imagery with historical data. Int J Remote Sens 20(3):561–584

Doka D (2005) Regionale und lokale Entwicklungen in Albanien – ausgewählte Beispiele. Potsdam – Germany

Doka D (2015) Gjeografia e Shqipërisë – Popullsia dhe Ekonomia. Tiranë

Heller W, Doka Dh, Berxholi A: Hoffnungsträger Tirana. In: Geographische Rundschau 56(2004) Heft 1. S. 50–58. Westermann – Gjermani

INSTAT: Demographic yearbook von 1989, 2000, 2005, 2010, 2015, 2019

Louis H (1925) Albanien – eine Landeskunde. Vornehmlich auf Grund eigener Reisen, Berlin

Qiriazi P (2019) Gjeografia fizike e Shqipërisë. Mediaprint

Qiriazi (2020) Trashëgimia Natyrore e Shqipërisë (Ribotim). Tiranët

Qiriazi P, Doka D (2020) Gjeografia e Shqipërisë. IDEART

Xhomo A, Qirici V, Kodra B, Pashko P, Meço S (1991) Stili tektonik mbulesor i zones së Korabit. Bul Shk Gjeol 1:205–212

Index